Recent Advances in

MICROBIOLOGY

Recent Advances in
MICROBIOLOGY

Edited By
Dana M. Santos, PhD
Professor, Department of Anthropology,
Binghamton University, State University of New York, U.S.A.

Apple Academic Press

TORONTO   NEW YORK

CRC Press
Taylor & Francis Group
6000 Broken Sound Parkway NW, Suite 300
Boca Raton, FL 33487-2742

Apple Academic Press, Inc
3333 Mistwell Crescent
Oakville, ON L6L 0A2
Canada

© 2012 by Apple Academic Press, Inc.
Exclusive worldwide distribution by CRC Press an imprint of Taylor & Francis Group, an Informa
business

First issued in paperback 2021

No claim to original U.S. Government works

Version Date: 20120530

ISBN 13: 978-1-77463-190-4 (pbk)
ISBN 13: 978-1-926692-71-5 (hbk)

Visit the Taylor & Francis Web site at
http://www.taylorandfrancis.com

and the CRC Press Web site at
http://www.crcpress.com

For information about Apple Academic Press product
http://www.appleacademicpress.com

Preface

Microbiology is a broad term for research including any type of microorganism such as bacteria, viruses, fungi, parasites and even prions. Current areas of particular interest to researchers include:

- antibiotics and antivirals
- antibiotic resistance
- bioremediation
- biotechnology

A degree in microbiology is ideal for work in various fields such as disease research, biotechnology, pharmaceuticals, or microbial genetics and systematics.

Understanding the genetics and behavior of disease-causing pathogens plays a crucial role in discovering effective pharmaceuticals, particularly in the face of increasing global antibiotic resistance. Microbiology is also important in present environmental concerns such as the use of microorganisms in bioremediation or energy production.

New diagnostics, pharmaceuticals, and understanding of the basis of disease will continue to be at the forefront of microbiology research. This is particularly relevant in the face of increasing antibiotic resistance. As concern for the state of our environment grows, there is increasing interest in using microorganisms to clean pollution or generate energy.

— Dana M. Santos, PhD

List of Contributors

N. A. Affara
Department of Pathology, University of Cambridge, Cambridge CB2,1QP, UK.

Misagh Alipour
The Novel Drug and Vaccine Delivery Systems Facility, Department of Chemistry and Biochemistry, Laurentian University, Sudbury, Ontario, Canada.

Ali O. Azghani
Department of Biology, University of Texas at Tyler, Tyler, TX, USA.

Valerio Berardi
Dipartimento di Genetica e Biologia Molecolare, Sapienza Università di Roma, Roma, Italy.

Agata Bielecka
Synthetic and Systems Biology Group, Helmholtz Center for Infection Research (HZI), Braunschweig, Germany.

Marcos Bilen
Laboratorio Ingeniería Genética, Biología Celular y Molecular- LIGBCM, Universidad Nacional de Quilmes, Roque Saenz Peña 352, Bernal, Buenos Aires, Argentina.

Ju Cao
Key Laboratory of Laboratory Medical Diagnostics, Ministry of Education, Faculty of Laboratory Medicine, Chongqing Medical University, Chongqing 400016, P.R. China.

Mauro Castelli
Regina Elena Cancer Institute, Roma, Italy.

O. E. Chausiaux
Department of Pathology, University of Cambridge, Cambridge CB2,1QP, UK.

Younghak Cho
School of Mechanical Design and Automation Engineering, Seoul National University of Technology, Seoul, Korea.

Ricardo S. Corral
Servicio de Parasitología y Enfermedad de Chagas. Hospital de Niños Dr. Ricardo Gutiérrez. Gallo 1330, Ciudad Autónoma de Buenos Aires, Argentina.

Romina A. Cutrullis
Servicio de Parasitología y Enfermedad de Chagas. Hospital de Niños Dr. Ricardo Gutiérrez. Gallo 1330, Ciudad Autónoma de Buenos Aires, Argentina.

Tamás Czárán
Ecology and Theoretical Biology Research Group of the Hungarian Academy of Science and Eötvös University, Budapest, Hungary.

Genevieve Di Bartolo
Genomics Division, DOE Joint Genome Institute, Walnut Creek, CA 94598, USA.
Boards 'N More, Brentwood, CA 94513, USA.

Paul de Figueiredo
Department of Plant Pathology and Microbiology, Texas A&M University, College Station, TX, USA.
Department of Veterinary Pathobiology, Texas A&M University, College Station, TX, USA.

Helene Feil
Department of Plant and Microbial Biology, University of California, Berkeley, CA 94720, USA.
Land for Urban Wildlife, Concord, CA 94527, USA.

William S. Feil
Department of Plant and Microbial Biology, University of California, Berkeley, CA 94720, USA.
Land for Urban Wildlife, Concord, CA 94527, USA.

R. A. Furlong
Department of Pathology, University of Cambridge, Cambridge CB2,1QP, UK.

Gaspare Galati
Dipartimento di Chirurgia "Pietro Valdoni", Roma, Italy.

Miguel Godinho
Synthetic and Systems Biology Group, Helmholtz Center for Infection Research (HZI), Braunschweig, Germany.

Raul O. Gonzalez
Programa de Nanomedicinas, Universidad Nacional de Quilmes, Roque Saenz Peña 352, Bernal, Buenos Aires, Argentina.

Murray Hackett
Department of Chemical Engineering, University of Washington, Seattle, WA, USA.

Majed Halwani
The Novel Drug and Vaccine Delivery Systems Facility, Department of Chemistry and Biochemistry, Laurentian University, Sudbury, Ontario, Canada.

Arum Han
Department of Electrical and Computer Engineering, Texas A&M University, College Station, TX, USA.

Sharifah Syed Hassan
School of Medicine and Health Sciences, Monash University, Sunway Campus, Kuala Lumpur, Malaysia.

Erik L. Hendrickson
Department of Chemical Engineering, University of Washington, Seattle, WA, USA.

Leticia H. Higa
Programa de Nanomedicinas, Universidad Nacional de Quilmes, Roque Saenz Peña 352, Bernal, Buenos Aires, Argentina.

Piotr Hildebrandt
Department of Microbiology, Chemical Faculty, Gdańsk University of Technology, Narutowicza 11/12, 80-952 Gdańsk, Poland.

Rolf F. Hoekstra
Laboratory of Genetics, Wageningen University, Wageningen, The Netherlands.

Huijie Hou
Department of Electrical and Computer Engineering, Texas A&M University, College Station, TX, USA.

Aini Ideris
Institute of Bioscience, University Putra Malaysia, UPM Serdang, Selangor, 43400, Malaysia.
Faculty of Veterinary Medicine, Universiti Putra Malaysia, UPM Serdang, Selangor, 43400, Malaysia.

Alfred Orina Isaac
The Department of Pathology, Case Western Reserve University, Cleveland, OH, USA.

O. Jafer
Department of Pathology, University of Cambridge, Cambridge CB2,1QP, UK.

Fatemeh Jahanshiri
Department of Microbiology, Faculty of Biotechnology and Biomolecular Sciences, University Putra Malaysia, UPM Serdang, Selangor, 43400, Malaysia.

Keith Keller
Physical Biosciences Division, Lawrence Berkeley National Laboratories, Berkeley, CA 94710, USA.

Masae Kuboniwa
Department of Preventive Dentistry, Osaka University Graduate School of Dentistry, Osaka, Japan.

Józef Kur
Department of Microbiology, Chemical Faculty, Gdańsk University of Technology, Narutowicza 11/12, 80-952 Gdańsk, Poland.

Richard J. Lamont
Department of Oral Biology, University of Florida, Gainesville, FL, USA.

Alla Lapidus
Genomics Division, DOE Joint Genome Institute, Walnut Creek, CA 94598, USA.

Lei Li
Department of Plant Pathology and Microbiology, Texas A&M University, College Station, TX, USA.

Nan Li
Key Laboratory of Laboratory Medical Diagnostics, Ministry of Education, Faculty of Laboratory Medicine, Chongqing Medical University, Chongqing 400016, P.R. China.

Youqiang Li
Key Laboratory of Laboratory Medical Diagnostics, Ministry of Education, Faculty of Laboratory Medicine, Chongqing Medical University, Chongqing 400016, P.R. China.

Xiu Luo
The Department of Pathology, Case Western Reserve University, Cleveland, OH, USA.

Vítor A. P. Martins dos Santos
Synthetic and Systems Biology Group, Helmholtz Center for Infection Research (HZI), Braunschweig, Germany.

Maradumane L. Mohan
The Department of Pathology, Case Western Reserve University, Cleveland, OH, USA.

Irma Morelli
Centro de Investigación y Desarrollo en Fermentaciones Industriales-CINDEFI, Facultad de Ciencia Exactas, Universidad Nacional de La Plata, 50 y 115 La Plata, Buenos Aires, Argentina.

Maria Jose Morilla
Programa de Nanomedicinas, Universidad Nacional de Quilmes, Roque Saenz Peña 352, Bernal, Buenos Aires, Argentina.

Siqiang Niu
Key Laboratory of Laboratory Medical Diagnostics, Ministry of Education, Faculty of Laboratory Medicine, Chongqing Medical University, Chongqing 400016, P.R. China.

Matthew A. Oberhardt
Department of Biomedical Engineering, University of Virginia, Health System, Charlottesville, VA, USA.

Abdul Rahman Omar
Institute of Bioscience, University Putra Malaysia, UPM Serdang, Selangor, 43400, Malaysia.
Faculty of Veterinary Medicine, Universiti Putra Malaysia, UPM Serdang, Selangor, 43400, Malaysia.

Abdelwahab Omri
The Novel Drug and Vaccine Delivery Systems Facility, Department of Chemistry and Biochemistry, Laurentian University, Sudbury, Ontario, Canada.

Jason A. Papin
Department of Biomedical Engineering, University of Virginia, Health System, Charlottesville, VA, USA.

Jiri Petrak
Department of Pathological Physiology, First Faculty of Medicine, Charles University in Prague, Prague, Czech Republic.

Patricia B. Petray
Servicio de Parasitología y Enfermedad de Chagas. Hospital de Niños Dr. Ricardo Gutiérrez. Gallo 1330, Ciudad Autónoma de Buenos Aires, Argentina.

Jacek Puchałka
Synthetic and Systems Biology Group, Helmholtz Center for Infection Research (HZI), Braunschweig, Germany.

Mohamed Rajik
Department of Microbiology, Faculty of Biotechnology and Biomolecular Sciences, University Putra Malaysia, UPM Serdang, Selangor, 43400, Malaysia.

Daniela Regenhardt
Environmental Microbiology Group, Helmholtz Center for Infection Research (HZI), Braunschweig, Germany.

Francesca Ricci
Dipartimento di Genetica e Biologia Molecolare, Sapienza Università di Roma, Roma, Italy.

Gianfranco Risuleo
Dipartimento di Genetica e Biologia Molecolare, Sapienza Università di Roma, Roma, Italy.

Eder L. Romero
Programa de Nanomedicinas, Universidad Nacional de Quilmes, Roque Saenz Peña 352, Bernal, Buenos Aires, Argentina.

Diana I. Roncaglia
Programa de Nanomedicinas, Universidad Nacional de Quilmes, Roque Saenz Peña 352, Bernal, Buenos Aires, Argentina.

Kennan Kellaris Salinero
Department of Plant and Microbial Biology, University of California, Berkeley, CA 94720, USA.
Yámana Science and Technology, Washington, DC 20009, USA.

C. A. Sargent
Department of Pathology, University of Cambridge, Cambridge CB2,1QP, UK.

Ajay Singh
The Department of Pathology, Case Western Reserve University, Cleveland, OH, USA.

Zacharias E. Suntres
The Novel Drug and Vaccine Delivery Systems Facility, Department of Chemistry and Biochemistry, Laurentian University, Sudbury, Ontario, Canada.
Medical Sciences Division, Northern Ontario School of Medicine, Lakehead University, Thunder Bay, Ontario, Canada.

Kenneth N. Timmis
Environmental Microbiology Group, Helmholtz Center for Infection Research (HZI), Braunschweig, Germany.

Stephan Trong
Genomics Division, DOE Joint Genome Institute, Walnut Creek, CA 94598, USA.

Marta Wanarska
Department of Microbiology, Chemical Faculty, Gdańsk University of Technology, Narutowicza 11/12, 80-952 Gdańsk, Poland.

Fei Wang
Drug Discovery and Design Centre, State Key Laboratory of Drug Research, Shanghai Institute of Material Medical, Graduate School of Chinese Academy of Sciences, Shanghai 201203, P.R. China.

Tiansong Wang
Department of Chemical Engineering, University of Washington, Seattle, WA, USA.
Department of Microbiology, University of Washington, Seattle, WA, USA.

Kaifeng Wu
Key Laboratory of Laboratory Medical Diagnostics, Ministry of Education, Faculty of Laboratory Medicine, Chongqing Medical University, Chongqing 400016, P.R. China

Qiangwei Xia
Department of Chemical Engineering, University of Washington, Seattle, WA, USA.
Department of Microbiology, University of Washington, Seattle, WA, USA.
University of Wisconsin-Madison, Department of Chemistry, Madison, WI, USA.

Hua Xie
Department of Oral Biology, University of Florida, Gainesville, FL, USA.

Nanlin Yin
Key Laboratory of Laboratory Medical Diagnostics, Ministry of Education, Faculty of Laboratory Medicine, Chongqing Medical University, Chongqing 400016, P.R. China.

Yibing Yin
Key Laboratory of Laboratory Medical Diagnostics, Ministry of Education, Faculty of Laboratory Medicine, Chongqing Medical University, Chongqing 400016, P.R. China.

J. F. Yuan
Key Lab of Agricultural Animal Genetics, Breeding and Reproduction of Ministry of Education, HuaZhong Agricultural University, 430070, P.R. China.
Department of Pathology, University of Cambridge, Cambridge CB2,1QP, UK.
College of Veterinary, South China Agricultural University, 510642, P.R. China.

Khatijah Yusoff
Department of Microbiology, Faculty of Biotechnology and Biomolecular Sciences, University Putra Malaysia, UPM Serdang, Selangor, 43400, Malaysia.
Institute of Bioscience, University Putra Malaysia, UPM Serdang, Selangor, 43400, Malaysia.

G. H. Zhang
College of Veterinary, South China Agricultural University, 510642, P.R. China.

S. J. Zhang
Key Lab of Agricultural Animal Genetics, Breeding and Reproduction of Ministry of Education, HuaZhong Agricultural University, 430070, P.R. China.
Department of Pathology, University of Cambridge, Cambridge CB2,1QP, UK.

Xuemei Zhang
Key Laboratory of Laboratory Medical Diagnostics, Ministry of Education, Faculty of Laboratory Medicine, Chongqing Medical University, Chongqing 400016, P.R. China.

Weiliang Zhu
Drug Discovery and Design Centre, State Key Laboratory of Drug Research, Shanghai Institute of Material Medical, Graduate School of Chinese Academy of Sciences, Shanghai 201203, P.R. China.

List of Abbreviations

AcCoA	Acetyl coenzyme A
AD	Activation domain
AIV	Avian influenza viruses
AKT1	v-Akt murine thymoma viral oncogene homolog 1
ALS	Amyotrophic lateral sclerosis
APLP1	Amyloid beta (A4) precursor-like protein 1
ARC	Archaeosomes
ATCC	American Type Culture Collection
BD	Binding domain
Bhp	Biphenyl/polychlorinated biphenyl
BHV-1	Bovine herpesvirus 1
BM	Black mud
Bp	Basepair(s)
BPG	Bisphosphatidyl glycerol
BSA	Bovine serum albumin
Bss	Benzylsuccinate synthase
BTEX	Benzene, toluene, ethylbenzene, and xylene compounds
BUB1	Budding uninhibited by benzimidazoles 1
CA	Cellular automaton
CALM3	Calmodulin 3
CAMH	Cation-adjusted Mueller-Hinton
CB	Constraint-based
CDC42	Cell division cycle 42
CDK4	Cyclin-dependent kinase 4
CDK7	Cyclin-dependent kinase 7
cDNA	Complementary deoxyribonucleic acid
CF	Cystic fibrosis
Chol	Cholesterol
CLSI	Clinical and Laboratory Standards Institute
CLSM	Confocal laser scanning microscopy
CLTB	Clathrin light chain
CLTC	Clathrin heavy chain
CNS	Central nervous system
COG	Clusters of orthologous gene
cRBCs	Chicken red blood cells

Ct	Threshold cycle
DAVID	Database for Annotation, Visualization and Integrated Discovery
DFO	Desferrioxamine
DMPC	1,2-Dimyristoyl-sn-glycero-3-phosphocholine
DNA	Deoxyribonucleic acid
DPPC	1,2-Dipalmitoyl-sn-glycero-3-phosphocholine
DSMZ	Deutsche Sammlung von Mikroorganismen and Zellkulturen
ED	Entner-Doudoroff
EE	Encapsulation efficiency
ESI-MS	Electrospray ionization-mass spectrometry
EST	Expressed sequence tag
FAC	Ferric ammonium citrate
FBA	Flux balance analysis
FBS	Fetal bovine serum
FBXW7	F-Box and WD repeat domain containing 7
FDR	False discovery rate
FGF	Fibroblast growth factor
FNR	False negative rate
FOS	v-Fos FBJ murine osteosarcoma viral oncogene homolog
F-PMB	Free polymyxin B
F-TOB	Free tobramycin
FVA	Flux variability analysis
GADD45	Growth arrest and DNA-damage-inducible alpha
GCs	Gray crystals
GPI	Glycosylphosphatidyl inositol
GPR	Gene-protein-reaction
GSS	Gerstmann-Straussler-Scheinker disease
HA	Hemagglutinin
HI	Hemagglutination inhibition
HK	Histidine kinase
HMM	Hidden Markov models
HPP	Hydroxyphenyl propionate
HPTS	8-Hydroxypyrene-1,3,6-trisulfonic acid
HS3ST5	Heparan sulfate (glucosamine) 3-O-sulfotransferase 5
HSPB2	Heat shock 27 kDa protein 2
HSPC	Hydrogenated phosphatidylcholine from soybean
HSPD1	Heat shock 60 kDa protein 1
HSV-1	Herpes simplex virus1
HTVS	High throughput virtual screening

HveC (PVRL1)	Herpesvirus entry mediator C (poliovirus receptor-related 1)
HveD (PVR)	Herpesvirus entry mediator D (poliovirus receptor)
ICL	Isocitrate lyase
ID2	Inhibitor of DNA binding 2
ID3	Inhibitor of DNA binding 3
ID4	Inhibitor of DNA binding 4
IPTG	Isopropyl-1-thio-β-D-galactopyranoside
IRPs	Iron regulatory proteins
KEGG	Kyoto Encyclopedia of Genes and Genomes
LANL	Los Alamos National Laboratory
LB	Luria–Bertani
LCP2	Lymphocyte cytosolic protein 2
LGA	Lamarchian genetic algorithm
LIP	Labile iron pool
LOWESS	Locally weighted scatterplot smoothing
LP	Linear programming
L-PMB	Liposomal polymyxin B
LPS	Lipopolysaccharide
LTA	Lipoteichoic acid
L-TOB	Liposomal tobramycin
MAPK	Mitogen-activated protein kinase
MBC	Minimal bactericidal concentration
mc	Metabolic cost
MCM7	Minichromosome maintenance complex component 7
MDR	Multi-drug resistant
MFC	Microbial fuel cells
MIC	Minimal inhibitory concentration
μL	Microliter(s)
MILP	Mixed-integer linear programming
MOSTI	Ministry of Science, Technology, and Innovation
mRNA	Messenger ribonucleic acid
MS	Mass spectrometry
MS^1	First stage of tandem mass spectrometry
MS^2	Second stage of tandem mass spectrometry
mTOR	Mechanistic target of rapamycin
NA	Neuraminidase
Na	Sodium
NAI	Neuraminidase inhibitors
NCBI	National Center for Biotechnology Information

NDV	Newcastle disease virus
NEFH	Neurofilament, heavy polypeptide
NEFL	Neurofilament, light polypeptide
NFE2L2	Nuclear factor (erythroid-derived 2)-like 2
NO	Nitrous oxide
OCV	Open circuit voltage
ONPGlu	*o*-Nitrophenyl-β-D-glucopyranoside
Orfs	Open reading frames
pAb	Polyclonal antibody
PBS	Phosphate buffered saline
PCNA	Proliferating cell nuclear antigen
PCR	Polymerase chain reaction
PDB	Protein Data Bank
PDCD8	Programmed cell death 8
PDH	Pyruvate dehydrogenase
PDMS	Polydimeyhylsiloxane
PEM	Proton exchange membrane
PG	Phosphatidylglycerol
PGD	Pseudomonas Genome Database
PGP-Me	Phosphatidylglycerophosphate-methyl ester
PHAs	Polyhydroxyalkanoates
PI	Post-inoculation
PIK3R1	Phosphoinositide-3-kinase regulatory subunit 1
PNPG	*p*-Nitrophenyl-β-D-galactopyranoside
PP	Pentose phosphate
PPP	Pentose phosphate pathway
PPP2CA	Protein phosphatase 2 catalytic subunit alpha isoform
PPP2CB	Protein phosphatase 2 catalytic subunit beta isoform
PPP3CA	Protein phosphatase 3 catalytic subunit alpha isoform
PRKACA	Protein kinase, cAMP-dependent, catalytic, alpha
PrP^C	Prion protein
PRRS	Porcine reproductive and respiratory syndrome
PRRSV	Porcine Reproductive, and Respiratory Syndrome Virus
PRV	Pseudorabies virus
PSSA	Pseudo steady-state assumption
PVRL3	Poliovirus receptor-related 3
Py	Polyomavirus
qRT	Quantitative real time

qRT-PCR	Quantitative real-time reverse transcriptase polymerase chain reaction
QS	Quorum sensing
QTL	Quantitative trait locus
rhDNase	Recombinant human DNase
RMSD	Root-mean-square deviation
ROS	Reactive oxygen species
RR	Response regulator
rRNA	Ribosomal ribonucleic acid
RSP	Rock-scissors-paper
RT-PCR	Reverse transcription-polymerase chain reaction
RV	Resveratrol
SBVS	Structure-based virtual screening
sCJD	Sporadic Cruetzfeldt–Jakob disease
SCX	Strong cation exchange
S-DGD	Sulfated diglycosyl diphytanylglycerol diether
SERPINE1	Plasminogen activator inhibitor, type I
SIH	Salicylaldehyde isonicotinoyhydrazone
SPP1	Secreted phosphoprotein 1
TCSs	Two-component systems
TEMED	N,N,N',N'-Tetramethylethylenediamine
TGFβ	Transforming growth factor, beta
THBS4	Thrombospondin 4
TIGR-CMR	The Institute for Genomic Research Comprehensive Microbial Resource
TM	Transmembrane
Tmo	Toluene mono-oxygenase
TNFRSF	Tumor necrosis factor receptor superfamily
ToBiN	Toolbox for Biochemical Networks
TPL	Total polar lipids
tRNA	Transfer ribonucleic acid
VIMSS	Virtual Institute for Microbial Stress and Survival
VZV	Varicellovirus

Contents

Chapter 1

Microfabricated Microbial Fuel Cell Arrays

Huijie Hou, Lei Li, Younghak Cho, Paul de Figueiredo, and Arum Han

INTRODUCTION

Microbial fuel cells (MFCs) are remarkable "green energy" devices that exploit microbes to generate electricity from organic compounds. The MFC devices currently being used and studied do not generate sufficient power to support widespread and cost-effective applications. Hence, research has focused on strategies to enhance the power output of the MFC devices, including exploring more electrochemically active microbes to expand the few already known electricigen families. However, most of the MFC devices are not compatible with high throughput screening for finding microbes with higher electricity generation capabilities. Here, we describe the development of a microfabricated MFC array, a compact and user-friendly platform for the identification and characterization of electrochemically active microbes. The MFC array consists of 24 integrated anode and cathode chambers, which function as 24 independent miniature MFCs and support direct and parallel comparisons of microbial electrochemical activities. The electricity generation profiles of spatially distinct MFC chambers on the array loaded with *Shewanella oneidensis* MR-1 differed by less than 8%. A screen of environmental microbes using the array identified an isolate that was related to *Shewanella putrefaciens* IR-1 and *Shewanella* sp. MR-7, and displayed 2.3-fold higher power output than the *S. oneidensis* MR-1 reference strain. Therefore, the utility of the MFC array was demonstrated.

The MFCs are devices that generate electricity from organic compounds through microbial catabolism [1-3]. A typical MFC contains an anaerobic anode chamber and an aerobic cathode chamber separated by a proton exchange membrane (PEM), and an external circuit connects the anode and the cathode [4, 5]. Electrochemically active microbes ("electricigens") reside within the anaerobic anode chamber. Electrons, generated during microbial oxidization of organic compounds, are delivered to the MFC anode via microbial membrane-associated components [3, 6], soluble electron shuttles [7, 8], or nanowires [9, 10]. Biofilms that support close physical interactions between microbial membranes and anode surfaces are also important for MFC power output [11]. Electrons flow from the anode to the cathode through the external electrical circuit. In parallel, protons generated at the anode diffuse through the PEM and join the electrons released to the catholyte (e.g., oxygen, ferricyanide) at the cathode chamber [1]. This electron transfer event completes the circuit.

The MFCs have generated significant excitement in the bioenergy community because of their potential for powering diverse technologies, including wastewater treatment, and bioremediation devices [12, 13], autonomous sensors for long-term operations in low accessibility regions [14, 15], mobile robot/sensor platforms [16], microscopic drug-delivery systems [17] and renewable energy systems [18]. In addition, MFCs hold significant promise for supporting civilian and combat operations in hostile environments [16]. Therefore, the development of efficient MFCs that are capable of producing high power densities remains an area of intense research interest. However, economical applications of existing MFCs are limited due to their low power output [19], which ranges from 100 to 1,000 W/m^3 [20, 21].

Important strategies for enhancing MFC performance include engineering optimized microbes (and microbial communities) for use in these devices [22] and improving cultivation practices for these organisms [23, 24]. To date, detailed description of individual microbe performance in MFCs has been limited to a surprisingly small number of organisms [25]. The MFCs that are fed by sediment and wastewater nutrient sources and that exploit mixed microbial consortia for electricity generation have been described [26, 27]. However, with the conventional two-bottle MFCs, characterization of the electrochemical activities of the microbial species in these consortia has not been possible because these conventional MFCs are not suitable for parallel analyses due to their bulkiness. To address this issue, MFC systems that support parallel, low cost, and reproducible analysis of the electrochemical activities of diverse microbes are required. High throughput microarrays, including DNA microarrays, protein microarrays, and cell arrays, are powerful platforms for screening and analyzing diverse biological phenomena [28]. Various MFC platforms, including miniature MFC devices that enable parallel comparison of electricity generation in MFCs, are emerging [29, 30]. However, state of the art microfabrication and highly integrated parallel measurement approaches [31, 32] have not yet been exploited to construct an MFC array with highly consistent architecture and performance.

Here we describe our development of a compact and user-friendly MFC array prototype capable of examining and comparing the electricity generation ability of environmental microbes in parallel. The parallel analyses platform can greatly speed up research on electricigens. Importantly, the array was fabricated using advanced microfabrication approaches that can accommodate scale-up to massively parallel systems. The MFC array consisted of 24 integrated cathode and anode pairs as well as 24 cathode and anode chambers, which functioned as 24 independent miniature MFCs. We validated the utility of our MFC array by screening environmental microbes for isolates with enhanced electrochemical activities. Our highly compact MFC array enabled parallel analyses of power generation of various microbes with 380 times less reagents, and was 24-fold more efficient than conventional MFC configurations. This effort identified a *Shewanella* isolate that generates more than twice as much power as the reference strain when tested in both conventional and microfabricated array formats.

MATERIALS AND METHODS

Twenty-four Well MFC Array Design

Figure 2A shows the schematic illustration of the MFC array. The array was microfabricated using micromachining and soft lithography techniques. The 24-well device was composed of layered functional compartments in which microbe culture wells were embedded. Each microliter-scale microbe culture well was combined with individually addressable anode and cathode electrodes and functioned as a separate MFC. The layers included anode electrodes, anaerobic microbe culture chambers (anode wells), a PEM, cathode chambers, and cathode electrodes (Figure 2A). The assembled device (Figure 2B–E) with acrylic supporting frames was coupled to a load resistor circuit board and a digital multimeter through a computer controlled switch box module and a data acquisition system (Figure 2F).

Twenty-four Well MFC Array Electrode Layer Microfabrication

An acrylic master mold having 4 × 6 arrays of pillars (diameter: 7 mm, height: 4 mm) was fabricated with a rapid prototyping machine (MDX-40, Roland Inc., Los Angeles, CA). PDMS precursor solution (Sylgard 184, Dow Corning, Auburn, MI) prepared by mixing base and curing agent at 10:1 ratio (v/v) was poured onto the acrylic master mold. After curing for 30 min at 85°C, the resulting polymerized polydimeyhylsiloxane (PDMS) slab was peeled off, creating an inverse replica of the acrylic master mold. The cathode layer was prepared by aligning and permanently bonding a PDMS well layer onto a patterned electrode layer. Platinum loaded carbon cloth (10%, A1STD ECC, BASF Fuel Cell, Inc., Somerset, NJ) was cut to the size of a well (diameter: 7 mm) and bonded on top of the Au electrode pads in the cathode electrode layer using silver paste (Structure probe, Inc., West Chester, PA). Cathode and anode electrode layers of the 24-well MFC array were fabricated using standard microfabrication techniques. The fabricated cathode and anode electrode substrates had 24 individually addressable electrodes, each having an 8 mm diameter disk pattern. Wires were then soldered to all contact pads to provide electrical interconnects between the MFC arrays and the voltage measurement setup.

Assembly of the MFC Array System

The 24-well MFC array system consisted of a 24-cathode array layer, a cathode well layer, a PEM, an anode well layer, and a 24-anode array layer. Cathode layer consisted of a PDMS well layer permanently bonded on an electrode layer. Platinum (Pt) loaded carbon cloth was cut to the size of the well (diameter: 7 mm) and bonded on top of the Au electrode pads. The anode layer consisted of three layers: two PDMS layers fabricated as described above and an acrylic layer (8 mm thick) having 4 × 6 arrays of through-holes (7 mm diameter) fabricated by a rapid prototyping machine. The two PDMS layers were placed on both sides of the acrylic layer. The rigid acrylic layer served as a support layer that could be clamped tightly in the subsequent assembly step. The cathode layer, activated PEM, and the anode layer were then assembled together. To assemble these layers together, a top and bottom acrylic support frame that could be used to tightly clamp all layers together in between was cut out using a rapid prototyping machine. The sequence of images (Figure 2B–F) shows how all parts of a 24-well MFC array system was assembled.

Organisms, Media and, Growth Conditions

Shewanella oneidensis MR-1 was obtained from American Type Culture Collection (ATCC, Manassas, VA). Environmental bacteria used for screening were isolated from eight different samples (soil and water) obtained from Lake Somerville (N30°30′09″ and E96°64′28″), Brazos River (N30°55′84″ and E96°42′24″; N30°62′64″ and E96°55′13″) and Lake Finheather (N30°64′93″ and E96°37′54″) around College Station, Texas.

Isolation and Pre-screening of Environmental Microbes

We performed a pre-screening for electrochemically active microbes. Each diluted sample was plated on nutrient agar and incubated under anaerobic conditions. The resulting 50–100 microbial colonies per plate were then used for plating on nutrient agar containing 100 μM Reaction Black 5, an azo dye that resulted in dark blue color of the media. After 3 days of incubation, a total of 26 colonies formed discoloration halos out of about 1,500 colonies plated for each of the eight environmental samples. The discoloration of the dye indicated reduction capability of the microbes. A total of 13 isolates were selected for MFC array screening. Un-inoculated medium was used as the negative control.

The 16s rDNA Amplification and Phylogenetic Analyses for Environmental Isolates

Colony PCRs were performed using different environmental isolates as the templates. The PCR products were then purified and sequenced with primers 11F and 1492R. The 16S rDNA sequences were BLAST searched against the GenBank database and the top hit for each isolate were used for alignment and phylogenetic tree generation. Sequences of the 16S rDNA of 15 members of genus *Shewanella* similar to 7Ca were aligned and phylogenetic tree was constructed among selected *Shewanella*. A matrix of pairwise genetic distances by the maximum-parsimony algorithm and the neighbor-joining method was used to generate phylogenetic trees.

The MFC Array Characterization and Data Acquisition

Two characterization methods were used to evaluate electricity generation from each of the 24 MFC wells. First, 24 1 MΩ fixed load resistors were connected to each of the MFC wells and voltage across these resistors were recorded. Load resistance of 1 MΩ was selected for MFC characterization because power output of *S. oneidensis* MR-1 at this resistance was close to maximum and the fabricated MFC array showed a steady state current output much faster than using resistors with lower resistances. A switch box module having an integrated digital multimeter (PXI-2575, PXI-4065, National Instruments, Austin, TX) and controlled through LabView™ (National Instruments, Austin, TX) was used to continuously measure voltages across the 24 load resistors that were connected individually to the 24 MFC array. The measured voltages were converted to current densities (mA/m^2, electrode area: 0.385 cm^2) using Ohm's law, and power densities were calculated using $P = VI/A$ (V: voltage, I: current, A: electrode area).

Full characterization of an MFC requires a current density versus power density plot, which can be obtained when measuring voltages across varying resistors. In this second characterization method, twenty-four 100 KΩ variable resistors (652-3296Y-1-104LF, Mouser Electronics, Mansfield, TX) were connected in series with twenty-four

2 MΩ variable resistors (652-3296Y-1-205LF, Mouser Electronics, Mansfield, TX) in pairs on a circuit board, connected correspondingly to the 24 MFC wells. For environmental screening of microbes using the MFC array, both the primary screening and the secondary confirmation started 1,000 min after loading microbes into the MFC array loaded with 1 MΩ resistors. One, 10, 20, 50, 100, 200, 500, 1,000, and 2,000 KΩ loading resistors were used and voltages across these resistors were continuously recorded.

RESULTS AND DISCUSSION

We recognized that selection of an appropriate anode material would be critical to the successful development of an MFC array and therefore initiated our studies by examining the performance of a commonly used anode material, carbon cloth in comparison to gold in a conventional MFC device (Figure 1). We used the model facultative anaerobe *S. oneidensis* MR-1 for these experiments because this organism had previously proven useful for the development of MFC applications [37]. The MFC power output was monitored for 5 hr after bacteria were introduced into the device. When the MFC was loaded with a 20 KΩ resistor, the gold electrode supported maximal power density of 3.77 ± 0.02 mW/m² at a current density of 16.47 ± 0.04 mA/m². A high standard deviation was observed (10% deviation). However, the MFC with gold anode displayed greater reproducibility (3.1% deviation). The open circuit voltage (OCV) of MFCs containing gold and carbon cloth anodes was 514 ± 12 mV (mean ± SE, n = 3) and 538 ± 51 mV (mean ± SE, n = 4), respectively. These results indicated that the OCV of the MFC with the gold anode was comparable to the corresponding OCV with the carbon cloth anode. However, OCV measurements with the carbon cloth anode displayed greater variance.

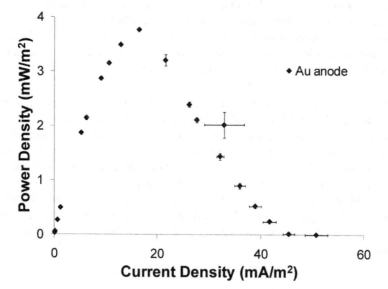

Figure 1. Au working as anode of MFC. Power density versus current density from an MFC with gold anode (n = 3).

An important hurdle to overcome in the development of MFC array systems is the identification of an electrode material that is durable, conductive, biocompatible, and easily fabricated [33]. Graphite, in the form of carbon cloth or graphite felt, has typically been the material of choice for the construction of MFC anodes, and conductive elements such as manganese, iron, quinines, and neutral red have been incorporated in graphite electrodes to significantly increase power output [33, 34]. However, graphite is not suitable for microfabricated MFC array systems [35]. The surfaces of graphite electrodes are non-uniform and difficult to pattern in small-scale devices. This non-uniformity thwarts efforts to compare performances between individual miniaturized MFCs. In addition, graphite materials are not compatible with most microfabrication technologies. Recently, gold has been identified as a potential material for MFC anode development [35]. Gold is highly conductive, can be vapor deposited, and is compatible with a wide array of conventional microfabrication modalities [36]. Thus, gold is a very attractive anode candidate for the development of an MFC screening platform. Our result showed that the MFC using gold as the anode material gave more reproducible results than its carbon cloth counterpart, a critical feature for side-by-side comparison in the MFC array. We therefore used gold as the anode material to develop the MFC array prototype.

Biofilms, when established on the anode of MFCs, enhance MFC performance when some microbial systems (including *S. oneidensis* MR-1) are employed. The enhanced performance has been suggested to result from the enhanced ability of biofilms to exploit close physical contacts between microbial membranes and the anode surface for electron transfer [11]. To investigate whether biofilms form on the surface of gold electrodes, light and fluorescence microscopy images of the electrode were captured 1 hr and 5 hr post-inoculation (PI). One hour PI, microbes started attaching to the gold electrode surface. Five hours later, an attached *S. oneidensis* biofilm was observed. Scanning electron micrographs of the electrode surface confirmed microbial attachment. Therefore, gold electrode supports *S. oneidensis* biofilm formation, and moreover, enables reproducible and consistent electrochemical activity to be measured when this model organism is used.

We were encouraged by our finding that gold electrodes can be employed in MFC devices, and exploited this material to develop an MFC array (Figure 2A). The array was successfully microfabricated using micromachining and soft lithography techniques. Performance and reproducibility of the MFC array were initially assessed by loading *S. oneidensis* MR-1 into the device and then measuring the electrical output. The current densities for negative control (un-inoculated medium) and *S. oneidensis* MR-1 chambers were 0.40 ± 0.01 mA/m^2 (mean $\pm$ SE, n = 4) and 1.80 ± 0.24 mA/m^2 (mean $\pm$ SE, n = 4), respectively (Figure 3A). Therefore the MFC array reproducibly measured the electrochemical activities of this microbial system (less than 14% of variance).

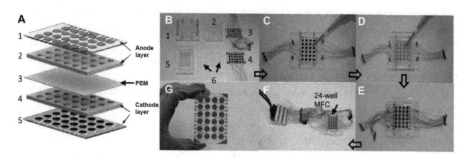

Figure 2. Illustration of the MFC array and its assembly steps. (A) Illustration of 24-well microbial fuel cell (MFC) array. Composed of an anode layer (1: anode electrode layer, 2: anode well layer), a proton exchange membrane (3: PEM), and a cathode layer (4: cathode well layer, 5: cathode electrode layer). (B)–(F) Microbial fuel cell array assembly. (B) Individual layers of the MFC array: acrylic support frames (1 and 2), cathode layer (3), anode electrode substrate (4), anode well layer (5 sandwiched by 6). (C) Assembly of the acrylic frame (1) with the cathode layer (3), followed by cathode solution loading. (D) Anode well layer and PEM assembly followed by microbe loading. (E) A fully assembled MFC array with the anode electrode layer (4) and acrylic frame (2) capping the anode wells. (F) Fully assembled MFC array connected to load resistors and a data acquisition system. (G) An MFC array device with no acrylic frame.

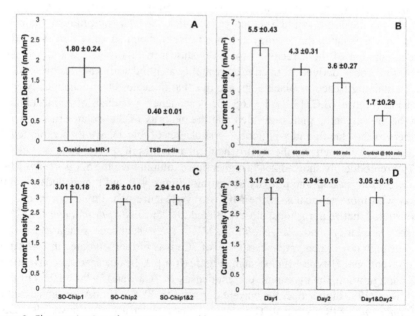

Figure 3. Characterization of current generated by *S. oneidensis* MR-1 in an MFC array. (A) Current densities generated by *S. oneidensis* MR-1 with PBS as the cathode solution at 350 min, TSB medium was used as the negative control (n = 4). (B) Repeatability of current densities generated by *S. oneidensis* MR-1 with ferricyanide as the cathode solution at different times after loading. TSB medium was used as the negative control (n = 5). (C) Chip-to-chip repeatability of current densities generated by *S. oneidensis* MR-1 with ferricyanide (100 mM) as the cathode solution at 1,000 min (n = 4 for each chip). (D) Batch-to-batch repeatability of current densities generated by *S. oneidensis* MR-1 with ferricyanide (100 mM) as the cathode solution at 1,000 min (n = 8 for each day). Means and standard errors were indicated above the bars (mean ± SE).

To increase the current, we used 100 mM ferricyanide at the cathode as the electrolyte due to the higher concentration of the electron acceptor in the cathode solution [4]. Under this condition, *S. oneidensis* MR-1 produced a current density of 5.54 ± 0.43 mA/m^2 (mean $\pm$ SE, n = 5) 100 min after loading microbes into the device (Figure 3B). Over time, the current density gradually dropped, but remained higher than that of the negative control wells containing medium only. The electricity generation profiles of spatially distinct wells measured at multiple time points during 15 hr of operation differed by less than 8%, demonstrating that individual wells of the MFC array displayed comparable performance characteristics.

Performances of the same microbial culture were also compared in different MFC arrays. Two arrays with the same configuration were tested with the same microbial culture (OD$_{600}$ = 0.8) and showed current density of 2.94 ± 0.16 mA/m^2 (mean $\pm$ SE, n = 8) (Figure 3C). Thus the MFC arrays showed chip-to-chip variances of less than 5.4%. Performances of the same chip with microbial cultures of the same concentration (OD$_{600}$ = 0.8) prepared on different days were also examined (Figure 3D). The current densities on two different days were 3.05 ± 0.18 mA/m^2 (mean $\pm$ SE, n = 16), showing a 5.9% variance. Therefore, the MFC array provided a platform for reproducible experimentation.

Encouraged by the performance and reproducibility of the MFC array, we examined whether the device could be employed to quickly screen environmental microbes for individual isolates that display enhanced electrochemical activities. A schematic representation of the screening process is shown in Figure 4A. We pre-screened ~12,000 microbes derived from environmental (water and soil) samples on solid medium containing Reaction Black 5, an azo dye that indicates electrochemically active organisms (Figure 4B,C) [39]. The 16S rDNA sequencing analysis of 26 hits obtained from the pre-screening plates revealed that the majority of the isolates (n = 10) were members of the Bacilli and γ-proteobacteria classes (Table 1). We then exploited the MFC arrays to characterize the electrochemical activities of several isolates. One isolate 7Ca reproducibly showed 266% higher power output than the *S. oneidensis* MR-1 reference strain in both the primary screening (Figure 4E) and the secondary confirmation with more replicates in the MFC arrays (Figure 4F). Phylogenetic analysis demonstrated that 7Ca was most closely related to *Shewanella putrefaciens* IR-1 (98% sequence similarity) and *Shewanella* sp. MR-7 (98% sequence similarity) (Figure 4D). The high power generation capability of 7Ca was further validated in 24-hr trials in a conventional H-type MFC system (Figure 4G) [6]. In our specific conventional MFC configuration, the maximum current density of 7Ca was 169.00 ± 10.60 mA/m^2, which was 217% higher than the current density (78.00 ± 7.30 mA/m^2) generated by the *S. oneidensis* MR-1 control. The maximum power density of 7Ca was also 233% higher than this reference strain. Although we used gold as the anode material in the MFC array and carbon cloth as the anode material in the H-type MFC system, the power density increases of 266% in the MFC array and 233% increase in the H-type MFC system showed that findings in our MFC array system can be translated to larger scale conventional systems. Thus, insights garnered using gold anodes in miniaturized MFCs can be transferred to conventional MFC formats using carbon anodes.

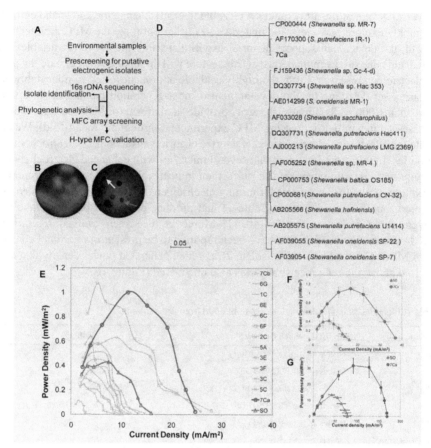

Figure 4. Environmental sample screening with the MFC array. (A) A schematic representation of the screening process for the environmental microbes with enhanced electricity generation capacities. (B) and (C) Electrochemically active microbes cause discoloration of an azo dye, reactive black 5 in the nutrient agar screening plate. (B) *S. oneidensis* MR-1 was used as the control; (C) A representative plate with a putative and non-putative microbe isolate indicated by arrow. (D) Phylogenetic tree based on 16S rDNA sequences indicating the relationship of various *Shewanella* species. (E)–(G) Screening of environmental isolates using the 24-well MFC array. (E) Screening of 13 environmental isolates with *S. oneidensis* MR-1 as the positive control (SO) using two 24-well MFC arrays in parallel. The average power density of two replicates was shown for each isolate. (F) The power density of 7Ca compared to the *S. oneidensis* MR-1 reference strain (n = 8). (G) Validation of current generation by 7Ca and *S. oneidensis* MR-1 in conventional MFCs (n = 4).

We have demonstrated that a microfabricated MFC array system can be exploited to rapidly screen and characterize microbial electrochemical activity. The universal design of our system has several attractive features. First, the microbe culture chamber was easy to assemble and reusable. The PDMS and electrode could be used at least ten times without degradation and the acryl anode chamber could be used more than ten times with proper cleaning. Therefore, the device has the potential for widespread adoption. Second, because the individual MFC chambers in the array hold a small volume, 380-fold fewer reagents are required than conventional MFC devices. Third, the universal design of the array makes it possible to easily change the anode and cathode to quickly explore new electrode materials to further optimize the MFC architecture. Fourth, the device can support factorial experiments in which several variables are tested and compared simultaneously. This feature will dramatically accelerate the pace of electricigen research. Fifth, the highly scalable approaches used to microfabricate the array set the stage for the development of next-generation parallel devices with more than 1,000 wells. Finally, the array provides a platform for MFC performance to be assessed in parallel (increasing MFC experimental throughput by 24-fold). We exploited this feature to identify and characterize electricigens with high electrochemical activities. In this regard, the fact that several microbes with enhanced electrochemical activity were rapidly uncovered in our screen indicates that the natural environment constitutes a plentiful reservoir for mining electricigens. For example, our screen uncovered four *Pseudomonas* sp. (Table 1) that produce phenazine-based metabolites that can serve as electron shuttles [40, 41]. Moreover, the screen resolved *Bacillus* sp. and *Aeromonas hydrophila* that have been reported to be present in microbial consortia of MFCs [26]. A *Shewanella* isolate 7Ca, which generated power density that was 2.3-fold above the reference strain, was also uncovered (Figure 4E).

Table 1. Identities of environmental isolates obtained from pre-screening.

Isolate	Top Hit (Genbank Number)	Identity%
1B	Bacillus sp. RC33 (F J263036.1)	100
1C	Bacillus sp. Wl-17 (FJ560473.1)	100
1D	Bacillus niacini strain YM 1 C7 (EU221338. 1)	97.93
2A#	Enterobacter sp. CTSP29 (EU855207 .1)	99
2B	Bacillus sp. SXSl (00227355.1)	
2C, 3B, 4A, 5A,5B	Bacillus thuringiensis serovar tenebrionis (EU429671.1)	100
3A	Bacterium 3A13 (00298760.1)	99
3C	Arthrobacter sp. FB24 (EU147009.1)	100
3E, 3F	Bacillus sp. A 1 (2008) (F J535468. 1)	100
5C	Bacillus sp. BMl-4 (FJ528077.1)	100
6A,6C*	Pseudomonas putida strain J3 1 2 (EF2032 1 0.1)	98.50
6B*· #	Stenotrophomonas maftophilia strain CMG3098 (EU048328.1)	98.99
6E*	Pseudomonas pfecogfossicida strain 519 (00095907.1)	98.52
6F*	Pseudomonas sp. lm 1 0 (EU240462. 1)	98.72
6G*	Pseudomonas sp. GNE25 (AM397659. 1)	95
7A*· #	Aeromonas sp. LOl 51 (AM913921.1)	97.34
7Ca*	Shewanella sp. Hac353 (00307734. 1)	99
7Cb	Bacillus pumifus strain TPRl 8 (EU373436. 1)	99
8A*· #	Paenibacillus sp. oral clone CA007 (AF385540. 1)	92
8B*·#	Aeromonas hydrophi/a strain IB343 (EU770277.1)	99

*isolates belonging to y-proteobacteria (1 0 out of 26 isolates).
#putative pathogenic isolates not subject to electricity generation screening using MFC array.

It is intriguing to speculate on the molecular mechanisms that contribute to the enhanced electrochemical activity of 7Ca. For example, the presence of gain of function mutations affecting biofilm formation, nanowire formation, cell membrane electron transfer, metabolic, and respiration capacities, regulatory components for these functions, and/or combinations of all the above, could contribute to the observed power output. In this regard, it is notable that previous genetic studies in *S. oneidensis* MR-1 revealed several genes that are directly involved in electricity generation, including *mtrA*, *mtrB*, *omcA/mtrC*, *cymA*, *fur*, and *crp*. Deletion of these genes caused severe reductions in current production [42]. Gain of function mutations at these candidate loci may therefore confer enhanced MFC power generating properties. Similarly, an engineered strain of *Geobacter sulfurreducens* generated by Izallalen et al. displayed enhanced electrochemical activity due to increased respiration rates [22]. Despite these molecular insights, there has been a paucity of functional studies that directly measure the electrochemical activities of environmental microbes. In fact, the electrochemical activities of only a handful of microbial species have been characterized in MFC systems [3]. Although MFCs using wastewater treatment and sediment nutrient sources have defined electrochemically active microbial consortia [43], the electrochemical activities of individual species within these consortia remain largely unexplored. It is likely that the MFC array system reported here will facilitate and accelerate these kinds of studies.

CONCLUSION

A microfabricated MFC array, a compact and user-friendly platform for the identification and characterization of electrochemically active microbes, was developed. The MFC array consisted of 24 integrated anode and cathode chambers, which functioned as 24 independent miniature MFCs. The electricity generation profiles of spatially distinct MFC chambers on the array loaded with *Shewanella oneidensis* MR-1 differed by less than 8%. The utility of the MFC array was demonstrated by screening environmental microbes and resulted in the identification of a microbe that displayed 2.3-fold higher power output than the *S. oneidensis* MR-1 reference strain. The MFC array consumed 380 fold less samples and reagents compared to a single H-type MFC, and 24 parallel analyses could be conducted simultaneously. We expect to further scale up the MFC array into a 96-well MFC array.

KEYWORDS

- **Microfabrication techniques**
- **Proton exchange membrane**
- ***Shewanella* sp.**

ACKNOWLEDGMENTS

The authors would like to thank for the insightful comments from Drs. Martin B. Dickman and Steven M. Wright (Texas A&M University).

AUTHORS' CONTRIBUTIONS

Conceived and designed the experiments: Huijie Hou, Lei Li, Younghak Cho, Paul de Figueiredo, and Arum Han. Performed the experiments: Huijie Hou, Lei Li, and Younghak Cho. Analyzed the data: Huijie Hou, Lei Li, Paul de Figueiredo, and Arum Han. Wrote thechapter: Huijie Hou, Lei Li, Paul de Figueiredo, and Arum Han.

Chapter 2

Quorum Sensing Drives the Evolution of Cooperation in Bacteria

Tamás Czárán and Rolf F. Hoekstra

INTRODUCTION

An increasing body of empirical evidence suggests that cooperation among clone-mates is common in bacteria. Bacterial cooperation may take the form of the excretion of "public goods": exoproducts such as virulence factors, exoenzymes, or components of the matrix in biofilms, to yield significant benefit for individuals joining in the common effort of producing them. Supposedly in order to spare unnecessary costs when the population is too sparse to supply the sufficient exoproduct level, many bacteria have evolved a simple chemical communication system called quorum sensing (QS), to "measure" the population density of clone-mates in their close neighborhood. Cooperation genes are expressed only above a threshold rate of QS signal molecule re-capture, that is, above the local quorum of cooperators. The cooperative population is exposed to exploitation by cheaters, that is mutants who contribute less or nil to the effort but fully enjoy the benefits of cooperation. The communication system is also vulnerable to a different type of cheaters ("Liars") who may produce the QS signal but not the exoproduct, thus ruining the reliability of the signal. Since there is no reason to assume that such cheaters cannot evolve and invade the populations of honestly signaling cooperators, the empirical fact of the existence of both bacterial cooperation and the associated QS communication system seems puzzling. Using a stochastic cellular automaton (CA) approach and allowing mutations in an initially non-cooperating, non-communicating strain we show that both cooperation and the associated communication system can evolve, spread and remain persistent. The *QS* genes help cooperative behavior to invade the population, and *vice versa*; cooperation and communication might have evolved synergistically in bacteria. Moreover, in good agreement with the empirical data recently available, this synergism opens up a remarkably rich repertoire of social interactions in which cheating and exploitation are commonplace.

Cooperation behavior that benefits other individualsis not easy to explain from an evolutionary perspective, because of its potential vulnerability to selfish cheating. A classic example is formed by the so-called tragedy of the commons [1]. A commons pasture is used by many herders, and the best strategy for an individual herder is to add as many cattle as possible, even if this eventually causes degradation of the pasture. The unfortunate outcome follows from the fact that the division of the costs and benefits of adding additional animals is unequalthe individual herder gains all of the advantage, but the disadvantage is shared among all herders using the pasture. Therefore, although cooperation (involving restraint in the input of animals) among the herders would yield the highest benefit for them as a group, each individual herder will be tempted to cheat by adding additional animals, causing the cooperation to break down.

The basis for evolutionary explanations of cooperation is provided by Hamilton's inclusive fitness (kin selection) theory [2]. Individuals gain inclusive fitness through their impact on their own reproduction (direct fitness effects) as well as through their impact on the reproduction of related individuals (indirect fitness effects) (see also [3]). Altruistic cooperative behavior (costly to the actor and beneficial to the recipient) can only be explained by indirect fitness effects. By helping a close relative reproduce, an individual is indirectly passing copies of its genes on to the next generation.

Another theoretical approach considers the evolution of cooperation in terms of two-level selection, namely between and within groups, rather than partitioning individual fitness into direct and indirect components. Cooperation is favored when the response to between-group selection is greater than the response to within-group selection. From yet another perspective, altruism will be favored by natural selection if carriers of altruistic genotypes are sufficiently overcompensated for their altruistic sacrifice by benefits they receive from others. In other words, there should be assortment between altruists and the helping behaviors of others [4]. Perhaps the most likely mechanism for such assortment is "population viscosity" (limited dispersal), causing the offspring of cooperators to remain spatially associated. These different theoretical approaches do not contradict each other but emphasize different aspects of altruistic behavior [4-6].

Although most studies of cooperation have been done on animals, there is a fast growing new field of socio-microbiology studying cooperative behaviors performed by microorganisms [7-10]. Consider a population of bacteria, in which individual cells are producing some public good. Public goods are costly to produce and provide a benefit to all the individuals in the local group. Examples of public goods are exoproducts like virulence factors damaging the host, enzymes for the digestion of food sources, surfactants for facilitating movement, and nutrient scavenging molecules such as siderophores. In many instances microbial cooperation is regulated by QS.

The QS involves the secretion by individual cells of "signaling" molecules. When the local concentration of these molecules has reached a threshold, the cells respond by switching on particular genes. In this way individual cells can sense the local density of bacteria, so that the population as a whole can make a coordinated response. In many situations bacterial activities, such as the production of the mentioned public goods, are only worthwhile as a joint activity by a sufficient number of collaborators. Regulation by QS would allow the cells to express appropriate behavior only when it is effective, thus saving resources under low density conditions. Therefore, QS has been interpreted as a bacterial communication system to coordinate behaviors at the population level [11, 12]. However, its evolutionary stability is somewhat problematic, since cooperative communication is vulnerable to cheating. For example, a signal-negative (mute) strain does not have to pay the metabolic cost of signal production, and a signal-blind (deaf) strain does not pay the cost of responding. Both type of mutants may still benefit from public goods produced in their neighborhood and have actually been observed among environmental and clinical isolates [13, 14] The question then is, under what conditions cheating strains will increase to such an extent that QS breaks down as a regulatory system of cooperative behaviorperhaps with the consequence that the cooperative behavior itself cannot be maintained.

Brookfield [15] and Brown and Johnstone [16] have analyzed models of the evolution of bacterial QS. Although differing in modeling approach, both have studied the evolution of QS in the context of explicit 2-level selection, where selection at the individual level operates against cooperation, while selection at the group level favors QS. These studies conclude that under fairly broad conditions either stable polymorphism may arise in bacterial populations between strains that exhibit QS and strains that do not [15] or the average resource investment into quorum signaling takes positive values, the actual investment depending on group size and within-group relatedness [16]. Since kin selection appears to be central for the evolution of altruistic cooperation, it is required that cooperation preferentially takes place among related individuals. As Hamilton [2] suggested, this could be brought about either by kin discrimination or by limited dispersal. The first mechanism may play some role in microbial communities, for example if a public good produced by a specific strain can only be utilized by clonemates [10]. However, limited dispersal is probably much more important in microbes because due to the clonal reproduction mode it would tend to keep close relatives together. This implies that the spatial population structure plays a key role in the evolution of bacterial cooperation.

In a previous work [17] we have analyzed the evolutionary stability of QS using a CA approach, which is eminently suitable to investigate the role of spatial population structure. There we asked whether QS could be stable as a regulatory mechanism of bacteriocin (anti-competitor toxin) production, and concluded that it could be maintained only when the competing strains were unrelated, and not when the bacteriocin is aimed at related strains which can share the signaling and responding genes involved in QS.

Here, we analyze a much more general model of the evolution of QS regulated cooperation, again using the CA approach. In fact, QS regulated cooperation can be viewed as a superposition and interaction between two cooperative behaviors: the cooperative QS communication system which coordinates another cooperative behavior (e.g., production of a public good). Both forms of cooperation are potentially vulnerable to being parasitized by cheating strains. We allow the reward and the cost of cooperation, the level of dispersal and the sensitivity of the QS system (the signal strength required to induce production of a public good) to vary, and ask for which parameter combinations cooperation and QS will evolve and be maintained, to what extent the presence of a QS system affects the evolution and maintenance of cooperation, how vulnerable the system is for social cheating and how equilibrium levels of QS and cooperation depend on the parameter values.

MATERIALS AND METHODS

The model we use is a two-dimensional CA of toroidal lattice topology. Each of the 300×300 grid-points of the square lattice represent a site for a single bacterium; all the sites are always occupied, that is bacteria may replace each other, but may not leave empty sites. The inhabitants of the sites may differ at three genetic loci: locus C for cooperation (production of a public good), the other two for QS: locus S for producing the signal molecule and locus R for signal response, which includes the signal receptor and the signal transduction machinery that triggers the cooperative behavior when the critical signal concentration has been reached. Each of these loci can harbor either a functional allele denoted by a capital letter (C, S, and R), or an inactive allele denoted by a small letter (c, s, and r). Thus the bacteria can have $2^3 = 8$ different genotypes, each paying its own metabolic cost of allele expression on the 3 loci (Table 1) besides the basic metabolic burden M_0 that is carried by all individuals.

Table 1. The eight possible genotypes of the cooperation-quorum sensing system and the corresponding total metabolic costs me of gene expression.

GENOTYPE	PHENOTYPE	Total cost m_e (with $m_c = 30.0$)	Total cost m_e (with $m_c = 10.0$)
csr	"Ignorant"	0.0	0.0
csR	"Voyeur"	1.0	1.0
cSr	"Liar"	3.0	3.0
cSR	"Lame"	4.0	4.0
Csr	"Blunt"	30.0	10.0
CsR	"Shy"	31.0	11.0
CSr	"Vain"	33.0	13.0
CSR	"Honest"	34.0	14.0

Cooperation can be costly (m_c= 30.0; left column) or relatively cheap (m_c= 1 0.0; right column). Cost of QS signalling: m_s= 3.0; Cost of QS re s onse: m_c= 1.0.

Fitness Effects of Cooperation

The product of the cooperating C allele is supposed to be an excreted "public good" molecule such as an exoenzyme for extracellular food digestion. It may increase the fitness of a bacterium, provided there are at least n_q bacteria (possibly, but not necessarily, including itself) expressing the C allele as well within its 3 × 3-cell neighborhood; n_q is the quorum threshold of cooperation. An individual can only obtain a fitness benefit from cooperative behavior in its neighborhood if at least nq cooperators are present in that neighborhood. On the other hand, cooperation carries a fitness cost which is always paid by the cooperator whether or not it enjoys the benefits of cooperation. The cost of cooperation is the metabolic burden associated with the production of the public good. That is, cooperation (expressing C) carries an inevitable fitness cost and a conditional fitness benefit. We study the effects of a high as well as a low cost of cooperation. Of course for cooperation to be feasible at all the benefit has to outweigh the cost.

Fitness Effects of Quorum Sensing

Cells carrying genotype S. (for the genotype notation, see Table 1) produce the quorum signal molecule, whereas R genotypes will respond to a sufficient amount of signal in their immediate environment. Both the expression of S and of R imply a fitness cost as well, because producing the signal and running the response machinery takes metabolic resources, although less than cooperation itself [13, 18]. The fitness benefit of a QS system is an indirect one: communication using a signaling system may spare unnecessary costs of futile attempts to cooperate whenever the local density of potential cooperators is lower than the quorum n_q. For this communication benefit to be feasible, the QS machinery altogether has to be much cheaper (in terms of metabolic costs) than cooperation itself, otherwise constitutive (unconditional and permanent) cooperation would be a better option for the bacterium, and resources invested into QS would be wasted. Thus the ordering of the metabolic fitness costs of cooperation and QS are assumed to be $m_C \gg m_S \geq m_R$. The inactive alleles c, s, and r carry no metabolic cost: $m_c = m_s = m_r = 0$.

The Effect of the Quorum Sensing Genes on the Cooperation Gene

The quorum signal is supposed to be the regulator of cooperation: bacteria with a C.R genome (i.e., those carrying a functional cooperation allele C and a working response module R) will actually express the C gene (i.e., cooperate) only if there is a sufficient quorum n_q of signalers (S. individuals) within their neighborhood. That is, C.R cells wait for a number of "promises" of cooperation in their 3 × 3-cell neighborhood before they switch to cooperating mode (produce the public good) themselves. The C.r genotypes do not have a functioning response module, therefore they produce the public good constitutively.

Selection

Individuals compete for sites. Competition is played out between randomly chosen pairs of neighboring cells, on the basis of the actual net metabolic burdens M(1) and M(2) they carry. The net metabolic burden M(i) of an individual i is calculated as the sum of the basic metabolic load M_0 carried by all individuals and the total metabolic cost $m_e(i)$ of the actual gene expressions at the three loci concerned (see Table 1), multiplied by the unit complement of the cooperation reward parameter (1–r) if it is surrounded by a sufficient quorum of cooperators:

$$M(i) = [M_0 + m_e(i)] \quad \textit{if \# of cooperators in neighborhood}$$
$$\textit{is below quorum threshold } n_q$$
$$m(i) = [M_0 + m_e(i)](1 - r) \quad \textit{otherwise} \quad (0 < r < 1).$$

Thus, successful cooperation reduces the total metabolic burden in a multiplicative fashion. The relative fitness of individual i is defined as its net metabolic burden relative to the basic metabolic load as $M_0/M(i)$. In practice, the outcome of competition is determined by a random draw, with chances of winning weighted in proportion to the relative fitness. The winner takes the site of the loser, replacing it by a copy of itself.

Mutations

During the takeover of a site by the winner of the competition the invading cell, that is the copy of the winner occupying the site of the loser, can change one of its three alleles (chosen at random) from functional to inactive or vice versa. We call these allele changes "mutations," but in fact they can be due to either mutation or some other process like transformation or even the immigration of individuals carrying the "mutant" allele. The point in allowing allele changes both ways (losing and obtaining them) is to maintain the presence of all six different genes (C, c, S, s, R, r) in the population so that the system does not get stuck in any particular genetic state because of the lack of alternative alleles. Thus, each of the six possible allele changes may have a positive probability. Mutations are independent at the three locifor example, the quorum signal gene S can be lost without losing the response module R at the same time; the resulting mutant will be "mute" yet still able to respond to quorum signals.

Diffusion

Each competition step may be followed by a number (D) of diffusion steps. One diffusion step consists of the random choice of a site, and the 90° rotation of the 2 × 2 subgrid with the randomly chosen site in its upper left corner. Rotation occurs in clockwise or anticlockwise direction with equal probability [19]. The D is the diffusion parameter of the model: it is proportional to the average number of diffusion steps taken by a cell per each competitive interaction it is engaged in. Larger D means faster mixing in the population. Since one diffusion move involves four cells, $D = 1.0$ amounts to an expected number of four diffusion steps per interaction per cell. In the simulations we use the range $0.0 \leq D \leq 1.0$ of the diffusion parameter, and occasionally much higher values ($D = 15.0$) as well.

Initial States and Output

At $t = 0$ the lattice is "seeded" either by the "Ignorant" (csr) genotype on all sites, or the initial state is a random pattern of all the eight possible genotypes present at equal proportions. We simulate pairwise competitive interactions, mutations, and diffusive movements for N generations. One generation consists of a number of competition steps equal to the number of sites in the lattice, so that each site is updated once per generation on average. In the majority of simulations we have applied mutation rates of 10^{-4} both ways at each locus, which is equivalent to an average of nine mutation events per generation within the whole habitat. The three functional alleles have a positive cost of expression, constrained by the relation $m_C \gg m_S > m_R$ (the actual values used throughout the simulations are given in Table 1).

Simulations

With the initial conditions specified above we follow the evolution (the change in allele frequencies) for both cooperation and the two components of QS. We investigate the qualitative or quantitative effects on the evolution of cooperation and QS of the crucial parameters of the model: the fitness reward of cooperation (r), the metabolic cost of cooperation (m_C), the intensity of diffusive mixing (D) and the quorum threshold (n_q). The simulations have been run until the relative frequencies of the three focal alleles (C, S, and R) approached their quasi-stationary values. This could be achieved within 10,000 generations in most cases. The first few simulations have been repeated three times with each parameter setting, using different random number arrays, but since variation in the results was very small at a lattice size of 300 × 300 in all cases, and each run took a long time to finish, we stopped producing replicate runs.

During the simulations we record and plot the time series of the eight different genotype frequencies, from which the frequencies of the three functional alleles can be calculated and plotted against time.

Evaluation of the Model Outputs

The simulation results are recorded as 10.000-generation time series of genotype frequencies and spatial patterns of the genotypes. With regard to allele frequencies we asked the following question: are the genes for cooperation (C) and QS (S and R) selected for beyond their respective mutation-selection equilibria based on the metabolic selection coefficients $s_C = (M_C-M_0)/M_C$, $s_S = (M_S-M_0)/M_S$ and $s_R = (M_R-M_0)/M_R$, and the (uniform) mutation rate μ. For example, relative frequencies of the cooperating allele above its mutation-selection equilibrium $\hat{q}_C = \mu / (s_C + \mu)$ indicate a net fitness benefit of cooperation and thus positive selection for the C allele. $\hat{q}_S$ and $\hat{q}_R$ can be calculated the same way.

DISCUSSION

Microorganisms display a wide range of social behaviors, such as swarming motility, virulence, biofilm formation, foraging, and "chemical warfare" [7-10, 21]. These social behaviors involve cooperation and communication. Cooperation often takes the form of a coordinated production and excretion of molecules like enzymes, toxins, and virulence factors. In bacteria, this cooperative behavior is typically regulated by QS, a chemical communication system in which cells produce diffusible molecules and can assess the cell density by sensing the local concentration of these signaling molecules [11, 12, 22, 23]. In fact, QS can be viewed itself as a cooperative behavior to optimize other forms of social behavior. An important issue is the evolutionary stability of cooperation because of its potential vulnerability to social cheating: the occurrence and selection of individuals who gain the benefit of cooperation without paying their share of the costs [2, 10]. We studied the evolution of cooperative behavior in bacteria (e.g., production of a public good) and of a QS system which coordinates this cooperative behavior, using CA modeling. This approach allows a fairly precise evaluation of the role played by the spatial population structure, because all bacterial interactions are supposed to occur between neighboring cells.

Our results allow three main conclusions, which we discuss in turn.

1. Cooperation Only Evolves Under Conditions of Limited Dispersal

The simulations in which we analyzed the evolution of cooperation without QS suggest that cooperation can only evolve when the degree of spatial mixing in the population is low, which implies a high relatedness between neighboring cells. Our model thus confirms the importance of the level of relatedness between interacting individuals and the evolutionary stability of cooperation, as first hypothesized by Hamilton (1964), and demonstrated experimentally in microbial populations [24-26].

The level of exploitation of cooperative behavior by non-cooperating strains is lowest when the required quorum of cooperators is relatively high and the dispersal rate is low (Figure 1).

2. The Presence of Cooperative Strains in a Population Always Selects for QS and Cooperation Becomes More Common as a Consequence of QS

The simulations in which we allow the simultaneous evolution of cooperation and QS suggest that whenever the gene for cooperation is selected, also one or both of the communication genes of the QS system are selected. Moreover, the presence of QS (either partial or complete) allows stable levels of cooperation in regions of the explored parameter space where cooperation without QS cannot invade (compare the corresponding columns in Figure 1 and Figure 3). Thus a communication system helps to establish stable cooperation. Of course, communication about willingness to cooperate will only be selected if at least part of the population is able to cooperate, so evolution of QS is not expected in a completely non-cooperating population. But it is not self-evident that QS always should enhance the frequency of cooperating strains in the population. Clearly, QS by cooperative strains is selected if the advantage derived from limiting the actual cooperative behavior to when it is most profitable outweighs its costs. In this way QS causes the gene for cooperation to increase. But *QS* genes may also be selected in non-cooperators, allowing exploitation of cooperative strains and lowering the frequency of cooperation. This applies in particular to "Liar" strains, non-cooperators which signal willingness to cooperate, which may manipulate fully QS "Honest" strains to cooperate when actually the number of local cooperators falls below the quorum n_q. As a consequence, these "Honest" cells pay the cost of cooperation but cannot enjoy its benefit.

3. The Communication Cooperation System as Modeled in This Study Displays a Remarkably Rich and Complex Pattern of Social Interactions in Which Cheating and Exploitation Play a Significant Role

The QS not only leads to a higher equilibrium frequency of the cooperation gene, but also allows a striking diversity of social interactions. Of the eight possible genotypes in our model, defined by the presence/absence of the three functional genes for respective cooperation, signaling, and responding, six genotypes may reach appreciable equilibrium frequencies, depending on the precise parameter combinations. Only two mutant types play an insignificant role in the system: "Voyeur" which responds to the signal but is unable to signal and cooperate itself, and "Lame," which is fully QS (signaling and responding) but cannot cooperate. Inspection of Figure 3 reveals the possibility of five different stable polymorhisms characterized by domination of two genotypes: [Blunt,Ignorant], [Blunt,Honest], [Shy,Liar], [Honest,Ignorant] [Honest,Vain]; five polymorphisms with three dominating genotypes: [Honest, Blunt, Ignorant], [Honest, Blunt, Liar], [Honest, Ignorant, Liar], [Honest, Vain, Blunt], [Shy, Liar, Blunt], and one in which four genotypes reach an appreciable frequency: [Honest,

Blunt, Liar, Ignorant]. It is important to note that in all cases some degree of social cheating occurs in the form of exploitation or parasitism. Thus our analysis predicts the large-scale occurrence of social cheating in microbial populations.

Two of the polymorphisms mentioned above merit more elaborate discussion.

The Janus Head of QS

In cases where the cooperation gene is (almost) fixed, one might at first sight expect a monomorphic unconditionally cooperating ("Blunt") population, because Blunt is the cooperator with the lowest metabolic costs, and in a fully cooperating population QS is not needed to obtain information about the potential level of cooperation. However, we find next to "Blunt" appreciable frequencies of fully QS ("Honest") and partially QS ("Vain") cells. The reason appears to be that here QS is selected because it allows exploitation of Blunt strains, which unconditionally cooperate. As soon as a quorum of Blunts is present, the other cells need not cooperate themselves in order to profit from the cooperation benefit. Adoption of the QS machinery allows them to do precisely this, since in such cases the level of signal is too low to trigger their cooperative behavior. This phenomenon is an unexpected and novel result, showing that QS not only prevents wasting resources when too few potential cooperators are around, but also allows cells to parasitize on unconditionally cooperating neighbors, when a sufficient number of those are present. It may occur at a large scale when the gene for cooperation is (almost) fixed in the population, due to a favorable benefit/cost ratio of the cooperative behavior and the quorum threshold relatively high. As explained more fully in the Results section, 100% cooperating populations seem to evolve in most cases to a [Blunt, Honest, Vain] mixture, characterized by a cyclic interaction pattern (Blunt>Vain>Honest>Blunt) reminiscent of the rock-scissors-paper (RSP) game [20, 21].

Spiteful Behavior

The second polymorphism we want to call attention to is the coexistence of Honest, Liar, and Ignorant, which occurs for example, at $n_q = 4$ and $n_q = 5$ for certain diffusion values. Clearly, the non-cooperative Ignorant and Liar cells exploit the Honest cells which provide cooperation benefits. Here the selective advantage of Liar is at first sight remarkable, since it pays a higher metabolic cost than Ignorant, and can only expect the same share as Ignorant from the cooperation benefit made available by the Honest cells. The only effect of Liar is to sometimes induce Honest cells to cooperate when actually less than the quorum of Honest is present. Nothing is gained, except that in such cases Honest is paying the cooperation cost without getting the benefit. Thus this coexistence is a clear example of spiteful behavior. Liar lowers the relative fitness of Honest, but also pays a fitness penalty (the cost of signaling) itself.

How do our results relate to previous theoretical and empirical work on the evolution of QS and cooperation? They confirm the basic result from earlier theoretical analyses of the evolution of QS [15], which predicted a stable polymorphism in microbial populations between QS and non-QS strains. The conclusion of Brown and Johnstone that the highest level of QS signal expression is expected for intermediate levels of relatedness between interacting strains [16] is confirmed in our study only for cases when the cooperation cost is low and the required quorum threshold is also low. Then the benefit of cooperation is relatively easy to obtain, and the QS machinery is too costly to operate. Only when the spatial population mixing becomes more intensive, causing the predictability of the neighborhood composition to go down, QS becomes profitable. The situation is different for costly cooperation and/or a high quorum threshold. Apparently, then QS is profitable even at a very low rate of dispersal (i.e., at high relatedness between interacting cells) because of the lower level of cooperation in the population and/or the greater sensitiveness to increased dispersal.

The available empirical observations on natural occurrence of QS cheats are mainly from work on *Pseudomonas aeruginosa* [13, 27-31]. Although these experimental studies cannot yet be informative with respect to the full spectrum of possible mutants and only focus on one or two QS mutants, they report a considerable level of social cheating, which is in agreement with our study.

Finally, we mention an experimental result that may be of relevance with respect to QS evolution but is not included in this model. It is related to the feedback-regulation of QS signal production: "signal deaf" signaler mutants (in our notation:.Sr genotypes) are shown to produce an excess of signal molecules compared to signal responsive ones, because in the latter signal production is downregulated by the extracellular concentration of the signal itself, which response-deficient mutants cannot sense [32, 33]. The effect of signal over-expression on the dynamics of QS evolution require further theoretical work.

In conclusion, we predict that the evolution of QS as a communication system regulating cooperative behavior such as the production of a public good has two striking effects. First, it enables the cooperative behavior to attain a higher frequency in the population, and second, it opens up a remarkably rich repertoire of social interactions in which cheating and exploitation are common place.

RESULTS

The Evolution of Cooperation Without Quorum Sensing

We first have performed simulations with the QS functions disabled (mutation rates in both ways set to 0.0 at the S and the R loci). Without QS allowed, the only possible genotypes are the "Ignorant" (no cooperation) and the "Blunt" (unconditional cooperation), of course.

(1) Cooperation is Relatively Costly ($m_c = 30$)

The left column in Figure 1 summarizes the results. Cooperation is only selected under a very low degree of dispersal. This confirms the essential role played by kin selection in the evolution of cooperation, since low dispersal in a microbial population implies that most social interactions are among related individuals. With a low quorum threshold (only few cooperators are necessary to provide the benefit to all the immediate neighbors) there is much scope for non-cooperators to parasitize, because sufficiently often they can enjoy the benefit from cooperative neighbors without paying the cost of cooperation themselves; therefore, with $n_q = 2$ and $n_q = 3$, only a minority of the population will consist of cooperator genotypes. With $n_q = 4$ and $n_q = 5$ there is obviously less opportunity for parasitism, and cooperators reach higher frequencies. However, then the system becomes more sensitive to the effects of spatial mixing; with $n_q = 6$ even at $D = 0$ successfully cooperating neighborhoods are disintegrated more often than that they are formed or maintained, and cooperative behavior is no longer selected.

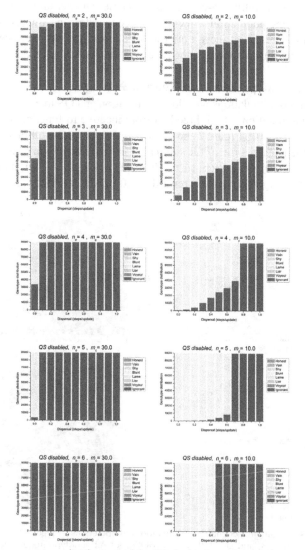

Figure 1. Stationary genotype distributions of the QS-disabled set of simulations. Fixed parameters: basic metabolic burden: $M_0 = 100.0$; metabolic cost of quorum signal production: $m_s = 3.0$; metabolic cost of quorum signal response: $m_r = 1.0$; fitness reward factor: $r = 0.9$; mutation rates: $\mu_s = \mu_r = 0.0$, $\mu_c = 10^{-4}$. Screened parameters: metabolic cost of cooperation (m_c), quorum threshold (ne) and dispersal (D). Simulations lasted for 10.000 generations and they were initiated with the "All-Ignorant" (csr) state.

(2) Cooperation is Relatively Cheap ($m_c = 10$)

Qualitatively the same trends are apparent when cooperation is less costly (Figure 1, right column). Cooperation is maintained over a broader range of diffusion rates, compared to the case of costly cooperation. Clearly, with less costly cooperation, occasional futile cooperation attempts (when the number of cooperators in a neighborhood is less than the quorum) are less deleterious. With increasing quorum threshold the scope for parasitism by non-cooperators becomes smaller and as a consequence a larger fraction of the population will consist of cooperators, as long as neighborhoods sufficiently often contain at least a quorum of cooperators. Above a certain level of population mixing this is no longer the case, and then cooperation does not evolve.

The Evolution of Cooperation and Quorum Sensing

In the next series of simulations we allow cooperation and QS to evolve simultaneously, and allowing mutations at all three loci from inactive to functional and vice versa with probability $\mu = 10^{-4}$. Figure 2 shows as an example the evolution of the genotype and allele frequencies in a run of the simulation model with a high cost of cooperation and a relatively cheap QS system ($m_C = 30.0$, $m_S = 3.0$, $m_R = 1.0$), medium quorum threshold ($n_e = 3$), high cooperation reward ($r = 0.9$), and no diffusion ($D = 0.0$).

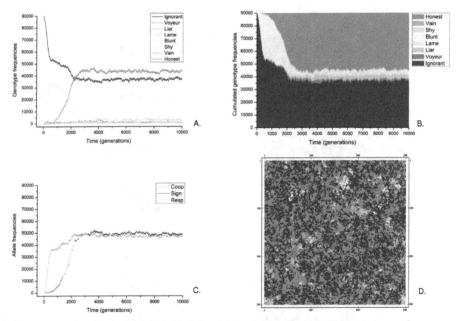

Figure 2. Details of a single QS-enabled simulation. Parameters as in Figure 1, except for $\mu_s = \mu_r = 10^{-4}$; $m_c = 30.0$, $n_e = 3$, and $D = 0.0$. Time evolution of A.: genotype frequencies; B.: genotype distribution; C.: allele frequencies. D.: The spatial pattern of genotypes at $T = 10.000$.

The first invading genotype is the "Blunt" one (Csr) which cooperates unconditionally but lacks QS. However, as soon as the "Blunt" type reaches a high frequency in the population, the adoption of QS genes obviously becomes profitable, because the "Honest" (CSR) genotype takes over, ultimately excluding the "Blunt" one. The "Honest" takeover renders the stationary population essentially dimorphic: the great majority of the individuals are either "Ignorant" or "Honest". The remaining six genotypes are present at very low frequencies, close to their metabolic mutation-selection equilibrium. What we end up with is thus the coexistence of cooperating-communicating cells ("Honest") and parasitic ones ("Ignorant").

The Effects of Changing the Quorum Threshold and Diffusion

(1) Cooperation is Costly (m$_C$ = 30.0).

Keeping the costs m$_C$, m$_S$, m$_R$ and the cooperation reward r constant, we have systematically screened the effects of the quorum threshold n$_q$ and the diffusion parameter D on the evolution of cooperation and QS (Figure 3, left column). Comparison with the corresponding cases without the possibility of QS (Figure 1, left column) immediately shows that the QS functions of signaling and responding are selected in a large part of the parameter space and that they have a positive effect on the relative frequency of cooperation in the population.

First consider the case of a low quorum threshold (n$_e$ = 2). If the population is not mixed at all (D = 0.0), cooperators do not need an intact QS machinery to have a reliable cue on the presence of cooperating neighbors: with a high chance at least one clone mate (mother or daughter) is always around, and that is sufficient for them to enjoy the cooperation reward during their next interaction. This is why the great majority of cooperators have disposed of one or both QS alleles (S and R) at D = 0.0. Most cooperators are of the "Shy" (CsR) genotype, which responds to quorum signals and cooperates accordingly, but does not itself produce the signal. Parasites capitalize on this feature by issuing the signal only, thereby persuading the "Shy" type to cooperate. This results in the parasite population to become, to an overwhelming majority, of the "Liar" (cSr) type which is the exact complement of the "Shy" one: it possesses the only functional allele that "Shy" is missing. Since the quorum signal is necessary for the onset of cooperation in "Shy" individuals, the interaction between these two dominant genotypes can be interpreted both as parasitism and as a peculiar type of "division of labor". The latter, less antagonistic component of the interaction immediately disappears with the introduction of the slightest diffusion into the system. At and above D = 0.1, the diffusion in the population creates already too many neighborhoods that do not contain the required two C and two S alleles distributed over separate genotypes (i.e., two "Shy" and two "Liar" types), and the presence of CSR ("Honest") is selected, guaranteeing successful cooperation as soon as two such genotypes are present in a neighborhood. This leaves ample space for csr ("Ignorant") parasites of course, which reach high frequencies. This will be true even at fairly high diffusion rates.

The n$_q$ = 3 case has already been described in some detail above (Figure 2). The special feature of this series of simulations is that the QS system is always adopted by the cooperators, even without diffusion. This might be accounted for by the fact that at about 50–60% (or less) of the population cooperating, the presence of three cooperators within a 9-individual neighborhood is far from guaranteed, making QS well worth its cost for the cooperators. Therefore both QS alleles (S and R) spread and become established within the cooperating population. Consequently parasites do not need to issue fake quorum signals to access the benefit. Increasing diffusion gradually reduces the likelihood of maintaining three cooperators in a neighborhood, resulting in a lower level of cooperation in the population. At very intensive diffusion (D = 6.0 in this parameter setting) both cooperation and QS disappear together abruptly, and the stage is left for the parasitic "Ignorant" type. Apparently then successful cooperation will be so rare that cooperators are losing more due to the cost of operating the QS machinery, than gaining from the cooperation benefit. Consequently, their relative fitness shrinks below that of the "Ignorant" type, and they vanish.

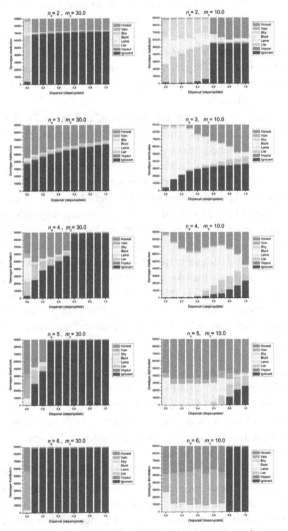

Figure 3. Stationary genotype distributions of the QS-disabled set of simulations. All parameters as in Figure 1. except $\mu_s = \mu_r = 10^{-4}$.

The n_q = 4, 5, and 6 cases are similar to the n_q = 3 case, except for two important aspects. One is that, due to the high quorum threshold, the system becomes more sensitive to spatial mixing. For n_q = 4, the upper limit of the diffusion parameter that still allows the cooperation and QS alleles to persist is D = 0.5; for n_q = 5 it is D = 0.2, and for n_q = 6 cooperation is only maintained in the absence of spatial mixing (D = 0). Above these D values, successfully cooperating clumps (neighborhoods with n_q or more cooperators) are disintegrated by too intensive mixing, at a rate faster than they are built by interactions. Second, at zero diffusion (D = 0.0), for n_q = 4, 5, and 6, cooperators increase in abundance and they tend to lose one or both components of the QS system, unlike in the n_q = 3 simulation. The reason for the loss of the communication device is that cooperators become so common, that QS is no longer needed to find out whether there is a sufficient number of cooperators present in the immediate neighborhood. Constitutively cooperating genotypes like "Blunt" and "Vain" increase in frequency because they do not pay the (complete) cost of QS. At n_q = 4 the "Honest" type is maintained at about 30%, because QS is still sufficiently often useful, with almost 30% of the population consisting of non-cooperators. Here most of the non-cooperating strains are of the "Liar" type: in neighborhoods with fewer than n_q = 4 "Honest" individuals, their signaling helps to persuade the latter to cooperate. At n_q = 5 and n_q = 6 with zero diffusion, the simulations bring an interesting strategic aspect of QS to the light. Although almost 100% of the population is cooperating, the fully QS "Honest" type is maintained at some 30–50% of the population. This is at first sight surprising, since the presence in local neighborhoods of a quorum of cooperators is practically guaranteed. However, here QS appears to function as a mechanism to avoid expression of the cooperative behavior when already a sufficient number of unconditionally cooperating ("Blunt") neighbors are producing the public good. Clearly, when less then n_q cells in a neighborhood are producing the quorum signal molecule, "Honest" types will not cooperate, thus saving the cost of cooperation while frequently enjoying the cooperation benefit thanks to their unconditionally cooperating neighbors. This explains the fairly high frequency of signaling unconditional cooperators ("Vains"). By enhancing the local concentration of the quorum signal they induce "Honest" cells to cooperate, thereby enhancing the likelihood that a quorum of cooperators is reached. Actually, in situations where cooperation is so attractive that the C allele is (almost) fixed, the three cooperating types: "Honest", "Blunt," and "Vain" display a cyclic interaction pattern (Blunt>Vain>Honest>Blunt) reminiscent of the RSP game [20, 21]. A population of "Blunt" is invaded by "Honest" because as explained above "Honest" parasitizes on the unconditional cooperation by the "Blunts". Conversely, an "Honest" population is invaded by "Blunt" and by "Vain", because they do not pay (part of) the costs of the QS machinery, and "Vain" invades a polymorphic ("Honest", "Blunt") population, enhancing the likelihood of a quorum of actual cooperators by inducing "Honest" cells. Figure 4 shows the evolutionary dynamics of such a population with the cooperating C allele fixed.

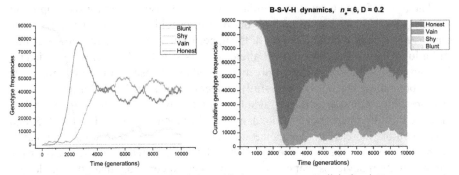

Figure 4. The evolution of QS in a population with the cooperating C allele fixed. Parameters as in Figure 3, with $n_e = 6$ and $D = 0.2$. The simulation was started from the "All-Blunt" initial state.

(2) Cooperation is Relatively Cheap ($m_c = 10.0$).
The right column in Figure 3 shows the simulation results for less costly cooperation. Just as in the case of costly cooperation, QS alleles are selected in a large part of the parameter space and boost the frequency of cooperators in the population (compare with Figure 1, right column). In particular, QS enables the maintenance of cooperative behavior at higher levels of population mixing. At low quorum thresholds ($n_q = 2$ and $n_q = 3$) and a low rate of diffusion, the QS machinery is too expensive because sufficiently often neighborhoods will contain a quorum of unconditional cooperators, and the QS alleles are not selected. When the rate of spatial mixing increases, the predictability of the local population composition goes down, and QS becomes profitable. At higher quorum thresholds ($n_q \geq 4$), we see again that if the population (almost) exclusively consists of (potential) cooperators, QS is selected because its machinery allows cells to avoid cooperating when the number of unconditionally cooperating neighbors is already equal to or higher than the quorum. The resulting cooperating population consists of a dynamical coexistence of fully QS ("Honest") genotypes, unconditionally cooperators ("Blunt" types), and signaling unconditionally cooperators ("Vains").

The Effect of Decreasing the Reward of Cooperation
At a lower cooperation reward of $r = 0.5$ neither cooperation nor QS evolves: the population becomes almost completely uniform "Ignorant" within the entire parameter space. This result is somewhat surprising, given that at $r = 0.5$, successful QS cooperators like the "Honest" genotype should still enjoy a substantial fitness advantage compared to the "Ignorant." The total metabolic burden of an "Honest" individual after getting the cooperation reward is $133.0 * 0.5 = 66.5$, whereas the "Ignorant" carries a burden of 100.0 units, that is the cooperator should have a fitness advantage of about 34% over the parasite. It cannot use it to the full, however, because nearby parasites may take the advantage as well without paying the costs, and those parasites which are successful in doing so carry an even lower (50.0 units) metabolic burden. Apparently a minimum threshold measure of fitness reward is necessary for cooperation to become an option. With the quorum threshold fixed at $n_q = 3$ and diffusion at $D = 0.0$, we looked for the critical value of the fitness reward by increasing r from 0.5 to 0.9, and found it to be $rc = 0.8$. This means that the kind of exploitable, broadcasted cooperation,

such as producing a public good needs to be highly rewarded for it to be worthwhile to adopt, otherwise parasitism prevails and ultimately eradicates cooperation.

AUTHORS' CONTRIBUTIONS

Conceived and designed the experiments: Tamás Czárán and Rolf F. Hoekstra. Performed the experiments: Tamás Czárán. Analyzed the data: Tamás Czárán and Rolf F. Hoekstra. Wrote the chapter: Tamás Czárán and Rolf F. Hoekstra.

KEYWORDS

- Allele
- Cellular automaton
- Cooperation
- Quorum sensing

Chapter 3

Resveratrol as an Antiviral Against Polyomavirus

Valerio Berardi, Francesca Ricci, Mauro Castelli, Gaspare Galati, and Gianfranco Risuleo

---

INTRODUCTION

Resveratrol (RV) is a non-flavonoid polyphenol compound present in many plants and fruits and, at especially high concentrations, in the grape berries of *Vitis vinifera*. This compound has a strong bioactivity and its cytoprotective action has been demonstrated, however at high concentrations the drug exhibits also an effective anti-proliferative action. We recently showed its ability to abolish the effects of oxidative stress in cultured cells. In this work we assayed the bioactivity of RV as antiproliferative and antiviral drug in cultured fibroblasts. Studies by other Authors showed that this natural compound inhibits the proliferation of different viruses such as herpes simplex, Varicella -zoster and influenza A. The results presented here show an evident toxic activity of the drug at high concentrations, on the other hand at sub-cytotoxic concentrations, RV can effectively inhibit the synthesis of polyomavirus DNA. A possible interpretation is that, due to the damage caused by RV to the plasma membrane, the transfer of the virus from the endoplasmic reticulum to the nucleus, may be hindered thus inhibiting the production of viral DNA.

The mouse fibroblast line 3T6 and the human tumor line HL60 were used throughout the work. Cell viability and vital cell count were assessed respectively, by the MTT assay and Trypan Blue staining. Cytotoxic properties and evaluation of viral DNA production by agarose gel electrophoresis were performed according to standard protocols.

Our results show a clear dose dependent both cytotoxic and antiviral effect of RV respectively at high and low concentrations. The cytotoxic action is exerted towards a stabilized cell-line (3T6) as well as a tumor-line (HL60). Furthermore the antiviral action is evident after the phase of virion entry, therefore data suggest that the drug acts during the synthesis of the viral progeny DNA.

The RV is cytotoxic and inhibits, in a dose dependent fashion, the synthesis of polyomavirus DNA in the infected cell. Furthermore, this inhibition is observed at non cytotoxic concentrations of the drug. Our data imply that cytotoxicity may be attributed to the membrane damage caused by the drug and that the transfer of polyomavirus from the endoplasmic reticulum to the cytoplasm may be hindered. In conclusion, the cytotoxic and antiviral properties of RV make it a potential candidate for the clinical control of proliferative as well as viral pathologies.

Murine polyomavirus (Py) is an ideal model system to investigate many different biological phenomena at cellular and molecular level. Polyomavirus is totally dependent on the metabolism of the infected cell: therefore, it has been used to study cellular and molecular functions. Classical works based on the study of the viral proliferation helped to elucidate the mechanisms of the regulation of DNA replication, RNA transcription, and translation as well as tumor transformation. Analogously to other polyomaviruses, with which it shares a high sequence homology, Py can very efficiently transform non permissive cells in culture and is able to cause tumors if injected in immuno-suppressed or singeneic animals (see: [1] for a compendium on polyomaviruses and [2-7] for more recent reviews on this subject).

In last decade we investigated the role of both natural and synthetic substances on Py DNA replication and RNA transcription [8-10]. Also, the cellular and metabolic response after exposure to these substances was studied [11-15]. We particularly focused our attention on a natural complex mixture, known as MEX, obtained by methanolic extraction of whole neem oil [13]. This oil is prepared from the seeds of *Azadirachta indica* and has been extensively used in Ayurveda, Unani, and Homoeopathic medicine possibly for centuries [16, 17]. In our laboratory MEX showed a significant and differential cytotoxic action, with the cancer cells being more sensitive than the normal ones [18]. The main target of MEX is the plasma membrane which, after treatment with this extract, becomes more fluid without a substantial loss of its structural properties [19]. In addition, preliminary experiments performed in our laboratory suggest that MEX has also an antiviral activity (Berardi et al., in preparation); in any case a similar activity of neem leaf extracts was reported in a model of Dengue virus [20].

In this work we assayed the action of RV, a natural compound raising an increasing interest on the proliferation of cultured cells that is : the murine fibroblast line 3T6 as well as in the tumor line HL60. In addition, we also investigated the action of this drug on the proliferation of the Py in the infected cell population.

The RV is a non-flavonoid polyphenol compound present in many plants and fruits, at especially high concentrations in the grape berries of *Vitis vinifera* [21]. This compound has a high bioactivity and its cytoprotective action has been demonstrated. As a matter of fact, possibly due to its polyphenol characteristics, RV was also shown to have antiviral action versus influenza A [22] and Varicella zoster virus in cultured cells [23]. Analogous properties of RV against Herpes virus simplex I were shown in animal models [24]. In this latter case, suppression of transcription factor NF-κ-B seems to be involved in its antiviral property [25].

The results presented here show that RV exhibits a cytotoxic activity and has an antiviral property since it efficiently inhibits the synthesis of Py DNA. The inhibition is observed at non cytotoxic concentrations of RV as shown by vital cell count and quantitative evaluation of the viral DNA synthesis after exposure to the drug. In addition, our results evidence a clear dose dependent antiviral effect of RV. Since this action appears after the phase of virion penetration, data suggest that RV exerts its antiviral properties during the synthesis of the viral DNA progeny. However, because of its cytotoxic properties, it may be envisaged an application of RV to control negatively the cell growth in proliferative diseases.

MATERIALS AND METHODS

Cell Cultures

The mouse fibroblast line 3T6 and the tumor line HL60 were used throughout the work. Cells were grown in high glucose DMEM, supplemented with newborn bovine serum (10% final concentration), glutamine (50 mM), and penicillin-streptomycin (10,000 U/ml). Growth temperature was 37°C in controlled humidity at 5% CO_2. Cells were routinely split and sub-cultured every third day.

Viral infection was performed at 4 pfu $\times$ cell^{-1}, for 2 hr at 37° with occasional rocking. Infection procedure and extraction (replication assays) of *de novo* synthesized DNA were described in detail in previous works, see for instance [9, 10, 26]. Viral DNA was visualized after agarose gel electrophoresis in the presence of ethidium bromide (0.5 µg/ml, final concentration). Evaluation of cell vitality: Cell viability was assessed by the colorimetric MTT assay [27]. Absorbance was measured at 570 nm to obtain a standard cell count. The number of cells surviving to the treatment with RV (20 µM) was also evaluated by vital cell count in Trypan Blue in a Burker chamber. The same approach was adopted to count the cell mortality consequent to Py infection [18].

All experiments were repeated at least three times. The error bars indicate the Standard Deviation of the Mean (±SEM).

DISCUSSION

In this work we report on cytotxicity versus two different cell lines: a normal mouse firbroblast line and tumoral one. The results clearly show that RV can exert a cytotoxic action both against a normal stabilized fibroblast cell line and human tumor cells. The human tumor line seems to be slightly more sensitive to the drug and this recalls results previously obtained in our laboratory with MEX: a partially purified natural mixture [18]. The antiviral activity of RV towards Py infection was also evaluated. The exposure to the drug was carried at a concentration of RV which did not show a significant cytotoxic effect. It is known that RV can exert anti-oxidant and anti-inflammatory activities and, also, it regulates multiple cellular events associated with carcinogenesis: for a relatively recent review see [28]. The cytotoxicity of RV is only apparently paradoxical; as a matter of fact this drug induces cell cycle arrest and stimulates the reactive oxygen species (ROS) activated mitochondrial pathway leading to apoptosis [29, 30]. An analogous paradoxical action has been described for another potent antioxidant: curcumin, which is able to induce apoptosis in human cervical cancer cells [31]. Therefore, an evaluation of possible cytotoxic effects of RV in our model system was a necessary pre-requisite. The Py productive cycle ends with the lysis of the infected cell: hence the actual number of cells dying as a consequence of viral proliferation had to be also assessed. The results of these experiments allowed us to find out the best conditions where the putative antiviral activity on murine Py could be investigated.

The results presented in this work show that like in the case of influenza A, HVS, and Varicella-zoster [22-25], the viral replication is severely inhibited by RV also in the case of murine Py. The inhibition is dose and time dependent and all experiments were carried out at 24 hr of infection time when the effects on the cell viability due to the exposure to the drug or to the viral proliferation are minimal. Similar results were obtained after 42 hr of infection but after such a prolonged time the significant cell mortality induced by RV and by the progression of the viral infection could overlap and/or mask the actual effects attributable to the drug (infection data non shown). Furthermore infection experiments performed in the presence of RV during the absorption phase gave essentially the same results obtained in infections experiments where drug was added after the viral penetration (not shown). This strongly suggests that virus entry is not the main target of RV whose action is therefore exerted during the phase of viral DNA synthesis. Furthermore, the presence of the drug for the whole duration of the infection is necessary to abrogate completely the viral DNA production. As a mater of fact exposure to the drug for shorter time has no effect on the overall yield of viral DNA. Incidentally, this data also shows that the intracellular "life time" of the viral DNA is fairly long, since removal of the drug after 12 hr exposure seems to have little effect on the amount of the progeny DNA. These data recall a similar observation made in our laboratory with a different natural substance [9].

At the moment the mechanism of action of RV remains to be elucidated; however in the case of influenza A virus, the translocation of viral ribo-nucleoprotein complexes to the endoplasmic reticulum is hindered and the expression of late viral proteins is reduced. These two phenomena could be related to the inhibition of protein kinase C activity and its dependent pathways [22]. Also, Py utilizes proteinprotein complexes associated to the endoplasmic reticulum and involving the viral capsid proteins VP2 and VP3 [32]. Therefore it could be speculated that the viral transfer to the nucleus, in the case of Py, follows an analogous process.

RESULTS

Evaluation of the Cytotoxicity of Resveratrol and of the Cell Death Consequent to Py Infection

In preliminary experiments we assessed the concentration at which RV may exert a putative cytotoxic activity. It should pointed out, as a matter of fact, that natural substances endowed of cytoprotective and antioxidant properties, may present a threshold effect above which they can paradoxically show cytotoxic properties. The phenomenon has been documented for RV and its analogues as well as for curcumin another potent antioxidant drug with cytoprotective features [28-31].

The cytotoxicity of RV on 3T6 cells has been evaluated by the Mossman assay [27] after treatment for 24 and 48 hr (Figure 1A and 1B, respectively), but in this latter case the treatment with 2 μM RV was omitted since this at this concentration the drug does not have a significant effect on cell mortality. The drug is dissolved in 0.02% DMSO (final concentration) in PBS but, at this low concentration, the organic solvent has no effects on cell survival, as shown by the second bar from the left. However cells exposed to RV at the concentrations of 20 and 40 μM show some sensitivity to the drug, since only about 60% of the cell population survives the treatment for 48 hr at the higher concentration. The vital cell count observed after trypan blue staining is in good agreement with the one obtained by the Mossman assay (Table 1, only the data for 20 μM RV at after 24 hr of treatment are shown).

Table 1. Vital cell count after trypan blue staining of cells treated with resveratrol (20 µM) for 24 hr.

	RV- Treated 3T6 Cells	
Sample	Vital	Non-Vital
1	27	3
2	32	2
3	28	3
4	30	3
	Average cell mortality 8.7 ± 2.6%	

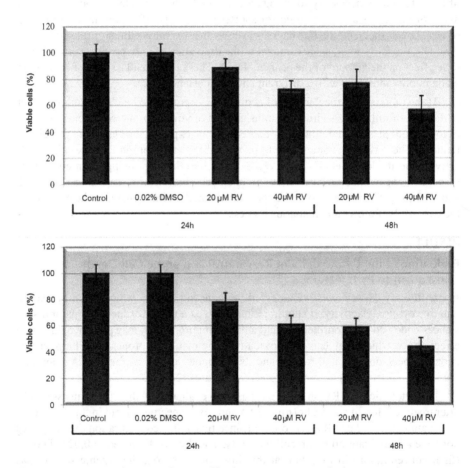

Figure 1. Cytotoxicity of resveratrol assessed by the Mossman assay. The bars report the percentage of viable cells after different times of exposure to the drug (24 hr: four bars to the left; 48 hr two bars to the right). The untreated control and the sample in DMSO at 48 hr are omitted since the data are virtually identical to the ones obtained at 24 hr. Data reported in upper panel refer to 3T6 cells while those shown in the lower panel refer to HL60 cells.

We also investigated the cytotoxic activity of RV on the tumor cells HL60: a human promyelocytic leukemia cell line. The results clearly show that RV can significantly inhibit the cell growth already at a concentration of 25 μM.

Subsequently we assessed the level of cell mortality induced by Py infection: in this case we used the method of vital cell staining only with trypan blue. As a matter of fact, the MTT assay is informative of cell death deriving from membrane damage and former data from our laboratory indicated that the plasma membrane is actually one of the targets of RV. On the contrary, trypan blue staining has a more general action ranging from a generic damage of cell membrane to severe problems in cell homeostasis. Table 2 reports the vital cell counts in control and Py infected cells.

Table 2. Assessment of the cell mortality rate due to Py proliferation.

Virus Py 24 h		
Sample	Vital	Non-Vital
1	44	1
2	45	2
3	41	2
4	52	1
Average cell mortality 3,4% ± 1,5%		
Virus Py 48 h		
Sample	Vital	Non-Vital
1	40	2
2	46	4
3	44	4
4	49	2
Average cell mortality 6,7% ± 2,5%		

The vital cell count was evalutated by trypan blue staining.

The reported data show that after 48 hr of infection the cell death rate is about as double as in controls: however the viral infection does not seem to cause extensive loss cell vitality.

In the light of these results the effect of RV on Py proliferation was evaluated at 24 hr post-infection in cells were treated with 20 μM RV or at the concentrations of drug reported in the legends to the figures.

Effect of Resveratrol on the Viral Proliferation

Semi-confluent cells were infected with Py and RV was added after the absorption phase at the indicated final concentrations. Infection was continued for 24 hr and progeny viral DNA was extracted according to the Hirt-procedure [26] (Figure 2A). The data clearly show that the viral replication is virtually abrogated at 20 μM RV. Infections performed in the presence of RV during the absorption phase gave essentially the same results (not shown).

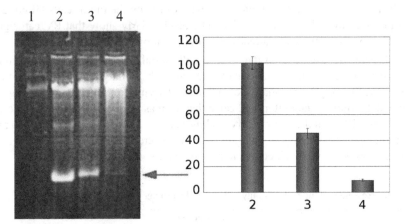

Figure 2. Viral DNA yield obtained at 24 hr post-infection. Left panel: Electropherogram of the *de novo* synthesized progeny viral DNA (form I) indicated by the arrow. Lane 1: Mock infected cells, Lane 2: Untreated control cells; Lane 3, and 4: Cells treated with RV 20 µM and 40, respectively. Right panel: Quantification of the fluorescence bands reported in the left panel. The yield of the viral DNA is normalized to the amount obtained in untreated control cells (Bar 1). Bar 3 and bar 4: viral DNA obtained after treatment with RV 20 µM and 40, respectively.

To assess whether the continuous presence of RV is necessary to inhibit the viral replication we removed the drug at different time points after the viral penetration into the cell (Figure 3). Therefore, the infection was carried out in 20 µM RV but the culture medium was changed to a drug-free fresh medium after different times of treatment and the incubation was continued for 24 hr. Results show that removal of RV after 4 hr incubation has little or no effect on the yield of viral progeny DNA (Lane 2). The drug must be present for the whole infection time to be effective and to cause the complete inhibition of the viral replication (Lanes 6 and 7).

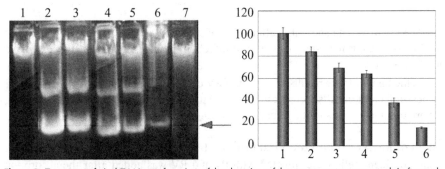

Figure 3. Decrease of viral DNA as a function of the duration of the exposure to resveratrol. Left panel: Progeny viral DNA (form I) is indicated by the arrow. In this case, the culture medium was changed to fresh drug-free medium at the following times post-infection. The incubation was continued for 24 hr. Lane 1: Mock infected cells; Lane 2: Untreated control cells; Lane 3 through 6: 4, 8, 12, and 16 hr, respectively; Lane 7: The medium was not changed and infection was carried permanently in the presence of RV (20 µM). Right panel: Quantification of the fluorescence bands reported in the left panel. The yield of the viral DNA is normalized to the amount obtained in untreated control cells (Bar 1). Withdrawal of RV is reported in the legend to left panel of this figure.

CONCLUSION

The results presented in this work demonstrate a clear, dose dependent cytotoxic and antiviral effect of RV: cytotoxicity at high concentration of the drug both on normal and tumor cells. On the other hand at low concentration, the continuous presence in the culture medium is necessary for the drug to be effective. The target of RV is the replication of viral DNA; however further studies are required for the full elucidation of the inhibitory mechanism mediated by RV leading to the abrogation of the viral DNA synthesis. This effect was demonstrated in the absence of significant cytotoxic effects induced by the drug. Removal of RV at short time after infection does not have a significant effect on the production of viral progeny DNA and this suggests that the viral penetration is not the main target of the drug. Therefore we may conclude that the RV dependent inhibition of the viral proliferation occurs at subsequent stages: possibly during translocation of the virion from cytoplasm to nucleus.

Finally this work gives a further support to the possibility that RV may find a potential clinical use for the control of proliferative pathologies and/or as an antiviral drug.

KEYWORDS

- **Cytotoxicity**
- **Murine polyomavirus**
- **Polyomavirus DNA**
- **Resveratrol**

AUTHORS' CONTRIBUTIONS

All Authors equally contributed to the completion of this work.

ACKNOWLEDGMENTS

Financial support by the Italian Ministry of Education and Sigma-Tau is acknowledged (grants to GR). The collaboration of Michela Di Nottia in performing some experiments is also acknowledged. The graphic elaboration of the figures by Riccardo Risuleo is also acknowledged.

Chapter 4

Novel Anti-viral Peptide Against Avian Influenza Virus H9N2

Mohamed Rajik, Fatemeh Jahanshiri, Abdul Rahman Omar, Aini Ideris, Sharifah Syed Hassan, and Khatijah Yusoff

INTRODUCTION

Avian influenza viruses (AIV) cause high morbidity and mortality among the poultry worldwide. Their highly mutative nature often results in the emergence of drug resistant strains, which have the potential of causing a pandemic. The virus has two immunologically important glycoproteins, hemagglutinin (HA), neuraminidase (NA), and one ion channel protein M2 which are the most important targets for drug discovery, on its surface. In order to identify a peptide-based virus inhibitor against any of these surface proteins, a disulfide constrained heptapeptide phage display library was biopanned against purified AIV sub-type H9N2 virus particles.

After four rounds of panning, four different fusion phages were identified. Among the four, the phage displaying the peptide NDFRSKT possessed good anti-viral properties *in vitro* and *in ovo*. Further, this peptide inhibited the hemagglutination activity of the viruses but showed very little and no effect on neuraminidase and hemolytic activities respectively. The phage-antibody competition assay proved that the peptide competed with anti-influenza H9N2 antibodies for the binding sites. Based on yeast two-hybrid assay, we observed that the peptide inhibited the viral replication by interacting with the HA protein and this observation was further confirmed by co-immunoprecipitation.

Our findings show that we have successfully identified a novel antiviral peptide against avian influenza virus H9N2 which act by binding with the hemagglutination protein of the virus. The broad spectrum activity of the peptide molecule against various subtypes of the avian and human influenza viruses and its comparative efficiency against currently available anti-influenza drugs are yet to be explored.

Avian influenza A viruses (AIV) are enveloped, segmented, and negative-stranded RNA viruses, that circulate worldwide and cause one of the most serious avian diseases called Bird Flu, with severe economic losses to the poultry industry [1]. They are divided into different subtypes based on two surface glycoproteins, hemagglutinin (HA) and neuraminidase (NA). Currently, there are 16 different types of HA and nine different types of NA circulating among aquatic birds [2]. Although, wild birds and domestic waterfowls are considered natural reservoirs for all subtypes, they usually do not show any symptoms of the disease. Domestic birds such as chickens are main victims of this virus especially of H5, H7, and H9 subtypes. The H9N2 viruses are endemic and highly prevalent in poultry of many Eurasian countries. These viruses cause severe morbidity and mortality in poultry as a result of co-infection with other pathogens [3, 4]. Recent studies have also shown that H9N2 prevalence in poultry pose a significant threat to humans [5-8].

Adamantane derivatives (amantadine and rimantadine) and neuraminidase inhibitors (NAIs; zanamivir and oseltamivir) are currently used for the chemoprophylaxis and treatment of influenza [9]. The drugs should be administered within 48 hr of infection to get the optimum results. Amantadine binds to and blocks the M2 ion channel proteins function and thereby inhibits viral replication within infected cells [10]. The NAIs inhibit the activity of neuraminidase enzymes and thus prevent the exit of virus from the infected cells [11].

In the last 15 years, the rate of amantadine resistant strains has risen from 2% during 1995–2000 to an alarming 92.3% in the early 2006 in the US alone for the H3N2 subtype [12] although none of the NAIs and adamantane resistant H5N1 viruses were reported in the south east Asian region from 2004 to 2006 [13]. Usually, these viruses are highly pathogenic and transmissible among animals [14, 15]. The viruses resistant to these drugs emerge due to mutations either at active sites of NA, altering its sensitivity to inhibition, or a mutation in the HA [9]. The mutations at HA reduce the affinity of the proteins to the cellular receptors and enable the virus to escape from infected cells without the need of NA. In several instances, strains which were resistant to both classes of antiviral drugs have been isolated from patients [16-18]. For these reasons, it has become necessary to identify novel drugs against the virus to control and treat infections.

Traditionally, the generation of new drugs involves screening hundreds of thousands of components against desired targets via *in vitro* screening and appropriate *in vivo* activity assays. Currently, new library methodologies have been developed with alternative and powerful strategies, which allow screening billions of components with a fast selection procedure to identify most interesting lead candidates. In this present study we used one of such methodologies called phage display technology to select novel peptides against avian influenza virus H9N2. The selected peptides were characterized for their anti-viral properties and their interaction site with the virus was identified by yeast two-hybrid assay and co-immunoprecipitation. The results showed that one of the peptides possesses good anti-viral property and inhibits the viral replication by binding with HA protein. The broad range anti-viral activity of the peptide against various subtypes of the virus is yet to be studied and if it turned positive, the peptide may serve as an alternative anti-viral agent to replace current potentially inefficient drugs.

Selection of Peptides that Interact with AIV

Peptides selected from phage display library have been used as effective anti-microbial agents in previous studies [19]. In this study, a 7-mer constrained phage displayed random peptide library containing about 3.7×10^9 different recombinant bacteriophages were used to select ligands that interact with the purified target molecule, AIV subtype H9N2. Four rounds of panning were carried out, each with a slight increase in stringency to isolate high-affinity peptide ligands.

Table 1 shows the heptapeptide sequences obtained from four rounds of panning the peptide library against AIV subtype H9N2. Seventeen out of 35 phages analyzed from the fourth round represented the sequence NDFRSKT and other major sequences found in the final round of panning were LPYAAKH and ILGDKVG. A new sequence carrying the peptide QHSTKWF emerged during the fourth round of panning represented 10% of the total phages sequenced.

Table 1. Heptapeptides binding to AIV subtype H9N2 and streptavidin selected from the phage display random peptide library.

Rounds of panning	Heptapeptide sequences	Frequency of sequences (%)
4<h round	NDFRSKT	47
	QHSTKWF	10.5
	LPYAAKH	5
	ILGDKVG	5
	Unrelated sequences	23
Panning of Streptavidin		
3rd round Streptavidin	HPQFLSL	55
	GLYNHPQ	27
	Unrelated sequences	18

Biopanning of the phage library against streptavidin (the positive control) gave a consensus sequences containing HPQ motif which totally represented 82% of the total phages screened from the third round of panning and these results are in good agreement to the reported findings [20-22]. No recognizable consensus sequence was observed with BSA, which served as a negative control. The peptide NDFRSKT was named as P1 (C-P1—cyclic form; L-P1—Linear form; FP-P1—fusion phage displaying this peptide).

Estimation of Binding Abilities of Selected Phage Clones

Recombinant phages selected from the fourth round of panning were further analyzed for their binding specificity by phage-ELISA which was carried out with all the four recombinant phages in varying phage concentrations against two different virus concentrations (5 μg and 10 μg/100 μl). The results (Figure 1) showed that all the phages selected from the biopanning were able to bind the virus efficiently and the higher the concentration of the recombinant phages, the higher the signal irrespective of the concentration of the virus.

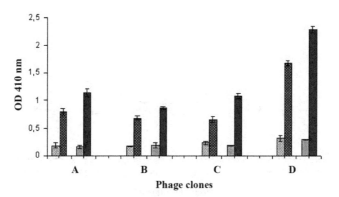

Figure 1. Binding ability of all four recombinant phages to AIV H9N2. Briefly, viruses were coated in the microwell plate at two different concentrations (5 µg and 10 µg/ml; 200 µl) and were detected by two different concentrations of recombinant phage molecules (1,012 pfu/ml and 1,011 pfu/ml). Dotted bars represent the 5 µg of target whereas solid bars represent 10 µg of target. All the four types of recombinant phage particles could able to detect the target AIV. Wild type phage M13 was used as control (Data not shown to avoid complexity of the graph). A—ILGDKVG (5%), B—NDFRSKT (47%), C—LPYAAKH (23%), D—QHSTKWF (5%), Blue Square—1,011 pfu/ml, Gray Square—012 pfu/ml.

Antiviral Activity of Peptides and Fusion Phages *In Vitro*

The fusion phage FP-P1 and the cyclic as well as linear peptides were evaluated for its ability to inhibit viral-induced cell death using a cytotoxicity assay as explained by Jones et al. (2005). Briefly, MDCK cells were mock inoculated (medium alone) or inoculated with different concentrations of phage or peptide treated AIV virus (MOI of 0.05 pfu/cell), and cell viability was evaluated at 48 hpi. If the FP-P1 phages were able to inactivate the AIV, then the AIV might not be able to induce the cell death and so the viability will increase. Interestingly, pre-treatment with increasing concentration of FP-P1 as well as the peptides increased the cell viability in dose dependent manner. More than 100% increase in viability was observed with the fusion phage and peptide treatment. In contrast, treatment with the wild type phage and control peptides did not show any significant increase in viability (Figures 2 and 3). This observation demonstrates that the fusion phage FP-P1 as well as the peptides (both in linear as well as cyclic form) was capable of inactivating the virus or inhibiting the viral replication *in vitro*.

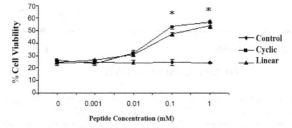

Figure 2. Antiviral activity of peptides *in vitro*. The MDCK cells were inoculated with untreated AIV H9N2 or treated with increasing concentration of linear, cyclic, and control peptides and the cell viability was determined by MTT assay. Results shown are the mean of three trials +/- SD. (*, statistical significance (P < 0.05).

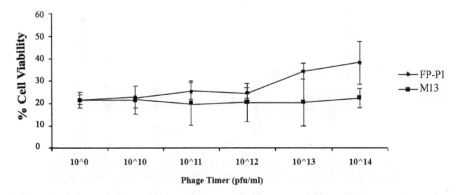

Figure 3. Antiviral activity of fusion phages *in vitro*. The MDCK cells were inoculated with untreated virus (AIV H9N2) or virus treated with increasing concentration of fusion phages and the cell viability was determined by MTT assay. Results shown are the mean of three trials +/- SD.

Antiviral Activity of Peptides and Fusion Phages *in Ovo*

Peptides were evaluated for their antiviral activity *in ovo* against AIV H9N2. Briefly, different concentrations of both cyclic and linear peptides (0.00, 0.001, 0.01, 0.1, and 1 mM) were mixed with constant amount of virus (8 HAU) and injected into allantoic cavity of embryonated chicken eggs. After 3 days, the allantoic fluid was harvested and the HA titer was determined. Complete inhibition was observed at the concentration 1 mM (Figure 4). The IC50 values of both cyclic and linear peptides were 48 µM and 71 µM respectively.

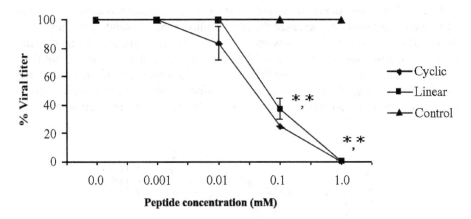

Figure 4. Antiviral activity of peptides *in ovo*. The peptide concentration needed to inhibit 50% of the virus growth was determined using different concentrations of peptides. Experiments were done in triplicates and the error bars represent the standard error of the mean. *, statistical significance ($P < 0.05$) (The SEM value is not shown for other values as there was little variation between repeated experiments).

To evaluate the efficacy of the fusion phage to inhibit the virus propagation *in ovo*, different pfu (10^8–10^{13}/100 µl) of recombinant fusion phages were mixed with constant amount of virus (16 HAU) and injected into the allantoic cavity of embryonated chicken eggs. After 3 days, the allantoic fluid was harvested and the HA titer was measured. The fusion phage FP-P1 reduced the viral titer in the allantoic fluid upto four fold at the concentration more than 10^{13} pfu/100 µl (Figure 5). Based on the dose response curve, the IC_{50} for FP-P1 was approximately 5×10^{11} pfu/100 µl.

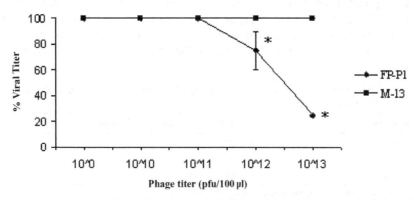

Figure 5. Antiviral activity of fusion phages *in ovo*. The fusion phage concentration needed to inhibit 50% of the virus growth was determined using different concentrations of recombinant phages FP-P1. Experiments were done in triplicates and the error bars represent the standard error of the mean. *, statistical significance (P < 0.05) (The SEM value is not shown for some data as there was no variation between repeated experiments).

Besides, to determine whether these peptides inhibit the virus replication specifically, these peptides (linear, cyclic and FP-P1) were tested for inhibitory effects against NDV strain AF2240. None of these molecules do not posses significant (ANOVA, p = 0.596) inhibitory effect against NDV replication (Figure 6).

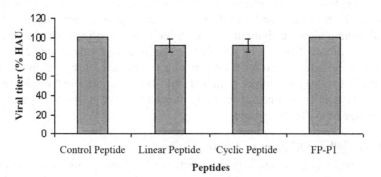

Figure 6. Effect of peptides against NDV. Cyclic, linear and FP-P1 at 100 µM concentrations were analyzed for their inhibitory ability against NDV in embryonated chicken eggs. Viral titers in the allantoic fluid were measured as HA units. Results are shown as the mean of three independent experiments and error bars represent the standard deviation of the mean. None of the peptides showed a statistically significant result (ANOVA, p = 0.596).

Inhibitory Effects of Peptides and Fusion Phages on Virus Adsorption onto Chicken Red Blood Cells (cRBCs)

Influenza A viruses, including AIV sub-type H9N2, have the ability to adsorb onto chicken RBCs, resulting in hemagglutination. So, inhibition of agglutination of blood cells was used to test the hypothesis that peptides C-P1, L-P1, and fusion phage FP-P1 inhibited viral attachment. Initially, the inhibition of viral-induced agglutination of cRBCs by the peptides and fusion phages were monitored. Twofold dilutions of untreated or peptide/phage treated virus were incubated with cRBCs, and agglutination was observed. All the three forms of peptides completely inhibited AIV sub-type H9N2 agglutination in a dose-dependent manner at concentrations of 100 μM or more (Table 2). In contrast, the control peptide CSWGEYDMC had no effect on agglutination.

Table 2. Inhibitory ability of the cyclic and linear peptides against the hemagglutination activity of the avian influenza virus H9N2.

Inhibitory Molecule	Minimum Inhibitory Concentration*
Cyclic Peptide	100 μM
Linear Peptide	100 μM
Fusion Phage	1013 pfu/100 μl

* Minimum concentration of peptides or phage required to inhibit the hemagglutination activity of 32 HAU of AIV

Inhibitory Effects of Peptides and Fusion Phages on Neuraminidase Activity

Based on the ability of the peptides and fusion phage to inhibit viral attachment, we hypothesized that the peptide interacted either with NA or HA since changes to either surface glycoproteins can alter fitness of the virus. Moreover, the biopanning experiment was carried out against the whole virus. As NA is one of the most abundant surface glycoproteins, the chances for binding of the peptides to this protein are relatively high. To determine if peptides or fusion phage inhibited enzymatic activity, untreated or peptide/fusion-phage—treated virus was tested for enzymatic activity. Untreated and cyclic peptide or fusion phage treated virus had similar enzymatic activity, suggesting that both of them had no effect on NA activity. But linear peptide showed reduced Neuraminidase activity at very very high concentrations. The 1,000 μM or more concentration of the linear peptide was required to reduce around 35% of the enzyme activity (data not shown). Considering the inability of cyclic and FP-P1 to inhibit the NA activity and the very limited ability of linear peptide it can be deduced that the linear peptide may non-specifically interact with the NA protein, perhaps taking advantage of its flexible nature.

Inhibition of Phage Binding to AIV by Antibody

Polyclonal antibody (pAb) and phage competition assay was performed to understand whether they both share common binding sites. Briefly, either fusion phages alone or fusion phage-antibody mixtures were added into wells coated with the virus and the eluted phages were titered. Figure 7 demonstrates that the fusion phages FP-P1 were able to compete with the pAb for binding sites on AIV. In the presence of the antibody, the number of phages bound to the AIV coated wells reduced dramatically as a result of the competition between these two molecules for the same binding site on AIV. For example, at input pfu of $1 \times 10^{12}/100$ µl, the output pfu for the FP-P1 phage alone was 1.8×10^4 plaques but in the presence of pAb, the output was reduced to 7.5×10^3 plaques, which is almost 2.4-fold reduction. This result clearly shows us that the phage molecules that display peptides on their surface can compete for the epitope binding sites on AIV with polyclonal antibodies.

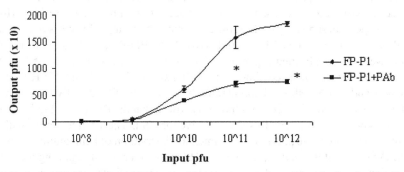

Figure 7. Antibody-phage competition assay. The phage competes with polyclonal antibodies for binding site on AIV, suggesting they may share common binding sites. Experiments were done in triplicates and the error bars represent the standard deviation of the mean. *, statistical significance ($P < 0.05$).

Peptide-phage Competition Assay

In order to identify whether the synthetic peptides and the phages (FP-P1) compete for the same binding sites on AIV H9N2, a peptide-phage competitive assay was performed. When the peptides (both linear and cyclic) were pre-incubated with the virus, the number of phages bound to the virus was reduced gradually in a dose-dependent manner. At 1 mM concentration of the peptides, the phage binding was almost completely inhibited (Figure 8). The control peptide does not possess any inhibitory effects on phage binding to AIV.

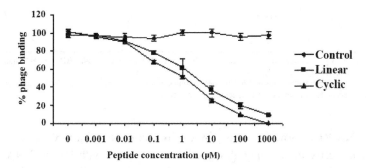

Figure 8. Peptide-phage competition assay. The peptides competes with the fusion phage FP-P1 for binding sites on AIV, suggesting that peptides displayed on the fusion phage FP-P1, and not other parts of the phage, binds to the AIV. Experiments were performed in triplicates and the error bars represent the standard deviation of the mean.

Interaction Between C-P1 Peptide and Hat/NA Protein by Yeast Two Hybrid Assay

The yeast two-hybrid assay was employed to validate the HA-P1 interaction and also to identify any interaction between NA-P1. To eliminate the false positive results (the possibility of binding domain (BD)-P1, activation domain (AD)-HA$_t$ and AD-NA fusion proteins themselves bringing about activation of the reporter genes), various combinations of the recombinant plasmids along with the parental vector were co-transformed into the yeast competent cells (Table 3). Three independent clones from each co-transformation were analyzed for the activation of the β-galactosidase (β-gal) reporter genes. As shown in Table 3, the co-transformed parental vectors did not show any β-gal activities. When BD-P1 and AD-HA$_t$ or BD-P1 and AD-NA fusion constructs were co-transformed separately along with their respective parental vectors, no β-gal activity was detected either. The co-transformed BD-P1 and AD-HA$_t$ as well as BD-P1 and AD-NA showed comparatively high level of β-gal activity (25 and 3.5 Miller Units respectively). This observation showed that the P1 peptide bind with both with HA glycoprotein as well as the NA glycoprotein. The P1 interaction with HA glycoprotein support the previous experimental observation of hemagglutination inhibition. As the yeast two-hybrid assay provided ambiguous result regarding the NA-P1 interaction, further experimental analysis (co-immunoprecipitation) was carried out.

Table 3. P1: HAt/NA interactions in the yeast two-hybrid system.

DBD Vectors[a]	AD Vectors[a]	β-gal activity[b]
Background		
BD	AD	0.05
BD-PI	AD	0.07
BD	AD-HA$_t$	0.05
BD	AD-NA	0.03
PI: HA/NA interactions		
BD-PI	AD-HA$_t$	25
BD-PI	AD-NA	3.5

[a]pHyblex/Zeo and pYESTryp2 are vectors encoding the LexA DNA binding domain (BD) and B42 transcriptional activation domain (AD), respectively.
[b]Average ~-gal activity (Miller Units of 3 independent colonies for each co-transformation (The SD value is not shown as there was very little variation between repeated experiment).

HAt-P1, NA-P1 Interaction Study by Co-immunoprecipitation

In order to verify the binding ability of the peptide P1 with HA$_t$ and NA proteins through Co-IP method, these three proteins were initially synthesized by *in vitro* transcription and translation methods. The P1 peptide was mixed either with the HA$_t$ protein or NA protein separately to allow the binding overnight or after incubation, the

HA or NA protein present in the mixture was immunoprecipitated by anti-AIV poly-clonal serum. After three rounds of washing, the bound P1 was detected by anti-His monoclonal antibodies (Novagen, USA). The P1 peptide was detected only in the HA complex (Figure 9). There was no P1 peptide visible in the NA complex (data not shown). This experiment confirmed the interaction of the P1 peptide to the HA protein.

Figure 9. Western blot analysis of immunoprecipitated HAt-P1 complex. *In vitro* translated NA protein or HA$_t$ protein was mixed with P1 peptide and the complex was co-immunoprecipitated using anti-AIV serum and the eluted complex was analyzed by SDS-15% PAGE, electrotransferred to a nitrocellulose membrane and probed with anti-His monoclonal antibody (Novagen, USA). Lane 1: HA$_t$ and P1 complex; Lane 2: NA and P1 complex; For control, *in vitro* translated NA or HA$_t$ mixed with control peptide SWGEYDM and detected using anti-His antibodies. Lane 3: HA$_t$ and Control peptide complex; Lane 4: NA and control peptide complex; Lane 5: *in vitro* translated P1 peptide (~12 kDa). The arrow indicates the precipitated P1 protein in the HA$_t$-P1 complex and the *in vitro* translated P1 peptide.

Peptide Toxicity

To analyze the cellular toxicity properties of the peptides and fusion phages, MDCK cells were exposed to 100 μM of cyclic, linear peptides or 10^{13} pfu/100 μl of FP-P1 for 24 hr and the cell viability was determined by MTT assay. There was no significant difference (Students t test, P > 0.05) observed in the cell viability of control and pep-tide treated cells (Figure 10).

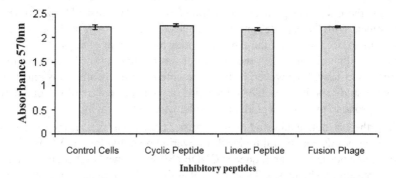

Figure 10. *In vitro* toxicity of inhibitory peptides. The MDCK cells were treated with 100 μM of C-P1 or L-P1 or 1,014 pfu/ml of FP-P1 and the cell viability was analyzed by MTT assay after 24 hrs of incubation (mean of three experiments +/- SD). No statistically significant differences in cell viability were observed (Students t test, P > 0.05).

DISCUSSION

Emerging and re-emerging infectious diseases remain to be one of the major causes of death worldwide. The current outbreak of avian influenza viruses is a major global concern due to the increasing number of fatalities among the poultry as well as human cases. Its highly mutative nature makes the current antiviral drugs not very effective. Therefore, there has been a constant need for broad-spectrum antiviral drugs against the currently circulating human as well as avian strains.

In this study, a phage displayed peptide library was used to select anti-viral peptides against the AIV H9N2. At the end of biopanning, four different peptide sequences were identified. Matching of these peptide sequences with protein sequences in the protein data banks (Swiss Prot and NCBI) showed no significant homology with any protein sequences. It is possible that these peptides might mimic a discontinuous binding site in which amino acids are brought from different positions of a protein to form an essential contact area with the virion [23, 24]. The lack of antiviral activity by the control peptide as well as the wild type phage suggest that the antiviral property of the peptides is specific to those peptides and neither a general property of any oligomeric peptide or wild type M13 bacteriophages nor based on charge or hydrophobic interactions. The peptide phage competition assay proved that the peptide displayed on the phage surface not the other parts of the phage binds to the virus.

Among the four different fusion phages isolated from the phage display library, the phage displaying the sequence NDFRSKT was selected for further analysis as it represented highest number of clones in the final round of biopanning. Besides, the peptides LPYAAKH, ILGDKVG, and QHSTKWF showed negligible or no anti-viral activity (data not shown); therefore, no further analyses on these peptides were carried out.

The *in ovo* model has been previously employed successfully by our group Ramanujam et al. [25] and Song et al. [26] to study the inhibitory effect of anti-viral molecules against the Newcastle disease virus and influenza virus respectively. Therefore, the antiviral activity of the synthetic peptides and the fusion phages themselves (or simply denoted as inhibitory peptides hereafter) were investigated in embryonated chicken eggs. All the peptides showed good anti-viral properties against AIV and interestingly there was no significant anti-viral effect found against NDV strain AF2240. Pre-treatment with the peptides or fusion phages reduced the AIV titre manifold (from two fold to six fold based on the type of peptide and number of days of treatment) in the infected allantoic fluid. But the post-infection treatment failed to protect the embryo (data not shown). However, it should be noted that the peptide was injected only once in the study and besides, the amino acids of the peptide were of L-isomers which are more prone to protease degradation inside the allantoic cavity.

Nevertheless, both cyclic and linear forms of peptides as well as the fusion phages proved their worth as antiviral molecules in varied potential levels. Among them, the cyclic peptide possessing the sequence CNDFRSKTC showed higher antiviral properties. The reason maybe its small size (only 9 amino acids in length for cyclic peptide) which helps its easy access to the respective binding site on the target molecule. Moreover, the cyclic peptides possess a stable structure due to the disulfide bond formed between the flanked cysteine residues which help to attain a stable interaction at a short time when compared to the linear peptides [27, 28]. Small peptide molecules have been used in the development of peptide based vaccines for melanoma [29], inhibitors against HIV [30], Dengue and West nile virus [31] and anti-angiogenic in the treatment of angiogenesis related diseases [32].

As whole virus particles were used in biopanning experiments, in principle, the selected peptides might interact with any of the three surface proteins such as HA, NA and M2. Since these inhibitory peptides possess strong anti-viral activity when used at pre-infection not at post-infection and also inhibit the hemagglutination, it can be deduced that the peptides (NDFRSKT and CNDFRSKTC) prevent the viral replication by inhibiting the attachment or entry of the virus into the target cells. There are many studies on the targeting of the conserved region of the HA protein. Recently, Jones et al. [33] identified that a well known cell-penetrating peptide, derived from the fibroblast growth factor 4 (FGF-4) signal sequence, possesses the broad-spectrum anti-influenza activity, which act by blocking the entry of virus through the HA protein interaction.

Neuraminidase (NA) is the second most abundant surface protein and responsible for the neuraminidase activity of the virus. It is important both for its biological activity in removing sialic acid from glycoproteins and as a major antigenic determinant that undergoes variation. At present, the NAIs such as zanamivir and oseltamivir are preferentially used for the treatment and prophylaxis of influenza [9], as the NA protein is less mutative when compared with HA. There are three receptor binding sites, two at the distal ends of both HA subunits and the third one in the NA protein [34] and changes in both HA and NA glycoproteins will affect the fitness of the virus [35]; therefore, the effect of peptide on the neuraminidase protein was assessed. Unfortunately, this experiment showed a negative result for the fusion phages and cyclic peptides and partial inhibition result at very high concentration of linear peptide (~35% inhibition at 1,000 µM). The latter inhibition may be nonspecific due to the increased ability of the linear molecules to attain a structure that facilitates the binding with NA molecule or merely based on hydrophobicity and charge.

The HA-P1 and NA-P1 interaction was further analyzed by the yeast two-hybrid system and co-immunoprecipitation. There has been a problem in amplifying the full length clone of *HA* gene for the past few years in our laboratory. The same problem has also been reported in few other laboratories working with the same strain in this region. The 3' end of the vRNA could not be amplified either by primer designed for conserved region or gene specific region based on other similar strain's sequence. The HA protein should be cleaved into two disulfide linked HA_1 and HA_2 in order to be infectious. The C-terminal HA_2 region is very important as it accounts for the entry of the virus into the host cell and thus serves as a fusion protein [36]. Therefore, the truncated HA protein representing C-terminal end (278 amino acids) of the full length HA protein was used for the yeast two-hybrid and co-immunoprecipitation experiments. The yeast hybrid assay turned positive for the both HA and NA proteins although the β-galactosidase activity for HA is nearly seven fold higher than the NA. Although, there was negligible or no interaction between NA and P1 as per the results of NA inhibition test and co-immunoprecipitation results, the yeast two-hybrid experiment showed a significant NA-P1 interaction which is almost 100 times higher than the control. So, NA-P1 interaction cannot be simply ignored and further investigations are required to analyze the kind of interaction between the NA glycoproteins and peptide P1. But, the HA and P1 interaction has been clearly demonstrated without any doubt in all the performed experiments.

Taking all together, this study has identified a novel antiviral molecule which inhibits the avian influenza virus infection by interacting with the surface glycoprotein HA and preventing its attachment to the host cell. To our knowledge, the selected peptide is the only antiviral peptide among the currently identified anti-viral peptides with 7 or 9 amino acids in length. This short sequence will be an added advantage for commercialization purpose as it can greatly reduce the cost of production. However, additional studies are required to define the broad-spectrum activity of the peptide against various strains including the currently circulating potential pandemic strains such as H1N1 and H5N1 as well as its diagnostic potential.

MATERIAL AND METHODS

Viruses, Cells, and Viral Purification

Avian influenza A/Chicken/Iran/16/2000(H9N2), a low pathogenic avian influenza virus and Newcastle disease virus (NDV) strain AF2240 was kindly provided by Abdul Rahman Omar. Viruses were propagated in 9-day old specific pathogen free embryonated chicken eggs. The allantoic fluid was clarified and the viruses were purified and concentrated as explained previously [25]. The virus titer was determined by hemagglutination test (HA) and the protein concentration of the purified virus was determined by Bradford assay [37].

Selection of Peptides Against AIV Sub-type H9N2

The virus (15 µg/ml; 100 µl) was coated onto a microtiter plate well with $NaHCO_3$ (0.1 M, pH 8.6) buffer overnight at 4°C. Streptavidin (0.1 mg/ml; 100 µl) was also coated and used as positive control. Phages from a disulfide constrained 7-mer phage display random peptide library (New England Biolabs, USA) were biopanned as explained by the manufacturer. The amplified phages from the first round of biopanning were used for the second round of biopanning. Totally four rounds of biopanning were carried out. Phage titration was carried out according to the method described by Sambrook et al. [38]. Phages were propagated in *Escherichia coli* (*E. coli*) host cells grown in LB broth (1 L). The phage particles were precipitated by PEG and purified through cesium chloride density gradient centrifugation as descried by Smith and Scott [39].

Sequence Analysis of Phagemids

The nucleotide sequence encoding the hypervariable heptapeptide region of pIII coat protein of M13 phage was sequenced by 1st Base Laboratories Sdn Bhd, Kuala Lumpur, with the -96 gIII sequencing primer 5' CCC TCA TAG TTA GCG TAA CG 3'. Sequence analyses such as comparison with wild type M13 phage pIII coat protein and prediction of amino acid sequences were performed with the free bioinformatics software package, SDSC Biology Workbench 3.2.

Estimation of Binding Abilities of Selected Phages

The avian influenza viruses were coated (5 or 10 µg/ml; 200 µl) on a microtiter plate with TBS buffer overnight at 4°C. The excess target was removed and blocked with blocking buffer (milk diluent KPL, USA) for 2 hr at 4°C. The plate was then washed with 1× TBST (TBS and 0.5% [v/v] Tween 20). Selected phages were added into the well at the concentration of either 10^{12} pfu/ml or 10^{11} pfu/ml and incubated for 2 hr at room temperature. The plate was again washed 6 times with 1× TBST. The HRP-conjugated anti-M13 antibody (Pharmacia, USA) was diluted into 1:5,000 with blocking buffer and added 200 µl into each well, incubated at room temperature for 1 hr with agitation. It was then washed six times with 1 × TBST as explained above. 200 µl substrate solutions (22 mg ABTS in 100 ml of 50 mM sodium citrate and 36 µl of 30% H_2O_2, pH 4.0) was added to each well and incubated for 60 min. Then the plate was read using a microplate reader (Model 550, BioRad, California, USA) at 405–415 nm.

Table 4. Peptides used in this study.

Name of the peptide	Sequence of the peptide
L-P I (Linear Peptide)	NDFRSKT
C-PI (Cyclic Peptide)	CNDFRSKTC
Control Peptide	CSWGEYDMC

Peptides

Peptides were synthesized at GL Biochem, Shanghai, China with more than 98% purity. The peptides contained the sequences as mentioned in Table 4.

Cytotoxicity Test by MTT Assay

MDCK cells (~5,000 cells/well) were grown on 96 well plates for 24 hr. The media was replaced by serially diluted peptides or fusion phages and incubated again for 48 hr. The culture medium was removed and 25 µl of MTT (3-(4,5-dimethylthiozol-2-yl)-3,5-dipheryl tetrazolium bromide) (Sigma) was added and incubated at 37°C for 5 hr. Then 50 µl of DMSO was added to solubilized the formazan crystals and incubated for 30 mn. The optical density was measured at 540 nm in an microplate reader (Model 550, BioRad, USA).

Virus Yield Reduction Assay in Egg Allantoic Fluid

The avian influenza A/Chicken/Iran/16/2000 (H9N2) virus suspension containing 8 or 16 HAU/50 µl was mixed with various concentrations of linear/cyclic peptides or fusion phages (50 µl) for 1 hr at room temperature. This mixture was then injected into the allantoic cavity of 9 day-old embryonated chicken eggs and incubated at 37°C for 3 days. After incubation, the eggs were chilled for 5 hr, the allantoic fluids were harvested and titrated by hemagglutination (HA) assay. As control, virus mixed with nonspecific peptides or wild phages were injected into the eggs.

Hemagglutination Inhibition Assay

The hemagglutination inhibition (HI) assay was carried out as originally explained by Ramanujam et al. (2002) with slight modifications to evaluate the ability of the peptides/fusion phages to inhibit the viral adsorption to target cells. Linear/Cyclic peptides or fusion phages (50 μl) in serial 2-fold dilutions in PBS were mixed with equal volume of influenza solution (8 HAU/50 μl) and incubated at room temperature for 1.5 hr. Subsequently, 50 μl of 0.8% red blood cells were added to the above mixture and further incubated at room temperature for 45 min.

Neuraminidase Inhibition Assay

The neuraminidase inhibition assay was carried out to test the ability of the peptide to inhibit the viral neuraminidase activity, as explained in Aymard-Hendry et al. [40] with slight modifications. The substrate used in this experiment was neuraminlactose rather than feutin.

Preparation of Anti-AIV Sera

Six month old New Zealand white rabbits were used for the production of polyclonal antibodies. Rabbits were pre-bleeded before injection. 50 μg of purified virus in PBS together with equal amount of Freund's adjuvant was injected into the rabbit subcutaneously. Subsequent booster injections were done with Freund's incomplete adjuvant. Injections were done for every 4 weeks, with bleeds 7–10 days after each injection. Antibodies were purified with Montage® antibody purification kits (Millipore, USA) as instructed by the manufacturer.

Antibody-phage Competition Assay

Wells were coated with AIV subtype H9N2 (20 μg/ml; 100 μl) as the aforesaid conditions of biopanning. A mixture of purified polyclonal antibodies (1:500 dilutions; 100 μl) raised against AIV sub-type H9N2 and a series of different concentrations of phage FP-P1 (10^8–10^{12} pfu; 100 μl) were prepared in Eppendorf tubes. After blocking the wells, these mixtures were added and incubated at room temperature for 1 hr. Wells were washed and bound phages were eluted and titrated. As for the positive control, AIV coated wells were incubated with the phage without the presence of the polyclonal antibodies.

Peptide-phage Competition Assay

The peptide-phage competition assay was performed to assay the inhibitory effects of synthetic peptides with its phage counterparts (FP-P1). The AIV H9N2 was coated on a multi-well plate at the aforesaid conditions of biopanning and incubated with different concentrations of either linear of cyclic peptides (0.0001–1,000 μM) in binding buffer for 1 hr at 4°C. After 1 hr incubation, phage FP-P1 (10^{10} pfu/100 μl) was added and incubated at 4°C for another 1 hr. Wells were then wash six times with TBST and the bound phages were eluted and titered. (Percentage of phage binding = (number of phage bound in the presence of peptide competitor/number of phage bound in the absence of peptide competitor) × 100).

In Vivo Study of Protein–Protein Interactions: Yeast Two-hybrid Assay

Cloning of Ha$_t$, NA snd P1 Genes Into Pyestrp2 and Phyblex/Zeo Vectors

The NA and truncated HA protein (*HA$_t$*) genes of AIV sub-type H9N2 were amplified by Reverse Transcription-Polymerase Chain Reaction (RT-PCR) from the viral RNA using the primers pY-HA$_t$-F and R and pY-NA-F and R, mentioned in Table 5. The *NA* gene carried the recognition sites for EcoRI and XhoI whereas the *HA$_t$* gene carried the recognition sites KpnI and XhoI restriction enzymes in their forward and reverse primers respectively. The peptide gene (*P1*) was amplified including the N1 domain of the P3 protein of the recombinant phage using the primer pH-P1-F and R (Table 5) from the ssDNA genome of the phage as the peptide is displayed as a fusion protein to this domain of the P3 protein. The *P1* gene carried the recognition sites for EcoRI and XhoI restriction enzymes in its forward and reverse primers respectively. The amplified *HA$_t$* and *NA* genes were ligated into pYESTrp2 vectors separately (Invitrogen, USA) and the *P1* gene was cloned into pHybLex/Zeo (Invitrogen, USA) vector. The resultant clones were named as pY-HA, pY-NA and pH-P1 respectively. The constructs were sequenced using the primers pYESTrp2-F and R and pHybLex/Zeo-F and R (Table 5) to check the reading frame and for the absence of mutations. The *Saccharomyces cerevisiae* strain L40 was then co-transformed with the recombinant plasmids using the lithium acetate method and the transformants were analyzed for their β-galactosidase activity as explained in Ausubel et al. [41].

Table 5. Oligonucleotides used to amplify the *NA*, *HA$_t$* and *P1* genes.

Primers	Sequence
pY-NA-F[a]	5' CAT AGAA TTCGCAAAAGCAGGAGT 3'
pY-NA-R	5' TATCGCTCGAGAGT AGAAACAAGGAG 3'
pY-HA-F	5' ATTTAAGGTACCGACAGCCATGGA 3'
pY-HAt-R	5' ATGCTGCTCGAGTATACAAATGTTGC 3'
pH-PI-F	5' AGCCTGGAATTCATGAAAAAATTA 3'
pH-PI-R	5' ATCGAACTCGAGA TTTTCAGGGAT 3'
pHyblex/Zeo-F	5' AGGGCTGGCGGTTGGGGGTT A TTCGC 3'
pHyblex/Zeo-R	5' GAGTCACTTTAAAATTTGTATACAC 3'
pYESTrp2-F	5' GATGTTAACGATACCAGCC 3'
pYESTrp2-R	5' GCGTGAA TGT AAGCGTGAC 3'
pC-HA-F	5'A TTT AAGGATCCGAGAGCCATGGA 3'
pC-HA-R	5'ATGCTGCTCGAGTTATATACAAA TGTTGC 3'
pC-NA-F	5'CATAGAA TTCGCAAAAGCAGGAGT 3'
pC-NA-R	5'T A TCGCTCGAGAGT AGAAACAAGGAG 3'
pC-PI-FP	5'AGCCTGGAA TTCATGAAAAAATT A 3'
pC-PI-RP	5'CTCACTCGAGACATTTTCAGGGAT 3'

• In all of the above mentioned oligonucleotides, the suffixes F and R refers Forward and Reverse primers respectively

In Vitro Study of Protein–protein Interactions

Construction of Recombinant Pc-Hat, Pc-NA, and Pc-P1 and In Vitro Transcription and Translation

The HA_t and *NA* gene of AIV strain H9N2 as well as the recombinant peptide gene *P1* was amplified from pY-HA, pY-NA and pH-P1 respectively as templates using the primers pC-HA-F and R, pC-NA-F and R and pC-P1-F and R respectively (Table 5) and cloned into the pCITE2a vector. The *in vitro* transcription and translation was performed in a single tube in a reaction mixture (15 µl) containing circular recombinant plasmid (1 µg), TNT® Quick Master Mix (12 µl; Promega, USA), Methionine (0.3 µl, 1 mM; Promega, USA). The above mixture was incubated at 30°C for 90 min. The translated products (3 µl) were electrophoresed on 15% SDS-PAGE and then transferred by electrophoresis for 1 hr onto a nitrocellulose membrane. They were detected with anti-His antibody for P1 protein and HAt/NA proteins were detected with the polyclonal antibodies raised against the AIV sub-type H9N2 in rabbit.

Co-immunoprecipitation

Co-immunoprecipitation was performed using the Pierce® Co-IP kit (Thermo Scientific, USA) as per the instructions given by the manufacturer. Briefly, the bait and pray complex was prepared separately by mixing the HA_t or NA with His-conjugated P1 peptide. The complex was precipitated using purified anti-AIV polyclonal antibodies, which were immobilised on antibody coupling resin. The peptide P1 in the eluted co-immunoprecipitated complex was analyzed by Western blotting using anti-His monoclonal antibodies (Novagen, USA) and detected with Amersham® ECL® Western blotting detection reagents (GE Healthcare, USA).

Statistical Analysis

All experiments were carried out in triplicate and are representative of at least three separate experiments. The results represent the means ± standard deviations or standard error means of triplicate determinations. Statistical significance of the data was determined by independent t test or one-way ANOVA method using SPSS software.

KEYWORDS

- **Co-immunoprecipitation**
- **Hemagglutination**
- **Hydrophobicity**

AUTHORS' CONTRIBUTIONS

Mohamed Rajik designed the study, carried out all of the experiments and drafted the manuscript. Fatemeh Jahanshiri participated in the design of yeast two hybrid assay experiments. Abdul Rahman Omar, Aini Ideris, and Sharifah Syed Hassan participated in the design of avian influenza virus related experiments. Khatijah Yusoff conceived the study, participated in its design and co-ordination and helped to draft the manuscript. All authors read and approved the final manuscript.

ACKNOWLEDGMENTS

This project is supported by the Ministry of Science, Technology and Innovation (MOSTI) of Government of Malaysia grant No.01-02-04-009 BTL/ER/38. Rajik is supported by the Universiti Putra Malaysia graduate research fellowship. The authors also acknowledge Ms. Hamidah for her help in statistical analysis.

COMPETING INTERESTS

Mohamed Rajik is a graduate student of Universiti Putra Malaysia (UPM). Fatemeh Jahanshiri, Abdul Rahman Omar, Aini Ideris, and Khatijah Yusoff are employees of the same institution. The university holds the rights for all the financial benefits that may result from this research. Neither Sharifah Syed Hassan nor her institutions do not have any competing interests with this study. UPM is financing this manuscript as well. The UPM is the owner of the patent for the peptides mentioned in this manuscript (Patent No.: PI20082061).

Chapter 5

Prion Disease Pathogenesis

Ajay Singh, Maradumane L. Mohan, Alfred Orina Isaac, Xiu Luo, Jiri Petrak, Daniel Vyoral, and Neena Singh

---

INTRODUCTION

Converging evidence leaves little doubt that a change in the conformation of prion protein (PrP^C) from a mainly α-helical to a β-sheet rich PrP-scrapie (PrP^{Sc}) form is the main event responsible for prion disease associated neurotoxicity. However, neither the mechanism of toxicity by PrP^{Sc}, nor the normal function of PrP^C is entirely clear. Recent reports suggest that imbalance of iron homeostasis is a common feature of prion infected cells and mouse models, implicating redox-iron in prion disease pathogenesis. In this report, we provide evidence that PrP^C mediates cellular iron uptake and transport, and mutant PrP forms alter cellular iron levels differentially. Using human neuroblastoma cells as models, we demonstrate that over-expression of PrP^C increases intracellular iron relative to non-transfected controls as indicated by an increase in total cellular iron, the cellular labile iron pool (LIP), and iron content of ferritin. As a result, the levels of iron uptake proteins transferrin (Tf) and transferrin receptor (TfR) are decreased, and expression of iron storage protein ferritin is increased. The positive effect of PrP^C on ferritin iron content is enhanced by stimulating PrP^C endocytosis, and reversed by cross-linking PrP^C on the plasma membrane. Expression of mutant PrP forms lacking the octapeptide-repeats, the membrane anchor, or carrying the pathogenic mutation PrP^{102L} decreases ferritin iron content significantly relative to PrP^C expressing cells, but the effect on cellular LIP and levels of Tf, TfR, and ferritin is complex, varying with the mutation. Neither PrP^C nor the mutant PrP forms influence the rate or amount of iron released into the medium, suggesting a functional role for PrP^C in cellular iron uptake and transport to ferritin, and dysfunction of PrP^C as a significant contributing factor of brain iron imbalance in prion disorders.

The PrP^C is an evolutionarily conserved cell surface glycoprotein expressed abundantly on neuronal cells. Despite its ubiquitous presence, the physiological function of PrP^C has remained ambiguous. The best characterized role for this protein remains its involvement in the pathogenesis of familial, infectious, and sporadic prion disorders, where a change in the conformation of PrP^C from a mainly α-helical to a β-sheet rich PrP^{Sc} form renders it infectious and pathogenic [1-5]. The mechanism by which PrP^{Sc} induces neurotoxicity, however, is not clear. Studies over the past decade have clarified several aspects of this process [1, 6, 7]. Prominent among these is the resistance of transgenic mice lacking neuronal PrPC expression to PrP^{Sc} induced toxicity, implicating PrP^C as the principal mediator of the neurotoxic signal [8, 9]. However, prion infected transgenic mice expressing PrP^C only on astrocytes accumulate PrP^{Sc} and succumb to disease [10], leaving the matter unresolved. Adding to the complexity is the development of prion specific neuropathology in mice over-expressing normal or mutant PrP in the wrong cellular compartment in the absence of detectable PrP^{Sc}, suggesting the presence of additional pathways of neurotoxicity [1, 7]. Although, brain homogenates from these animals are not infectious in bioassays, these models suggest that a disproportionate change in the physiological function of PrP^C is as neurotoxic as the gain of toxic function by PrP^{Sc}. Investigations on both fronts are therefore essential to uncover the underlying mechanism(s) of neurotoxicity in these disorders.

Efforts aimed at understanding the physiological function of PrP^C and pathological implications thereof have revealed several possibilities, varying with the model, the physiological state, and the extra- and intracellular milieu in a particular tissue. Some of the reported functions include a role in cell adhesion, signal transduction, and as an anti-oxidant and anti-apoptotic protein [7, 11, 12]. While the importance of these observations cannot be under-estimated, they fail to provide a direct link between PrP^C function and dysfunction to prion disease pathogenesis. In this context, it is interesting to note that PrP^C binds iron and copper, and is believed to play a functional role in neuronal iron and copper metabolism [13, 14]. Since both iron and copper are highly redox-active and neurotoxic if mis-managed, it is conceivable that dysfunction of PrP^C due to aggregation to the PrP^{Sc} form causes the reported accumulation of redox-active PrP^{Sc} complexes in prion infected cell and mouse models, inducing a state of iron imbalance [15-17]. A phenotype of iron deficiency in the presence of excess iron is noted in sporadic Cruetzfeldt–Jakob disease (sCJD) affected human and scrapie infected animal brain tissue, lending credence to this assumption [45].

To explore if PrP^C is involved in cellular iron metabolism, we investigated the influence of PrP^C and mutant PrP forms on cellular iron levels in human neuroblastoma cells expressing endogenous levels (M17) or transfected to express 6–7 fold higher levels of PrP^C or mutant PrP forms. The following parameters were evaluated: (1) total cellular iron, (2) intracellular LIP, (3) iron content of ferritin, and (4) levels of iron uptake proteins TfR and Tf and iron storage protein ferritin that respond to minor changes in the LIP [18, 19]. Our data demonstrate that PrP^C increases cellular iron levels and the cells demonstrate a state of mild overload, while pathogenic and non-pathogenic mutations of PrP alter cellular iron levels differentially, specific to the mutation.

Normal and Mutant Prp Forms Influence Cellular Iron Levels Differentially

The influence of PrP expression on cellular iron status was evaluated in M17 cells expressing endogenous PrP^C or stably transfected to express 6–7 fold higher levels of PrP^C or the following mutant PrP forms: (1) $PrP^{231stop}$ that lacks the glycosylphosphatidyl inositol (GPI) anchor and is secreted into the medium, (2) $PrP^{\Delta51-89}$ that lacks the copper binding octa-peptide repeat region, (3) $PrP^{\Delta23-89}$ that lacks the N-terminal 90 amino acids, and (4) PrP^{102L} associated with Gerstmann–Straussler–Scheinker disease (GSS), a familial prion disorder (Figure 1A). Expression of PrP in transfected cell lines was assessed by separating cell lysates on SDS-PAGE and probing transferred proteins with the PrP specific monoclonal antibody 3F4 [26]. As expected, the di-, mono-, and unglycosylated forms of PrP^C, $PrP^{\Delta51-89}$, $PrP^{\Delta23-89}$, and PrP^{102L} migrating between 20 and 37 kDa are detected (Figure 1B, lanes 2–5). Deletion mutations $PrP^{\Delta51-89}$ and $PrP^{\Delta23-89}$ migrate faster than PrP^C and PrP^{102L} as expected (Figure 1B, lanes 3 and 4).

M17 lysates show barely detectable levels of PrP^C, while transfected cell lines express significantly higher levels of PrP^C and mutant PrP forms (Figure 1B, lanes 1–5).

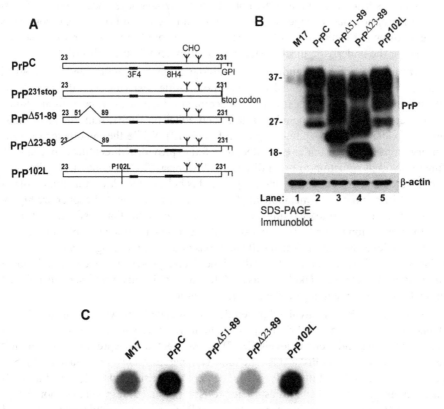

Figure 1. Cells expressing normal and mutant PrP forms incorporate different levels of iron. (A) Diagrammatic representation of PrPC and mutant PrP forms evaluated in this study. (B) Lysates of M17, PrP^C, PrP$^{\Delta51-89}$, PrP$^{\Delta23-89}$, and PrP102L were resolved by SDS-PAGE and immunoreacted for PrP and β-actin. All transfected cell lines express 6–7 fold higher levels of PrP relative to non-transfected M17 cells (lanes 1–5). (C) Cell lines in (B) were radiolabeled with ^{59}FeCl$_3$-citrate complex, washed with PBS supplemented with 100 µM DFO to chelate surface bound iron, and lysed. Equal amount of protein from each sample was spotted on a PVDF membrane, air dried, and exposed to an X-ray film.

To evaluate if PrP^C or mutant PrP forms influence cellular iron uptake, M17, PrP^C, PrP$^{\Delta51-89}$, PrP$^{\Delta23-89}$, and PrP^{102L} cells cultured in serum-free medium for 1 hr were radio-labeled with ^{59}FeCl$_3$-citrate complex for 4 hr in the same medium, washed with PBS supplemented with 100 µM desferrioxamine (DFO) to remove surface bound iron, and lysed in non-denaturing buffer. Equal amount of protein from lysates was spotted on a PVDF membrane, air-dried, and exposed to an X-ray film. Surprisingly, PrP^C and PrP^{102L} cells incorporate significantly more ^{59}Fe, while PrP$^{\Delta51-89}$, PrP$^{\Delta23-89}$ cells take up less ^{59}Fe than M17 controls (Figure 1C).

Major ^{59}Fe labeled proteins in these cells were identified by separating cell lysates prepared in non-denaturing buffer on a 3–20% native gel in duplicate. One part of the gel was dried and subjected to autoradiography (Figure 2, lanes 1–5), while the other was transferred to a PVDF membrane under native conditions and probed for ferritin and Tf using specific antibodies [19, 20] (Figure 2, lanes 6–15). Autoradiography shows a prominent iron labeled band consistent with ferritin (Figure 2, lanes 1–5 and 6–10, black arrow), and a faster migrating band representing Tf (Figure 2, lanes 1–5 and 11–15, open arrow) (the lower part of the autoradiograph is over-exposed to highlight the Tf band). Compared to M17 lysates, the amount of ^{59}Fe bound to ferritin is higher in PrP^C and PrP^{102L} lysates, and lower in PrP$^{\Delta51-89}$ and PrP$^{\Delta23-89}$ lysates (Figure 2, lanes 1–5). On the other hand, Tf bound iron is higher in M17 compared to PrP^C, PrP$^{\Delta51-89}$, and PrP$^{\Delta23-89}$ lysates, and equivalent to PrP^{102L} lysates (Figure 2, lanes 1–5 and 11–15). The slower migrating iron labeled bands (*) probably represent a complex of Tf and TfR (Figure 2, lanes 1–5) [20, 21]. Probing for ferritin shows a major band and minor slower migrating forms probably representing ferritin complexes (Figure 2, lanes 6–10, black arrow). Probing for Tf shows oligomers or glycosylation variants of Tf that correspond to ^{59}Fe labeled purified transferrin fractionated similarly (Figure 2, lanes 1–5, 11–15). The relative levels of ferritin and Tf proteins in the samples correspond to radioactive iron in labeled ferritin and Tf bands in all samples (Figure 2, lanes 1–15). Similar results were obtained when the cells were labeled with [59]FeCl$_3$-citrate complex for 16 hr or with purified ^{59}Fe-Tf for 4 and 16 hr (data not shown), indicating similar uptake of non-transferrin and Tf bound Fe by these cells. Silver staining of re-hydrated autoradiographed gel confirms equal loading of protein for all samples analyzed. Quantitative comparison of ferritin iron and levels of PrP, ferritin, and Tf between the cell lines is shown below in Figure 4.

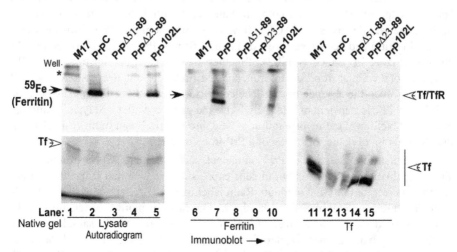

Figure 2. The PrP influences iron incorporation in cellular ferritin. Radiolabeled lysates were fractionated on a 3–20% native gradient gel in duplicate. One set was subjected to autoradiography (lanes 1–5) and the other was transblotted and probed for ferritin and transferrin under native conditions (lanes 6–15).

The identity of iron labeled bands in Figure 2 was further confirmed by cutting each band from fractionated PrP^C lysates and re-fractionating electro-eluted proteins on SDS-PAGE followed by immunoblotting. Lane 1 represents proteins eluted from the loading well that did not enter the running native gel. Lanes 2, 3, and 5 represent iron labeled bands that resolve adequately on native gels, and lane 4 represents unlabeled section of the gel that serves as a negative control. Sequential immunoreaction with specific antibodies confirms the presence of PrP in band 1, TfR in bands 1 and 2, ferritin in band 3, and Tf in band 5. Band 4 does not react with antibodies to known iron binding proteins. Silver staining shows co-migration of a few other un-identified proteins with bands 1–3, and almost none with bands 4 and 5.

To determine if PrP^C mediates iron uptake directly, a modified non-denaturing gel system with a 3–9% gradient was used to separate ^{59}Fe-labeled PrP effectively. Accordingly, M17 and PrP^C cells were radiolabeled with $^{59}FeCl_3$-citrate complex for 4 hr as above, and lysates were fractionated in duplicate under non-denaturing conditions. One part was dried and exposed to an X-ray film, while the other was transferred to a PVDF membrane and probed for PrP, ferritin, TfR, and Tf. As in Figure 2, the amount of ^{59}Fe incorporated by ferritin in PrP^C cells is significantly higher than M17 cells (Figure 3A, lanes 1 and 2, black arrow). A slower migrating ^{59}Fe labeled band corresponding to Tf/TfR complex is detected in M17 lysates (Figure 3A, lanes 1, 7, and 9, open arrow). Unlike Figure 2, PrP is resolved on this less concentrated gel system and is detected by PrP specific antibody 3F4 (Figure 3A, lane 4, arrow-head). However, a corresponding ^{59}Fe labeled band is not detected in lane 2, though pure ^{59}Fe-labeled recombinant PrP is readily detected by this method as demonstrated previously [17]. Evaluation of iron modulating proteins shows higher levels of ferritin and lower levels of TfR and Tf in PrP^C lysates relative to M17 as in Figure 1 above (Figure 3A, lanes 5–10, arrow-head). Ferritin and Tf/TfR complex show corresponding iron labeled bands as expected (Figure 3A, compare lanes 1, 2 with 5, 6, 9, 10). Fractionation of the same samples by SDS-PAGE followed by immunoblotting confirms increased levels of ferritin and decreased levels of Tf and TfR in PrP^C lysates compared to M17 controls (Figure 3B, lanes 1 and 2). Together, these results demonstrate that PrPC increases total cellular iron, ferritin iron, and ferritin levels, and decreases Tf and TfR levels. However, the absence of ^{59}Fe-labeled PrP^C indicates that either the association of PrP with ^{59}Fe is transient or relatively weak and disrupted after cell lysis, or alternatively, PrP facilitates the incorporation of ^{59}Fe into ferritin by an indirect mechanism that does not involve the formation of a PrP-iron complex.

To evaluate if expression of PrP^C on the cell surface is required for iron uptake, a similar evaluation was carried out in cells expressing $PrP^{231stop}$ that lacks the GPI anchor and is secreted into the medium. Radiolabeling of M17, PrP^C, and $PrP^{231stop}$ cells with $^{59}FeCl_3$-citrate complex for 4 hr shows significantly more ^{59}Fe-ferritin in PrP^C cells compared to M17 as above, and minimal change in $PrP^{231stop}$ samples (Figure 3C, lanes 1–3, black arrow). Western blotting of M17, PrP^C, and $PrP^{231stop}$ lysates and medium sample from $PrP^{231stop}$ cells cultured overnight in serum-free medium with 3F4 shows the expected glycoforms of PrP in PrP^C lysates, and undetectable reactivity in M17 and $PrP^{231stop}$ lysates as expected (Figure 3D, lanes 1–3). However, significant reactivity is detected in the medium of $PrP^{231stop}$ cells, demonstrating adequate expression and secretion of PrP[231stop] in transfected cells (Figure 3D, lane 4) [22, 23]. Re-probing of lysate samples for ferritin, Tf, and TfR shows increased levels of ferritin and decreased levels of Tf and TfR in PrP[C] samples compared to M17 lysates (Figure 3E, lanes 1 and 2). The PrP[231stop] lysates show minimal change in ferritin levels, and surprisingly, lower levels of Tf and TfR relative to M17 lysates (Figure 3E, lanes 1 and 3). This observation is surprising since ^{59}Fe-ferritin levels in PrP[231stop] cells are as low as M17, and yet the cells do not show increased levels of Tf and TfR as in M17-cells. Reaction for β-actin confirms equal loading of protein in all samples (Figure 3E, lanes 1–3).

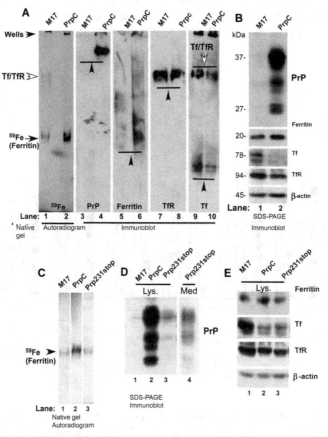

Figure 3. Expression of PrP on the plasma membrane is essential for iron incorporation in ferritin. (A) ^{59}Fe-labeled M17 and PrP[C] lysates were fractionated on a 3–9% native gradient gel and autoradiographed (lanes 1 and 2), or immunoblotted as above with antibodies specific to PrP, ferritin, TfR, and Tf (lanes 3–10). (B) Immunoblotting of the same samples following fractionation by SDS-PAGE shows similar differences in the levels of PrP, ferritin, Tf, and TfR as in (A) after normalization with actin (lanes 1 and 2). (C) ^{59}Fe-labeled M17, PrP[C], and PrP[231stop] lysates were fractionated by native gel electrophoresis and subjected to autoradiography (lanes 1–3). (D) Unlabeled lysates prepared from M17, PrP[C], and PrP[231stop] lysates, and methanol precipitated proteins from the medium sample of PrP[231stop] cells were fractionated by SDS-PAGE and immunoblotted for PrP using 3F4 (lanes 1–4). (E) Membrane from (D) was re-probed for ferritin, Tf, TfR, and β-actin (lanes 1–3).

Quantitative comparison of ferritin iron and levels of ferritin, Tf, and TfR shows significant differences between cell lines. Thus, relative to M17 cells, PrP^C cells show an increase in ferritin iron and ferritin levels to 570 and 565%, and a decrease in Tf and TfR levels to 70 and 75% respectively. A similar comparison of mutant cell lines relative to PrP^C cells shows the following: PrP$^{\Delta51-89}$ cells show a decrease in ferritin iron and ferritin to 7.0, 6.9%, and insignificant change in Tf and TfR levels. PrP$^{\Delta23-89}$ cells show a similar decrease in ferritin iron and ferritin levels to 7.5 and 7.2%, an increase in Tf to 120%, and insignificant change in TfR levels. PrP^{102L}-cells show a decrease in ferritin iron and ferritin levels to 89 and 90%, and an increase in Tf and TfR levels to 300 and 142% respectively. The PrP^{231stop} cells show a decrease in ferritin iron and ferritin to 27 and 16%, and a decrease in Tf and TfR levels to 89 and 67% respectively. Quantification of PrP expression relative to M17 shows levels of 650, 710, 750, 610, and 5% in PrP^C, PrP$^{\Delta51-89}$, PrP$^{\Delta23-89}$, PrP^{102L}, and PrP^{231stop} cells respectively (Figure 4).

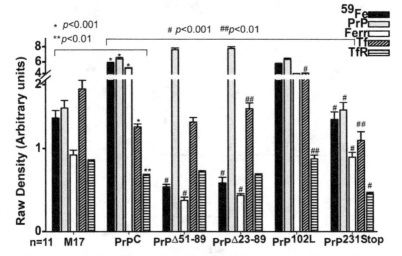

Figure 4. Quantitative analysis of the results in Figures 1–3. Quantitative evaluation after densitometry of ferritin iron and levels of PrP, ferritin, Tf, and TfR in PrP^C, PrP$^{\Delta51-89}$, PrP$^{\Delta23-89}$, PrP^{102L}, PrP^{231stop}-cells relative to non-transfected M17 controls. Values are mean±SEM of 11 independent experiments. The y-scale is linear but has been re-scaled after the break to illustrate the data clearly. For M17 vs. PrP^C *p<0.001, **p<0.01, and for PrP^C vs. mutant cell lines #p<0.001, ##p<0.01).

Considering the tightly orchestrated and coordinated balance between cellular iron levels and iron uptake and storage proteins [18, 21], these results indicate a mild iron overload in PrP^C-cells relative to M17-cells, and an indefinable phenotype in mutant cell lines since the iron uptake proteins Tf and TfR do not respond to ferritin iron levels as expected. Since Tf and TfR levels are reflective of the biologically available intracellular LIP that is maintained within the physiological range by ferritin, these results indicate a disconnect between ferritin iron and the cellular LIP, or a failure of the iron regulatory loop involving the LIP, iron binding proteins 1 and 2, TfR, and ferritin to induce appropriate response.

Mutant PrP Forms Influence the Uptake of Iron by Ferritin

The influence of normal and mutant PrP forms on intracellular LIP was evaluated in M17 and transfected cell lines cultured in complete medium under normal culture conditions. All cell lines were loaded with the iron binding dye calcein-AM, and the increase in fluorescence in response to salicylaldehyde isonicotinoyhydrazone (SIH), a cell permeable iron chelator, was measured (Figure 5A) [24]. Relative to M17 cells, PrP^C cells show an increase in LIP to 143%, an expected observation since the ferritin iron levels of these cells are also higher than M17 cells (compare Figures 5A and 4). A similar evaluation of mutant cell lines relative to PrP^C-cells shows a decrease in LIP to 95, 78, and 67% in PrP$^{\Delta 51-89}$, PrP$^{\Delta 23-89}$, PrP^{102L}-cells, and an increase to 155% in PrP^{231stop} cells respectively (Figure 5A). These results indicate that Tf and TfR levels in mutant cell lines observed in Figure 4 above respond to the LIP rather than ferritin iron content as expected. More importantly, these results indicate a block in uptake or increased uptake of iron by ferritin in specific cell lines, accounting for the disproportionate levels of ferritin iron and intracellular LIP, and the unexpected response of Tf and TfR to cellular iron content.

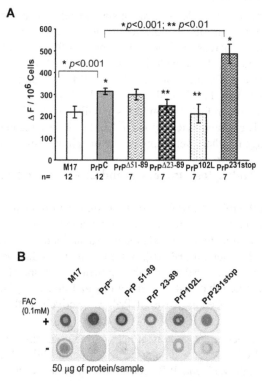

Figure 5. Cells expressing normal and mutant PrP forms show differential levels of LIP and uptake of extra-cellular iron. (A) Indicated cell lines were loaded with calcein and intracellular LIP was estimated by quantifying the SIH chelatable iron pool. Values are mean±SEM. n = 12 for M17 and PrP^C, and 7 for mutant cell lines. *p<0.001, **p<0.01, #p<0.001, ##p<0.01. (B) The same cell lines were exposed to 0.1 mM FAC for 16 hr and 50 µg of protein from cell homogenates was spotted on a PVDF membrane and reacted with Ferene-S, a dye that forms a blue reaction product with iron [25].

To evaluate if the difference in ferritin iron content of different cell lines is maintained in the presence of excess extra-cellular iron, M17, PrP^C, $PrP^{\Delta 51-89}$, $PrP^{\Delta 23-89}$, PrP^{102L}, and $PrP^{231stop}$-cells were cultured overnight in the presence of 0.1 mM ferric ammonium citrate (FAC). (This dose of FAC was found to cause <1% cell death after overnight exposure). After washing the cells with PBS supplemented with 100 µM DFO to remove surface bound iron, cells were disrupted with glacial acetic acid and equal amount of protein from each cell line was spotted on a PVDF membrane. Reaction with Ferene-S, a dye that forms a blue reaction product with iron [25], shows a marked increase in protein bound iron in all cell lines compared to unexposed controls (Figure 5B). More importantly, each cell line reflects cell-specific differences in protein bound iron as observed for ferritin iron above (Figure 5B). Fractation of lysates by SDS-PAGE followed by immunoblotting for PrP, ferritin, and TfR shows up-regulation of PrP and ferritin, and down-regulation of TfR to undetectable levels in FAC exposed lysates [17]. Up-regulation of PrP in response to FAC appears to be at the mRNA level. These results suggest a dominant role for PrP in the transport of extra-cellular iron to ferritin both under normal culture conditions and in the presence of excess extra-cellular iron.

Together, the above results demonstrate a state of relative iron overload in PrP^C-cells compared to M17 controls as indicated by an increase in intracellular LIP and iron content of ferritin, increase in iron storage protein ferritin, and decrease in iron uptake proteins Tf and TfR. Relative to PrP^C-cells, mutant PrP expressing cells show a substantial decrease in ferritin iron in $PrP^{\Delta 51-89}$, $PrP^{\Delta 23-89}$, and $PrP^{231stop}$-cells, and relatively less reduction in PrP^{102L}-cells. Intracellular LIP is reduced in $PrP^{\Delta 23-89}$ and PrP^{102L}, minimally altered in $PrP^{\Delta 51-89}$, and substantially increased in $PrP^{231stop}$-cells relative to PrP^C-cells. The Tf and TfR respond to LIP levels in some cell lines, but show an unexpected change in others, reflecting a state of cellular iron imbalance.

Stimulation of PrP Endocytosis Increases, and Cross-linking Decreases Ferritin Iron Content

Further support for the role of PrP in mediating cellular iron uptake was obtained by assessing iron incorporation into ferritin following stimulation or disruption of PrP^C endocytosis by 3F4, a well characterized monoclonal antibody specific for methionine residues 109 and 112 of human PrP [26]. A similar approach has been used successfully to down-regulate mouse PrP using Fab fragments of PrP specific antibodies [27]. Initial evaluation revealed that 3F4 concentrations of 1 and 12 µg/ml are optimal for stimulating and disrupting endocytosis of PrP^C respectively without compromising cell viability.

To evaluate the effect of antibody treatment morphologically, M17 and PrP^C-cells exposed to 1 µg/ml of 3F4 for 5 days were fixed, permeabilized, and reacted with anti-mouse-FITC. Both M17 and PrP^C-cells show minimal reactivity at the plasma membrane, but significant reactivity in endocytic vesicles that are more prominent in PrP^C cells (Figure 6A, panels 1 and 2, arrow-head). These observations suggest significant endocytosis of PrP^C along with 3F4. Untreated PrP^C-cells reacted with 3F4-anti-mouse-FITC show punctuate reaction at the plasma membrane and minimal intracellular reaction as expected for normal distribution of PrP^C (Figure 6A, panel 3, arrow).

Exposure to 12 µg/ml of 3F4, however, cross-links PrP^C at the plasma membrane and reduces its endocytosis significantly (Figure 6B, panels 1 and 2). As a control, mouse neuroblastoma cells (N2a) expressing mouse PrP that does not react with 3F4 were exposed to 3F4 and reacted with mouse PrP-specific antibody 8H4 followed by anti-mouse-FITC. Examination shows normal distribution of PrP^C at the plasma membrane and some reactivity in the Golgi region as expected (Figure 6B, panel 3) [26]. Exposure of PrP^C cells to anti-Thy-1, a monoclonal antibody to an irrelevant GPI-linked protein abundant on neuronal cells shows normal distribution of PrP^C when reacted with 8H4-anti-mouse-FITC (Figure 6B, panel 4), confirming the specificity of 3F4 mediated endocytosis and cross-linking of PrP^C.

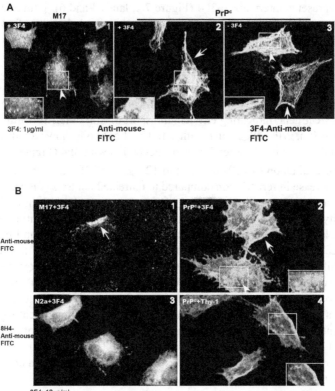

Figure 6. Exposure of PrP^C-cells to different concentrations of 3F4 induces endocytosis or cross-linking of PrP. (A) Immunostaining of M17 and PrP^C cells exposed to 1 µg/ml of 3F4 for 5 days shows a prominent reaction in vesicular structures in M17 and PrP^C cells (panels 1 and 2). Coalesced vesicles simulating aggregated PrP^C are evident near the Golgi region and in the cytosol of PrP^C cells (panel 2). Untreated PrP^C-cells reacted with 8H4-anti-mouse-FITC show a prominent reaction at the plasma membrane as expected (panel 3). (B) Reaction of M17 and PrP^C cells exposed to 12 µg/ml of 3F4 for 4 hr with anti-mouse FITC shows cross-linking of PrP on the plasma membrane of M17 and PrP^C cells (panels 1 and 2, arrow) and a slight increase of reactivity in vesicular structures in the latter (panel 2, arrow-head). Similar exposure of N2a-cells to 3F4 and PrP^C-cells to anti-Thy1 antibody followed by immunoreaction with 8H4-anti-mouse-FITC shows plasma membrane and Golgi reaction of endogenous PrP in N2a cells (panel 3) and plasma membrane distribution of PrP in Thy-1 exposed cells (panel 4). (Mouse PrP expressed by N2a cells does not react with 3F4).

The effect of increased endocytosis of PrP^C on ferritin iron content was evaluated by radiolabeling cells cultured in the presence of 1 µg/ml of 3F4 with $^{59}FeCl_3$ for the last 4 hr of the incubation, and analyzing radiolabeled lysates as in Figure 1 above. Fractionation by non-denaturing page shows a significant increase in ferritin iron in the 3F4 exposed lysate compared to untreated control (Figure 7A, lanes 1 and 2, open arrow). Analysis by SDS-PAGE and immunoblotting shows 2–3 fold increase in reactivity for all PrP glycoforms with anti-PrP antibodies 3F4 and 8H4 (Figure 7A, lanes 3–6). However, the 18 kDa fragment that results from recycling of PrPC from the plasma membrane is not increased in 3F4 exposed lysates, indicating stimulation of PrPC internalization and possible intracellular accumulation by 3F4 binding rather than increased recycling from the plasma membrane (Figure 7A, lanes 5 and 6) [28]. The 50 kDa band represents internalized 3F4 (Figure 7A, lanes 4 and 6). Immunoblotting for ferritin, Tf, and TfR shows an increase in TfR, and minimal change in ferritin and Tf levels (Figure 7A, lanes 7 and 8). Quantification by densitometry shows an increase in ferritin iron to 271%, and insignificant change in ferritin and Tf levels by 3F4 treatment. The increase in TfR levels to 175% is probably due to co-endocytosis with PrP-antibody complex (Figure 7B). Measurement of cellular LIP revealed insignificant difference between 3F4 exposed and untreated controls after 24 hr (data not shown) or 5 days of treatment, indicating efficient transport of iron to ferritin within this time frame (Figure 7C). PrP^C cells treated with anti-Thy-1 antibody, however, demonstrated a significant decrease in LIP after 5 days of incubation with 3F4 (Figure 7C).

A similar evaluation of cells exposed to 12 µg/ml of 3F4 for 4 hr shows significantly less increase in ferritin iron compared to untreated controls (Figure 8A, lanes 1 and 2, open arrow). Separation by SDS-PAGE and immunoblotting shows increase in PrP reactivity (Figure 8A, lane 4) and an increase in the levels of ferritin, Tf, and TfR (Figure 8A, lanes 5 and 6). Quantification shows an increase in ferritin iron to 148%, and an increase in the levels of ferritin and TfR to 153 and 146% respectively. Tf levels show insignificant change by this treatment (Figure 8B). A similar increase in ferritin iron is observed when M17 cells expressing endogenous levels of PrP are exposed to 3F4, ruling out the effect of over-expression of PrP^C on these observations. Exposure to equivalent amounts of anti-Thy-1 does not alter ferritin iron content significantly. Measurement of intracellular LIP after 4 hr of exposure to 12 µg/ml of 3F4 shows an increase to 170% in treated cells compared to untreated controls. Exposure to similar concentrations of anti-Thy-1 shows a decrease to 70% (Figure 8C), an unexpected effect that requires further evaluation.

The above results indicate that stimulation of PrP^C endocytosis over a prolonged period increases iron incorporation into ferritin, whereas cross-linking of PrP^C that is likely to result in its degradation following endocytosis has relatively less effect on ferritin iron. The increase in intracellular LIP by cross-linking PrP without any increase in ferritin iron probably reflects inefficient transport of iron to ferritin in the absence of PrP, as observed for certain mutant forms of PrP. The levels of ferritin, Tf, and TfR probably reflect an artifactual change due to membrane perturbation by antibody treatment rather than a response to intracellular LIP.

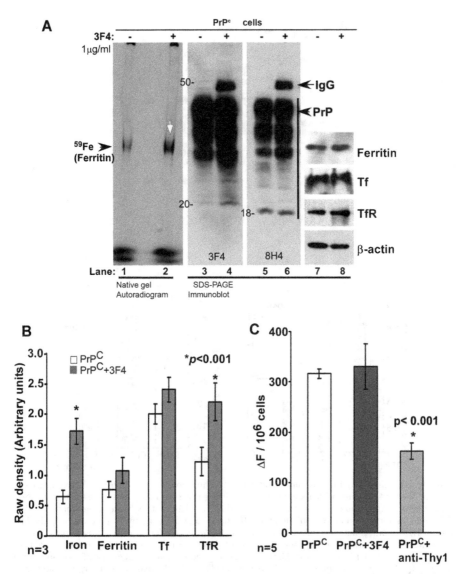

Figure 7. Endocytosis of PrP increases ferritin iron content. (A) PrP^C-cells exposed to 1 µg/ml of 3F4 for 5 days were radiolabeled with ^{59}FeCl$_3$ for 4 hr, and lysates were fractionated on a non-denaturing gel and auto-radiographed (lanes 1 and 2). Equal aliquots of the same samples were boiled in SDS-containing sample buffer and fractionated in duplicate by SDS-PAGE followed by immunoblotting with PrP specific antibodies 3F4 and 8H4 (lanes 3–6). Subsequently, the membranes were re-probed for ferritin, Tf, TfR, and β-actin (lanes 7 and 8). (B) Quantification by densitometry shows an increase in ferritin iron and TfR levels, and insignificant change in Tf levels in 3F4 exposed cells. Values are mean±SEM of three independent experiments. *p<0.001 compared to untreated cells. (C) Estimation of LIP after exposing the cells to 1 µg/ml of 3F4 or anti-Thy-1 antibody for 5 days shows insignificant difference between untreated and 3F4 treated PrP^C cells, and a decrease in anti-Thy-1 treated cells. *p<0.001. n = 5.

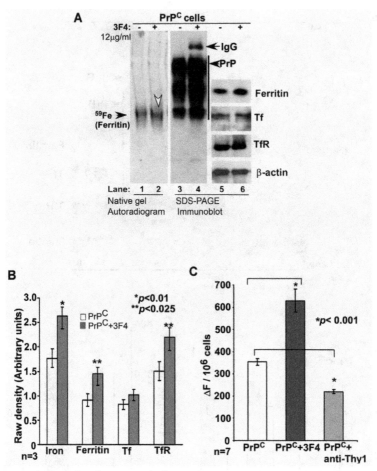

Figure 8. Cross-linking of PrP has minimal effect on ferritin iron content. (A) PrP^C-cells exposed to 12 µg/ml of 3F4 for 4 hr were radiolabeled with ^{59}FeCl$_3$ in the last 2 hr, and lysates were fractionated on a native gel followed by autoradiography (lanes 1 and 2). Equal aliquots of lysates were fractionated by SDS-PAGE as above and immunoblotted with 3F4 (lanes 3 and 4). The membrane was re-pobed for ferritin, Tf, TfR, and β-actin (lanes 5 and 6). (B) Quantification by densitometry shows an increase in ferritin iron, ferritin, and TfR levels, and insignificant change in Tf levels by 3F4 treatment. *p<0.001, **p<0.025. n = 3. (C) Estimation of LIP after exposing the cells to 12 µg/ml of 3F4 or anti-Thy-1 antibody for 4 hr shows an increase in 3F4 exposed cells, and a decrease in anti-Thy-1 treated cells. *p<0.001. n = 7.

Prp Does Not Modulate Release of Iron From Cells

To determine if the difference in cellular iron levels between cell lines is due to differential release into the medium, M17, PrP^C, PrP$^{\Delta 51-89}$, PrP$^{\Delta 23-89}$, and PrP^{102L} cells were cultured in the presence of 3H-thymidine overnight to monitor cell proliferation and radiolabeled with ^{59}FeCl$_3$ for 4 hr as above. Labeled cells were washed with PBS containing 100 µM DFO to remove surface bound ^{59}Fe, and chased in complete medium for 30 min to 16 hr. At the indicated times equal aliquots of medium were retrieved and released ^{59}Fe was quantified in a γ-counter. Kinetic analysis shows minimal difference in extra-cellular iron between cell lines after normalizing with 3H-thymidine (Figure 9A). Estimation of cell-associated ^{59}Fe after 16 hr of chase shows more ^{59}Fe in PrP^C and PrP^{102L}, and significantly less in PrP$^{\Delta51-89}$ and PrP$^{\Delta23-89}$ compared to M17 lysates as observed in Figure 1 above (Figure 9B). However, the fold difference in ferritin iron content between M17 and other cell lines is significantly less after 16 hr of chase, and represents steady state levels of iron content in each cell line. Evaluation of possible ferroxidase activity of recombinant PrP using plasma as a positive control yielded negative results (Figure 9C). Though informative, this result does not rule out possible ferroxidase activity of cell-associated PrP, a technically challenging assay that has yielded inconclusive results (data not shown).

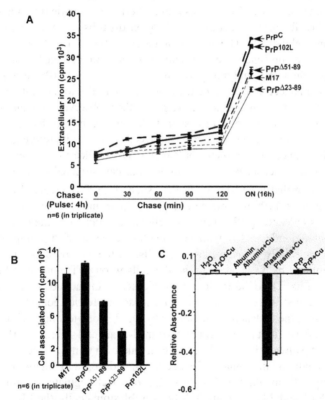

Figure 9. PrP is not involved in the export of iron from cells. (A) Cells expressing PrP^C, PrP$^{\Delta51-89}$, PrP$^{\Delta23-89}$, and PrP^{102L} were radiolabeled with ^{59}FeCl$_3$, washed with PBS supplemented with DFO, and chased in complete medium for 30, 60, 90, 120 min, and 16 hr. At the indicated time points equal aliquots of medium samples were quantified in a γ-counter. Estimation of released ^{59}Fe does not show a significant difference between the indicated cell lines at any time point. n = 6 experiments in triplicate. (B) Cell associated ^{59}Fe after 16 hr of chase reflects the ferritin iron content of each cell line noted in Figure 1 above, though the difference between cell lines is significantly less. (C) Possible ferroxidase activity of recombinant PrP was measured using the established colorimetric method [44] with modifications. Negative controls included water and albumin supplemented with copper, and positive controls included plasma in the absence or presence of copper. Recombinant PrP does not show detectable ferroxidase activity either in the absence or presence of copper, whereas plasma shows a robust reaction under similar conditions.

DISCUSSION

The results presented in this report demonstrate an unprecedented role of PrP in facilitating iron uptake by cells and its transport to cellular ferritin. Using a combination of neuroblastoma cell lines expressing normal and mutant PrP forms, we demonstrate that over-expression of PrP^C increases intracellular LIP and the amount of iron deposited in ferritin. Pathogenic and non-pathogenic mutations of PrP over-expressed to the same extent as PrP^C alter cellular LIP and ferritin iron content differentially, specific to the mutation. Certain cell lines, especially cells expressing anchorless PrP^{231stop}, demonstrate increased LIP in the presence of decreased ferritin iron, while PrP^{102L}-cells display low LIP in the presence of adequate ferritin iron. Furthermore, stimulation of endocytosis by PrP specific antibody increases ferritin iron, while cross-linking at the plasma membrane increases LIP but has minimal effect on ferritin iron, indicating that alteration of PrP function or cellular localization disturbs the homeostasis between ferritin iron and cellular LIP. The differential incorporation of iron by mutant cell lines is maintained in the presence of excess extra-cellular iron, demonstrating a dominant role of PrP^C in iron uptake and transport. The positive effect of PrP^C on cellular iron is mainly due to enhanced uptake since the amount released into the culture medium is not altered in any of the cell lines tested. Together, these observations suggest a role for PrP^C in mediating iron uptake and transport to ferritin directly, or by interacting with other iron modulating proteins. Below we discuss these data with reference to possible functions of PrP^C in cellular iron metabolism, and the implications thereof in inducing imbalance in iron homeostasis observed in prion disease affected brains [15, 16, 45].

It is surprising that a GPI-linked protein such as PrP^C is involved in iron transport to ferritin since PrP^C is a membrane protein that undergoes vesicular transport while ferritin is cytosolic [29]. Normally, cellular iron uptake is mediated by the Tf/TfR dependent and independent pathways, the former being most prominent and well characterized especially in neuroblastoma cells. In the Tf/TfR dependent pathway, ferric iron captured by Tf is taken up by the cells through TfR-mediated uptake via clathrin coated pits. Tf-bound ferric iron is released in the acidic environment of the endosomes, reduced to ferrous iron by an endosomal ferric reductase Steap3, and transported across the endosomal membrane by DMT1 to cytosolic ferritin where it is oxidized to the fairly inert ferric form by ferritin H-chain and stored [18, 19, 29]. In the Tf-independent pathway, iron is taken up by an unknown transport mechanism, possibly non-specifically by fluid phase of endocytosis, and stored in ferritin. Ferritin regulates the biologically available LIP in the cell, and is itself regulated by iron regulatory proteins (IRPs) 1 and 2 [18, 19, 30, 31]. In neuroblastoma cells, the LIP is a function of total cellular iron, and an increase in cellular iron is accompanied by increased ferritin content to maintain the LIP within safe limits [32, 33]. Where might PrP intersect with this tightly orchestrated mechanism of iron uptake, transport, and storage? Three potential mechanisms are plausible: (1) modulation of uptake at the plasma membrane independently or by interacting with the Tf/TfR dependent pathway, (2) facilitation of iron transport to cytosolic ferritin across the endosomal membrane by promoting ferric iron release from Tf and/or its reduction for transfer through DMT1 [19], or (3) assistance in deposition into ferritin by oxidizing ferrous iron to the ferric form.

It is unlikely that PrP facilitates export of iron from neuroblastoma cells based on our observations.

At the plasma membrane, PrP^C could take up iron directly from the extra-cellular milieu and deliver to an endosomal compartment as suggested for copper [34]. However, this seems unlikely for three reasons; (1) ^{59}Fe-labeled PrP^C could not be detected in radiolabeled cells although labeled recombinant PrP is easily detected using the same procedure [17], (2) ^{59}Fe-labeled recombinant PrP loses its label to Tf when added to cells, indicating lower affinity for iron relative to Tf (unpublished observations), and (3) intracellular LIP is high in cells expressing anchorless PrP^{231stop} despite low ferritin iron content, indicating efficient uptake of iron in the absence of cell surface PrP^C. It remains plausible, though, that PrP^C modulates iron uptake by the Tf/TfR pathway at the plasma membrane or in an endosomal compartment [35].

It is also possible that extra-cellular iron induces the movement of PrP^C from detergent insoluble membrane domains where it normally resides to the proximity of TfR in a similar manner as in the presence of copper [34]. Here, it may enhance the binding of iron loaded Tf to its receptor, or stimulate the endocytosis of Tf/TfR complex by a direct or an indirect interaction. In this context, it is interesting to note that PrP^C undergoes endocytosis through clathrin coated pits after associating with a transmembrane protein through its N-terminal domain [36], suggesting that the reported co-localization of PrP^C with Tf and TfR within endosomes may reflect a functional association rather than co-residence due to a common mode of endocytosis [37]. Assuming this scenario, the increase in TfR levels by stimulation of PrP endocytosis by 3F4 and the differential effect of mutant PrP forms on ferritin iron content may be explained by a change in the rate of endocytosis, or altered interaction of normal and mutant PrP forms with Tf or TfR due to misfolding [35-37]. We have previously reported increased endocytosis and defective recycling of mutant PrP^{102L} in neuroblastoma cells [38], a fact that may account for increased ferritin iron in these cells. Though attractive, this model fails to explain decreased ferritin iron in the presence of significantly high LIP in cells expressing anchorless PrP^{231stop} and by cross-linking PrP at the plasma membrane, indicating a role downstream from the plasma membrane. The up-regulation of PrP^C at the transcriptional and translational level when cells are exposed to excess extra-cellular iron (supporting information) perhaps reflects its function as an iron regulatory protein, though a protective response to oxidative stress cannot be ruled out under these experimental conditions [39]. However, since all cell lines display similar differences in ^{59}Fe-ferritin content when labeled with ^{59}FeCl$_3$ or purified ^{59}Fe-Tf (unpublished observations), it is likely that PrP^C functions downstream of the iron uptake pathways specific for free and Tf bound iron, perhaps in an endosomal compartment.

Keeping the above facts in mind, it is plausible that PrP^C functions as a ferric reductase along with Steap3 to facilitate the transport of ferric iron released from Tf across the endosomal membrane to cytosolic ferritin. This assumption is supported by the fact that PrP^C functions as a copper transport protein by reducing copper (II) prior to transfer to copper (I) specific trafficking proteins within cells [34]. Such a function would explain the low ferritin iron content in cells expressing mutant PrP lacking the octapeptide region responsible for reducing copper (II) [34], the observed up-regulation of PrP^C in response to exogenous iron, increase in ferritin iron by increased expression of PrP^C and stimulation of PrP endocytosis, and co-localization of PrP^C and ferritin in cells exposed to excess iron [17]. However, decreased ferritin iron despite high LIP levels in cells expressing anchor-less PrP and the opposite scenario in PrP^{102L}-cells suggests an additional role in iron transport between the LIP and cellular ferritin, a function that is hard to explain merely by the altered reductase activity of mutant proteins. Although, we could not detect measurable ferroxidase activity of recombinant PrP, such a function of cell associated PrP^C would explain the facilitative effect of PrP^C on iron incorporation into ferritin. Further studies are required to resolve this question.

Despite obvious shortcomings in our data in explaining the mechanistic details of cellular iron modulation by PrP, this report clearly shows the effect of PrP and its mutants on iron uptake and transport. We demonstrate a state of mild iron overload mediated by PrP^C, and mild iron deficiency or imbalance by pathogenic and non-pathogenic mutations of PrP. The positive role of PrP^C on cellular iron levels is further supported by a recent study where transgenic mice lacking PrP^C expression (PrP$^{-/-}$) recover slowly from experimentally induced hemolytic anemia [40], indicating a functional role for PrP^C in iron uptake by hematopoietic cells. These findings take on a greater significance since prion disease affected human and animal brains show signs of iron imbalance [45], a potentially neurotoxic state due to the highly redox-active nature of iron. It is conceivable that dysfunction of PrP due to aggregation combined with the formation of redox-active PrP^{Sc} aggregates [17] induces brain iron imbalance, contributing to prion disease associated neurotoxicity. Future studies are required to define the precise biochemical pathway of iron modulation by PrP, and develop therapeutic strategies to prevent iron induced neuronal death in prion disorders.

MATERIALS AND METHODS

Antibodies and Chemicals

Monoclonal anti-PrP antibodies 3F4 and 8H4 were obtained from Signet (Dedham, MA) and Drs. Man-Sun Sy (Case Western Reserve University) and Pierluigi Gambetti (National Prion Surveillance Center, Case Western Reserve University) respectively. Antibody against human ferritin was purchased from Sigma (St. Louis, MO), anti-transferrin from GeneTex (San Antonio, TX), anti transferrin receptor from Zymed Laboratories Inc (Carlsbad, CA), and anti-Thy 1.1 from eBioscience (SanDiego, CA). Secondary antibodies tagged with HRP or fluorophores FITC and TRITC were obtained from Amersham Biosciences (England) and Southern Biotechnology Associates (Birmingham, AL) respectively. Ferrous ammonium sulfate, Ferene S, and all other chemicals were purchased from Sigma. All cell culture supplies were obtained from Invitrogen. The ^{59}FeCl$_3$ was from Perkin-Elmer.

Cell Lines and Culture Conditions

Human neuroblastoma cells (M17) were obtained from J. Biedler (Memorial Sloan-Kattering Cancer Center, New York) and purchased from ATCC. The M17 cells expressing PrP^C, PrP^{231stop}, PrP$^{\Delta51-89}$, PrP$^{\Delta23-89}$, and PrP^{102L} were generated and cultured as described in previous reports [41, 42]. For this study M17 cells from two different sources were transfected at least three separate times and bulk transfected cells were used to avoid cloning artifacts. Similarly transfected cells from two different investigators and cells cultured in DMEM supplemented with 10% FBS and Opti-MEM supplemented with different lots of FCS were also tried to avoid errors due to culture conditions.

Radiolabeling with ^{59}FeCl$_3$

The M17, PrP^C, and mutant PrP$^{\Delta 51-89}$, PrP$^{\Delta 23-89}$, PrP^{102L}, and PrP^{231stop} cells cultured overnight to 80% confluency were serum starved for 1 hr and incubated with ^{59}FeCl$_3$-citrate complex (1 mM sodium citrate and 20–25 μCi of ^{59}FeCl$_3$ in serum free Opti-MEM; molar ratio of citrate to iron was maintained at 100:1) for 4 hr at 37°C in the incubator. At the end of the incubation cells were washed 3 times with ice cold PBS and lysed with native lysis buffer (0.14 M NaCl, 0.1 M HEPES, pH 7.4, 1.5% Triton X-100 and 1 mM PMSF). Aliquots of lysates were mixed with glycerol (to a final concentration of 5%) and traces of bromophenol blue, and equal amount of protein from each sample was resolved on 3–9% native gradient gel. For fractionation on SDS-PAGE, the same samples were mixed with 4× SDS-sample buffer, boiled for 10 min and resolved on SDS-PAGE followed by immunoblotting.

Native Gradient Gel Electrophoresis, Autoradiography, Immunoblotting, and Electroelution

Electrophoresis of lysates was performed using a Hoefer SE 600 vertical apparatus with a cooling system. Linear 3–20% (Figure 1) or 3–9% (Figure 3) gradient poly-acrylamide gels were prepared as described by Vyoral et al. [21] with modifications. The gel mixture contained 0.375 M Tris, pH 6.8, 1.5% Triton X-100, and 1.18 mM ammonium persulfate. N,N,N′,N′-Tetramethylethylenediamine (TEMED) was added to a final concentration of 5.38 mM. Radiolabeled lysates mixed with glycerol were subjected to electrophoresis using electrode/running buffer (25 mM Tris, 192 mM glycine pH 8.3, and 1.5% Triton X-100) under constant current (100 mA) for 4 hr at 4°C. Gels were either electroblotted or vacuum dried (BioRad) and exposed to X-ray film (Kodak BioMax XAR) fitted with intensifying screens. For Western Blotting, gels were washed thoroughly with electrode buffer without Triton X-100 for 2 hr (each wash of 200 ml, 10 min) on a slowly rocking platform to remove Triton. The gel was electroblotted to a PVDF membrane using BioRad semi-dry electroblotting system with anode buffer (25 mM Tris, pH 10.4) and cathode buffer (25 mM Tris, 39 mM glycine, pH 9.2) at 25 V for 90 min. Membranes were further processed for immunodetection as described below. To confirm the identity of iron labeled proteins, iron bands were excised from native gels and proteins were electro-eluted using Biorad electro-eluter at 60 mA for 4 hr. Eluted proteins were concentrated by methanol precipitation and analyzed by SDS-PAGE.

SDS-PAGE and Western Blotting

Cells cultured under different conditions were fractionated by SDS-PAGE and immunoblotted as described previously [41, 42]. The following antibody dilutions were used: 8H4 (1:3000), 3F4 (1:5000), ferritin (1:1000), Tf (1:6000), TfR (1:3000), actin (1:7500), secondary antibodies conjugated with horseradish peroxidase (1:6000). Immunoreactive bands were visualized by ECL (Amersham Biosciences Inc.).

Measurement of Intracellular Calcein-chelatable Iron

Cellular LIP was assayed as described by Tenopoulou et al. [43] using the iron sensitive fluorescent dye calcein. When incubated with cells as a lipophilic calcein-AM-ester (molecular probes), it enters the cells and is cleaved by cellular esterases to release calcein that binds iron and is quenched by this reaction. Upon addition of the cell permeable iron chelator SIH, iron is released from calcein that regains its fluorescence (recorded at λ_{ex} 488 nm and λ_{em} 518 nm). Briefly, 5×10^5 M17 cells or cell lines expressing PrP^C and mutant PrP forms plated in 35 mm Petri dishes were washed with PBS containing 1 mg/ml BSA and 20 mM Hepes, pH 7.3 and incubated with 0.25 μM calcein-AM for 20 min at 37°C in same buffer. After calcein loading, cells were trypsinized, washed and re-suspended in 1.0 ml of the above buffer without calcein-AM and placed in a 24 well micro-plate in a thermostatically controlled (37°C) fluorescence plate reader (Microtek). The fluorescence was monitored at λ_{ex} 488 nm and λ_{em} 518 nm. Iron-induced quenching of calcein was reduced by the addition of 20 μM SIH. Cell number and viability was checked by Trypan Blue dye exclusion and results were expressed as $\Delta F/10^6$ cells.

Detection of Iron with Ferene S

Cell lines cultured overnight in complete medium or in the presence of 0.1 mM FAC were washed with PBS supplemented with EDTA to chelate surface bound iron and pelleted. The pellet was dissolved in 50 μl of acetic acid and equal amount of protein (50 μg) was spotted on a PVDF membrane and immersed in a freshly prepared solution of Ferene S (0.75 mM 3-(2-pyridyl)-5, 6-bis(2-(-furyl sulfonic acid)-2, 4-triazine, 2% (v/v) acetic acid, 0.1% thioglycolic acid) (24) for 30 minutes at 37°C. Ferene reacts with iron in the presence of acetic acid and thioglycolic acid to form a dark blue complex. Stained membranes were de-stained with 2% acetic acid and scanned.

Stimulation of Endocytosis with 3F4 Antibody

The M17 and PrPC cells were cultured in DMEM supplemented with 5% FBS and 1% PSF at 37°C in a humidified atmosphere in absence or presence of 1 μg/ml of 3F4 for 5 days [26], [27]. Medium containing 3F4 was replaced every second day and care was taken to make sure that the cells did not achieve confluency. On the fifth day, cells were washed and incubated with serum free DMEM for 1 hr, followed by radiolabeling with $^{59}FeCl_3$-citrate complex in DMEM for 4 hr as above. In a separate experimental paradigm, N2a, M17, and PrP^C cells were radiolabeled as above in the presence of 12 μg/ml of 3F4 or Thy-1 4 hr. After labeling, cells were washed, lysed in native lysis buffer, and analyzed as above.

Immunostaining and Fluorescence Microscopy

Cell lines subjected to different experimental conditions were processed for immunostaining as described in a previous report [41].

Estimation of Iron Export from Cells

Cell lines expressing different PrP forms were radiolabeled with ^{59}FeCl$_3$-citrate complex as above. Cell surface bound iron was chelated with three washes of PBS supplemented with DFO (100 µM) and the cells were chased in complete medium for different time periods. A 50 µl aliquot of the medium was retrieved at each time point and counted in a γ-counter. After 16 hr, cells were lysed and cell associated iron was measured in a gamma counter.

Estimation of Ferroxidase Activity of Recombinant PrP

Ferroxidase activity of PrP was measured by the published colorimetric method using 3-(2-pyridyl)-5,6-bis(2-(5furylsulfonic acid))-1,2,4-triazine that forms a colored Fe^{2+} complex with ferrous iron (44) with the following modifications: Reagent A: 0,45 mol/l sodium acetate, pH 5.8, reagent B:130 mmol/l thiourea, 367 µM/l Fe(NH$_4$) (SO$_4$)$_2$×6H$_2$O, reagent C (chromogen): 18 mmol/l 3-(2-pyridyl)-5,6-bis(2-(5-furylsulfonic acid))-1,2,4-triazine in 0.01 M Tris pH 7.0. Each sample contained either 1 µl of water or 1 µl of 300 µM CuSO$_4$, 6 µL of the sample (undiluted human plasma, human serum albumin 70 g/l (Sigma A1653-5G) in PBS or recombinant prion protein (0.6 µg/ml) and 820 µl of reagent A. Multichannel pipette (Finnpipette) was used for therapid addition of the reagent B (substrate) to minimize the time difference in sample processing. Sample quadruplicates were incubated at 37°C for 4 min. Unoxidized Fe^{2+} was reacted with 60 µl of chromogen solution (reagent C) and absorbance was measured at 600 nm with Smart Spec Plus (BioRad) spectrophotometer. Copper was added to provide two copper ions per PrP molecule, and was also added to human albumin and plasma samples. The amount of PrP protein in PrP-containing samples (3.6 µg/sample) roughly corresponds to a known amount of ceruloplasmin in 6 µl of undiluted human plasma. As a control, purified 99% human serum albumin was used (70 g/l in PBS) to mimic the total protein concentration in plasma. As a blank samples were supplemented with 6 µl of de-ionized water instead of albumin solution, plasma or recombinant PrP solution.

RNA Isolation and Northern Blotting

The M17 and WT cells cultured in the absence or presence of 0.1 mM FAC for 24 hr were washed with cold PBS, trypsinized, and collected in 1.5 ml eppendorf tubes. Total RNA was isolated by using SV total RNA isolation kit (Promega, Madison, WI) and quantified. 15 µg of total RNA was fractionated on 0.8% formaldehyde agarose gel followed by blotting to positively charged Nylon membranes (Roche diagnostics). Membranes were hybridized with DIG-labeled PrP or β-actin probes and binding was detected by the CSPD reagent.

STATISTICAL ANALYSIS

Data are presented as the mean ± SEM values. Statistical evaluation of the data was performed by using Students t-test (unpaired).

KEYWORDS

- **Endocytosis**
- **Ferritin**
- **Ferroxidase**
- **Immunoblotting**

AUTHORS' CONTRIBUTIONS

Conceived and designed the experiments: Neena Singh. Performed the experiments: Ajay Singh, Maradumane L. Mohan, Alfred Orina Isaac, Xiu Luo, Jiri Petrak, and Neena Singh. Analyzed the data: Ajay Singh, Maradumane L. Mohan, Alfred Orina Isaac, Jiri Petrak, Daniel Vyoral, and Neena Singh. Contributed reagents/materials/analysis tools: Daniel Vyoral. Wrote the chapter: Neena Singh.

Chapter 6

Activity and Interactions of Liposomal Antibiotics in Presence of Polyanions and Sputum of Patients with Cystic Fibrosis

Misagh Alipour, Zacharias E. Suntres, Majed Halwani, Ali O. Azghani, and Abdelwahab Omri

INTRODUCTION

To compare the effectiveness of liposomal tobramycin or polymyxin B against *Pseudomonas aeruginosa* in the cystic fibrosis (CF) sputum and its inhibition by common polyanionic components such as DNA, F-actin, lipopolysaccharides (LPS), and lipoteichoic acid (LTA).

Liposomal formulations were prepared from a mixture of 1,2-dimyristoyl-sn-glycero-3-phosphocholine (DMPC) or 1,2-dipalmitoyl-sn-glycero-3-phosphocholine (DPPC), and cholesterol (Chol), respectively. Stability of the formulations in different biological milieus and antibacterial activities compared to conventional forms in the presence of the aforementioned inhibitory factors or CF sputum were evaluated.

The formulations were stable in all conditions tested with no significant differences compared to the controls. Inhibition of antibiotic formulations by DNA/F-actin and LPS/LTA was concentration dependent. The DNA/F-actin (1251,000 mg/l) and LPS/LTA (1 to 1,000 mg/l) inhibited conventional tobramycin bioactivity, whereas, liposome-entrapped tobramycin was inhibited at higher concentrationsDNA/F-actin (5001,000 mg/l) and LPS/LTA (1001,000 mg/l). Neither polymyxin B formulation was inactivated by DNA/F-actin, but LPS/LTA (11,000 mg/l) inhibited the drug in conventional form completely and higher concentrations of the inhibitors (1001,000 mg/l) was required to inhibit the liposome-entrapped polymyxin B. Co-incubation with inhibitory factors (1,000 mg/l) increased conventional (16-fold) and liposomal (4-fold) tobramycin minimum bactericidal concentrations (MBCs), while both polymyxin B formulations were inhibited 64-fold.

Liposome-entrapment reduced antibiotic inhibition up to 100-fold and the CFU of endogenous *P. aeruginosa* in sputum by 4-fold compared to the conventional antibiotic, suggesting their potential applications in CF lung infections.

Chronic bronchial infections caused by opportunistic pathogens in the lower respiratory tract are a major cause of health decline in the CF population [1]. These recurrent infections are mainly due to gram-negative bacteria, with *P. aeruginosa* being the most common species isolated [2-4]. Bacterial infections lead to biofilm formation and host inflammatory responses and the ultimate resistance to antibacterial therapies results in increased morbidity and mortality [5-10]. Presently, prophylactic anti-inflammatory and antibacterial chemotherapy have dramatically improved the life span of the CF population, albeit pathogenic resistance to commonly used antibiotics has raised the demand for the development of novel therapeutic modalities [11-13].

Antibiotics like aminoglycosides and polymyxins have been used for the treatment of acute or chronic exacerbations in response to multi-drug resistant (MDR) bacteria, particularly Gram-negative bacilli such as *P. aeruginosa* [14-17]. Aminoglycosides including tobramycin contain broad antibacterial and post-antibiotic effect, but due to their hydrophilic nature, they are not absorbed and have adverse effects (i.e., nephrotoxicity, ototoxicity) when parenterally administered [17-19]. Presently, intravenous administration of aminoglycosides is widely used by CF clinicians and limiting the dose to daily administration seems to reduce adverse effects [20, 21]. Polymyxins are cationic polypeptides that bind to lipopolysaccharide of the Gram-negative bacteria and increase their membrane permeability and cell death. Cytotoxicity issues and adaptive resistance by bacterial cell surface alterations have limited their application to cases where other antibiotics have failed [22-24].

Clinical studies have shown the success of antibiotics used in inhalation therapy, alone or in synergism, to combat MDR *P. aeruginosa* [25-31]. However, loss of innate immune response, the emergence of resistant mucoidal strains, and increase in biofilm production, and the buildup of thick polyanionic sputum have hampered complete eradication of these infections [7,32-35]. Although an antibiotic may display activity against planktonic bacteria *in vitro*, the harsh environment of sputum containing factors produced by host and the microbes reduce their potential interactions with the targeted pathogens [36, 37]. Clinical experiments have shown that in the presence of sputum, antibiotic potency is reduced mainly because of binding to sputum and its inhibitory components like glycoproteins [e.g., mucin (8–47 mg/ml)] [38], neutrophil derived DNA (0.6–6.6 mg/ml) [38], and actin filaments (0.1–5 mg/ml) [39], and bacterial endotoxins such as LPS and LTA [40-47].

Liposomes are biodegradable delivery vesicles made up of single or multiple phospholipids in the range of several nanometers to micrometers [48, 49]. It is clear that entrapment of the majority of antibacterial agents in liposomes tends to enhance bioactivity, bioavailability, and lower drug toxicity [50-52]. Liposomes may protect the entrapped agent from aggregation and inactivation with polyanionic components of the CF sputum, hence increasing its activity at the site, although the sputum may act as a barrier to larger liposomes [53-55]The present study was carried out to answer the following questions: (i) Are liposome-entrapped antibiotics stable in the environment of the sputum? (ii) Will the entrapment within liposomes reduce antibiotic interaction with the inhibitory factors present in the sputum? (iii) Will liposome-entrapped antibiotics reduce the number of live bacteria in sputum more effectively than the free antibiotics?

Our data demonstrate that liposomes are stable in presence of sputum and inhibitory factors. This data is encouraging as it displays the ability of lipid vesicles to protect the antibiotics from inactivation. The study shows that free tobramycin and polymyxin B, incubated with negatively charged inhibitory factors, is greatly inhibited compared to liposome-entrapped forms at higher concentrations. Liposome-entrapped antibiotics display higher reduction in CFU of endogenous *P. aeruginosa* in sputum compared to the free antibiotic suggesting its potency in CF lung infections.

MATERIALS AND METHODS

F-actin and Other Chemicals

Human placental DNA, G-actin, *Escherichia coli* (O111:B4) lipopolysaccharide (LPS), and *Staphylococcus aureus* LTA were purchased from Sigma Chemicals Co (St. Louis, MO, USA). Monomeric G-actin was prepared from an acetone powder of rabbit skeletal muscle in a non-polymerizing buffer (10 mM TRIS, pH 7.4, 0.2 mM $CaCl_2$, 0.2 mM ATP, 1 mM Dithiothreitol). The G-actin was then polymerized to F-actin with the addition of 2 mM $MgCl_2$ and 150 mM KCl and gently shaken for 1 hr at room temperature. Depending on the experiment, DNA, LPS, and LTA were dissolved in double distilled H_2O or in cation-adjusted MuellerHinton (CAMH) broth. Synthetic DMPC, Chol, and DPPC were obtained from Northern Lipids Inc (Burnaby, BC, Canada). Polymyxin B (Alexis Biochemicals, Burlington, NC, USA), and tobramycin (Sandoz Laboratories, Boucherville, QC, Canada), were diluted in Phosphate Buffered Saline solution (PBS: 160 mM NaCl, 10 mM KH_2PO_4, pH 7.4).

Organisms

Reference strains *P. aeruginosa* (ATCC 27853) were purchased from PML Microbiologicals (Mississauga, ON, Canada). Clinical isolate strains PA-48912-1, PA-48912-2, and PA-48913 were kindly obtained from the Clinical Microbiology Laboratory of Memorial Hospital (Sudbury, ON, Canada) and grown to form biofilm as described elsewhere [56]. The strains were inoculated onto CAMH agar plates and incubated for 18 hr at 37°C before any experiments. For any bactericidal experiment involving ATCC 27853, single colonies were suspended to a concentration of 1×10^6 cfu/ml in CAMH broth before addition to 96-well plates.

Preparation and Characterization of Liposomal Antibiotics

Liposome-entrapped tobramycin or polymyxin B was prepared from a lipid mixture of either DMPC or DPPC and Chol (molar ratio of 2:1), respectively, by dehydration-rehydration method as described previously with slight modifications [57, 58]. In brief, lipids were dissolved in chloroform and removed under vacuum at 53°C using a rotary evaporator (Buchi-Rotavapor R205, Brinkmann, Toronto, ON, Canada). 2 ml of an aqueous solution of tobramycin or polymyxin B at a concentration of 10 mg/ml were added to the thin dry lipid film and hand shaken in a warm water bath for 1 min. The lipid suspensions were sonicated in a round-bottom Erlenmeyer flask for 5 min (Sonic Dismembrator Model 500, Fischer Scientific, USA) while submerged in an ice-bath. The sonicator was not in direct contact with the liposome suspension at any time. The suspension was freeze-dried overnight for preservation and higher entrapment (Labconco model 77540, USA). At the time of experiment, dehydrated liposomes were rehydrated in PBS above the phase transition temperature of lipids (DMPC $T_c = 23°C$; DPPC $T_c = 41°C$), for 2 hr and unentrapped drug was washed off twice by ultracentrifugation at 62,000 g. This step ensures that the unentrapped drug (in the supernatant) is separated from the liposomal pellet and is aspirated from the formulation. The liposomal suspensions were diluted at room temperature and size and polydispersity index was automatically determined with the use of a NICOMP 270/autodilute Submicron Particle Sizer according to manufacturer instructions (Santa Barbara, CA, USA). The content of antibiotic entrapped in liposomes (after disruption with 0.2% Triton X-100) was measured by an established method as described previously for tobramycin and polymyxin B [51, 58]. Encapsulation efficiency (EE) was calculated as follows:

$$EE(\%) = (concentration\ of\ antibiotic\ released)\ /\ (concentration\ of\ initial\ antibiotic) \times 100\%$$

Stability of Liposomes Loaded with Antibiotics

The stability of antibiotics in the formulations was examined according to Mugabe et al. [59] at 37°C for 18 hr in the presence of PBS, CAMH broth, supernatant of biofilm forming *P. aeruginosa* (PA-48912-1, PA-48912-2, and PA-48913), a combination of DNA, F-actin, LPS, and LTA at a concentration of 1,000 mg/l, and intact or autoclaved sputum. In experiments involving sputum, pooled CF sputum was either kept intact and diluted 1:10 (w/v), or autoclaved for 10 min before mixing with CAMH broth. After incubation, aliquots of the mixtures were removed and centrifuged. Antibiotic presence in the pellet was assayed by the microbiological assay as described above, and the amount of antibiotic released from the liposomes was expressed as a percentage of the total antibiotic concentration at 0 hr.

Bacterial Killing Assays in Presence of Polyanions

Antibacterial activity of the formulations was measured in the presence or absence of polyanions found in the CF lung. *P. aeruginosa* (ATCC 27853) was grown on CAMH agar overnight at 37°C. Single colonies were diluted and suspended in CAMH broth alone or with 2-fold dilutions of LPS, LTA, DNA, and F-actin (125 to 1,000 mg/l); 2-fold dilutions of DNA and F-actin (125 to 1,000 mg/l); and 10-fold dilutions of LPS and LTA (1 to 1,000 mg/l). Equal volumes of 100 µl were added to a 96-well plate to a final concentration of 1×10^6 cfu/ml. To each well, 100 µl of the free or liposome-entrapped antibiotic (0.125–256 mg/l; final concentration) was added and the plates were incubated for 3 hr at 37°C. The incubation period and concentration chosen were adequate to allow liposome or free antibiotic-bacteria interaction and eradication. After incubation, the suspensions were kept cool on ice and bacterial suspensions were diluted 10–10,000 folds in PBS. Wells treated in the absence of polyanions were plated as is, that is without any dilutions. Aliquots (100 µl) of each dilution were plated on CAMH agar and incubated overnight at 37°C. The cfu/ml values were then determined for each of the three independent experiments.

To determine the ability of liposomes to retain their antibiotic activity, the MBC of the antibiotic formulations were determined in an 18 hr period by a standard microbroth dilution assay in CAMH broth alone or with a mixture of LPS, LTA, DNA, and F-actin at a fixed concentration of 1,000 mg/l. The MBC assay was performed as mentioned above with addition of 2-fold dilutions of the free or liposomal antibiotic formulations added to the 96-well plates. The final volume in each well was 200 µl, and PBS or CAMH alone were used as positive (no antibiotic) and negative (no bacteria) controls, respectively. Following incubation for 18 hr, aliquots (100 µl) were aspirated from each well and subcultured on CAMH agar plates overnight. The MBC was defined as the lowest concentration of the antibiotic that resulted in less than 30 cfu live bacteria/Petri dish.

Expectorated Sputum

The sputum samples were collected by spontaneous expectoration from nine CF patients following informed consent and a protocol approved by the Research Ethics Committee (Sudbury Regional Hospital, Sudbury, Ontario, Canada). Patients' age, name, treatments, and exacerbation records were kept confidential. Sputum samples colonized with moderate to heavy growth of *P. aeruginosa* were pooled and frozen at −80°C in aliquots. To measure the effects of the formulations on endogenous *P. aeruginosa*, aliquots of sputum samples were also stored at 4°C and used within 24 hr of collection. At the time of the experiment, the sputum samples were diluted 1:10 (w/v) in CAMH broth and mixed with the antibiotic formulations to achieve concentrations ranged 1 to 512 mg/l. The mixtures were then incubated for 18 hr at 37°C and the cfu/ml of live bacteria was determined according to the aforementioned protocol. The dilution of the samples should have affected the viscoelastic properties of the sputum, and this dilution was only done for easier handling and measurement of the sputum samples.

Data Analysis

All results were expressed as mean ± SEM obtained from three trials. Comparisons between free and liposomal formulations were made by ANOVA one-way post t-test, and P-values were considered significant when (*) $p<0.05$, (**) $p<0.01$, (***) $p<0.001$.

DISCUSSION

Polycationic antibacterial agents, like aminoglycosides and polymyxins, require self-promoted uptake pathways for entry and eradication of gram-negative bacteria [60]. The cationic antibiotics increase bacterial outer membrane permeability by displacing magnesium ions and binding to LPS [41, 60]. In the highly ionic CF sputum, however, the high affinity of excreted polyanionic bacterial endotoxins and glycoproteins from lysed white blood cells towards cationic antibiotics decreases their overall interaction with the bacteria in the lungs [46, 62]. Liposomes may create a protective environment for antibacterial agents to minimize such interactions and subsequently maintain a steady drug concentration in the lungs. Our data on the stability of the liposomal formulations displays that tobramycin leakage was at equilibrium after 3 hr, while polymyxin B leakage continued up to half its concentration over 18 hr. This suggests that these nanoparticles are effective in protecting the antibiotics in the CF sputum *in vitro*. The stability will ensure a continuous presence of the antibiotic at the site of infection, and improves antibiotic bioavailability and biodistribution *in vivo* [63].

Polyanions like DNA and F-actin have strong affinity for their multivalent counterions and tend to aggregate (form bundles) in the presence of cationic antibiotics which block their bioactivity [40, 46, 64, 65]. Our results demonstrate the capability of liposomes to reduce the antibiotics' contact with polyanionic factors in the sputum and enhance bacteria-antibiotic(s) interactions. The liposomal formulation protected tobramycin from the inhibitory actions of DNA/F-actin at low concentrations while neither polymyxin B formulations were inactivated. Our findings are in agreement with those reported by Hunt et al. [36] who found a reduction in tobramycin activity in the presence of DNA (within a 2 hr exposure) even when it was pretreated with recombinant human DNase (rhDNase). Weiner et al. [43] on the other hand, reported DNA and F-actin aggregation (within a 5 hr exposure) with increasing concentrations of tobramycin, yet bioactivity in a microbroth dilution assay (within an 18 hr exposure) was not hindered by the presence of either DNA or F-actin. The inconsistencies among the results of the different studies may be attributed to factors such as incubation time, co-incubation of DNA and F-actin, and that DNA/F-actin concentrations were increased as tobramycin concentration was kept constant. The protective effect of the liposomes at the lower DNA/F-actin concentrations may be attributed to the neutral nature of the phospholipids comprising the liposomes which would not favor electrostatic interactions between phospholipids with DNA or F-actin. The lack of effectiveness of tobramycin encapsulated within liposomes in the presence of the higher concentrations of polyanionic factors cannot be explained from the results of this study but it may be possible that a build up of F-actin/DNA aggregates leads to an increase in viscoelasticity, which ultimately hinders liposome-bacteria interaction [64, 65]. Reports from other studies have shown that DNA greatly hampers nanosphere diffusion through sputum and that the rhDNase improves its diffusion [53, 55, 56].

With regards to polymyxin B, reports from studies have shown G-actin polymerization in the presence of polymyxin B [37] and DNA and polymyxin B precipitation *in vitro* [67]. In our studies, there was no loss of bioactivity when DNA, F-actin, or both were incubated with polymyxin B (data not shown). The observation of consistent bacterial killing by polymyxin B can be attributed to the ability of the antibiotic to resist bundle formation, and having a higher affinity for polyanionic LPS of the bacterial outer wall than DNA or F-actin. Weiner et al. [43] reported no aggregation or reduction of bioactivity between colymycin, an anionic colistin form, and DNA or F-actin. However, the absence of aggregation may be due to the similar negative charges of the antibiotic and DNA or F-actin.

The binding of free bacterial surface components (e.g., LPS and LTA) to polycationic antibiotics like polymyxin B may be beneficial to the host in terms of suppressing inflammation however it will compromise the antibacterial effect of the antibiotic. Tobramycin and polymyxin B tend to interact with the bacterial lipid membranes as indicated by the results of this study where the bioactivity of both antibiotics was reduced when co-incubated with LPS/LTA. However, the bioactivity of the antibiotics within the liposomes fared better (Figure 3) although inhibited at the higher LPS/LTA concentrations. The mechanism of inactivation of liposomal antibiotics by the higher polyanionic LPS/LTA levels cannot be attributed to the release of antibiotics from liposomes and subsequent inactivation because results from the liposomal stability studies (Table 1) showed that the lipid bilayers were not lysed. This is consistent with results from another study reported by Davies et al. [68] where divalent anions entrapped in negative or positive charged liposomes when incubated with LPS were not significantly leaked from the liposomes which were not lysed. It is possible then that the higher concentrations of LPS/LTA may contribute to the stabilization of the liposomes, reduce antibiotic release, and thus prevent the leakage of the antibiotics leading to reduction of their interaction with bacteria.

If the lipid bilayers of liposomes can decrease antibiotic interactions with the poly-anionic components found in CF lungs and reduce bacterial growth within a 3 hr pe-riod much more strongly than free antibiotics, its long term advantage and presence in an 18 hr period would be advantageous (Table 2). Unfortunately, prolonged contact between polyanions and the formulations greatly increased the free and liposomal polymyxin B bactericidal concentrations, with liposomal tobramycin exhibiting bet-ter activity than free tobramycin. The dissimilar inhibitory effects on tobramycin and polymyxin B may be attributed to differences in their mechanisms of action, as tobra-mycin, a polar drug can enter the cell while polymyxin B a lipophilic agent interacts with LPS on the cellular surface. The interaction of polymyxin B with cell surface LPS, in addition to the interaction with the polyanions might leads to competition at the LPS binding site of bacteria, ultimately reducing antibiotic binding.

In light of the higher bactericidal activities and lower inactivation of liposomal antibiotics in the presence of polyanionic components *in vitro*, we sought to compare the bactericidal activity of these formulations against endogenous *P. aeruginosa* in CF sputa to that of the free drug. As shown in Figure 4, the antibacterial activity of liposomal antibiotics was more effective than the free antibiotics by 4-fold, although due to a large microbial population in the CF sputum, neither of the formulations fully eradicated bacterial growth. While liposomal tobramycin (128 mg/l) reduced growth, liposomal polymyxin B (8 mg/l) fell into clinically acceptable levels. The high con-centrations of antibiotics, tobramycin in particular, required to lower growth, may be primarily due to samples containing antibiotic resistant strains, or the sputum and its contents impeding antibiotic effects by acting as a physical barrier or inhibitor. Several studies have dealt with the inhibitory properties of sputum on antibiotics [34, 42, 69] while there have been a limited number of studies focused on liposomal penetration and interaction with sputum [53-55, 70]. The majority of these studies have focused on gene therapy and their transport across the sputum, but a recent work by Meers et al. [54] showed the ability of labeled neutral liposomes to penetrate sputum, and further-more, aminoglycosidic amikacin-entrapped liposomes were more efficacious than free amikacin in reducing bacterial growth in a rat *P. aeruginosa* infection model. In our study, due to issues of confidentiality, we did not have access as to the clinical status of the patients or their pathology laboratory reports. Nevertheless, delivery of antibiotics via a liposomal system enhanced their antibacterial activity in sputum.

Although liposome entrapment of antibiotics and their increased efficacy is not a novel finding, neutral liposome-entrapped antibiotics tended to be more bactericidal in sputum and in the presence of sputum components when compared to free antibiotics, but with reduced efficacies over a longer period of time *in vitro* (18 hr exposure). This decrease in efficacy appears to be the result of pro-longed interactions of the liposomes with the polyanionic factors found in sputum. As prophylactic and anti-inflammatory treatments are improving the lung function of CF patients, reduction in neutrophil inflammatory response and bacterial infections may reduce its lysis and the presence of charged macromolecules which tend to inactivate cationic antibiotics. As novel approaches proceed towards a cure for CF, research must also be directed on strategies that obstruct the presence and/or action of inhibitory factors associated with the disease. Future work in our laboratory will tend to focus on disruption of these negatively charged factors for increased liposomal penetration.

RESULTS

Liposome Entrapment and Sizing
The entrapment efficiency of tobramycin in liposomes composed of DMPC/Chol (35 mg:10 mg) was 2.47 ± 0.19 mg/ml with a mean size of 293.7 ± 41.1 nm, (polydispersity index of 0.70 ± 0.12). Liposomes containing polymyxin B in DPPC/Chol (38 mg:10 mg) had an entrapment efficiency of 0.4 ± 0.02 mg/ml, with a mean size of 445.1 ± 49.3 nm (polydispersity index of 0.91 ± 0.06).

Stability of Liposome-entrapped Antibiotics
Liposomal stability and antibiotic leakage in different environments including CF sputum at 3 or 18 hr post-exposure are shown in Table 1. The release rate of antibiotics in the presence of bacterial supernatant, polyanionic components, autoclaved, or intact sputum was comparable to PBS buffer or CAMH broth controls.

Table 1. Liposome-entrapped antibiotic stability assayed by microbiological assay.

Formulations	Conditions	Retention at 3 h	Retention at 18 h
Liposomal tobramycin	PBS buffer	72.7±3.2%	71.5±2.6%
	CAMH broth	74.6±2.1%	73.3±0.5%
	Bacterial Supernatant	72.9±1.6%	74.1±2.0%
	Polyanionic broth	74.3±2.2%	73.3±1.8%
	Sterile Sputum	72.1±3.0%	71.4±1.1%
	Intact Sputum	73.8±1.9%	71.7±1.4%
Liposomal polymyxin B	PBS buffer	67.5±1.3%	54.9±1.8%
	CAMH broth	65.2±1.2%	54.7±1.7%
	Bacterial Supernatant	67.5±2.1%	54.9±2.7%
	Polyanionic broth	69.4±2.7%	51.3±2.4%
	Sterile Sputum	67.2±2.6%	53.3±2.4%
	Intact Sputum	65.4±1.5%	53.3±2.4%

The stability of the liposomal formulations were examined at 37°(in an 18 h period in the presence of PBS, CAMH broth, supernatant of biofilm forming *P. aeruginosa*, a combination of DNA, F-actin LPS, and LTA and diluted intact or autoclaved s utum.

Effect of DNA, F-Actin, LPS, and LTA on Bactericidal Activity
To determine the inhibitory effects of DNA, F-actin, LPS, or LTA on the activity of antibiotics, different concentrations of these inhibitory factors were co-incubated with

the antibiotic formulations during a 3 hr pre-incubation first. The free and liposomal tobramycin at 2 mg/l killed all ATCC 27853 strain within 3 hr, while the liposomal and free polymyxin B eradicated bacteria at 1 mg/l and 2 mg/l, respectively. The bactericidal activity for both groups of antibiotics in the presence of DNA, F-actin, LPS, and LTA at the concentrations of 125 to 1,000 mg/l is shown in Figure 1. The activities of free and liposomal antibiotics were strongly inhibited and liposomal formulations tended to display lower antibiotic inhibition.

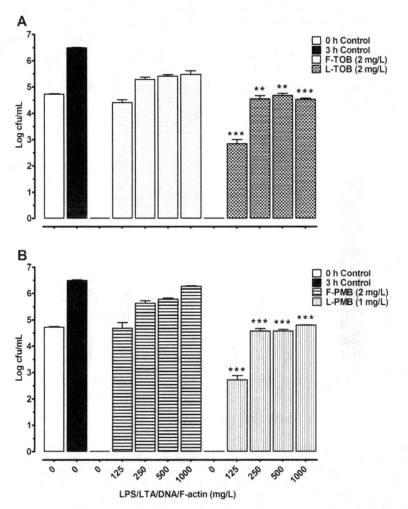

Figure 1. Bactericidal activity and inhibition of antibiotics by DNA, F-actin, LPS, and LTA. A) Bactericidal concentrations of free tobramycin (F-TOB) and liposomal tobramycin (L-TOB) were incubated in presence of LPS/LTA (1 to 1,000 mg/l). B) Bactericidal concentrations of free polymyxin B (F-PMB) and liposomal polymyxin B (L-PMB) were incubated in presence of DNA/F-actin/LPS/ LTA (125 to 1,000 mg/l). Growth controls are represented at 0 hr, and 3 hr. Comparisons between free and liposomal formulations were made by ANOVA one-way post t-test, and P-values were considered significant when (**) p<0.01, (***) p<0.001.

Separately, the inhibition of activity by DNA and F-actin, LPS, and LTA were assessed. When DNA and F-actin were co-incubated with the tobramycin formulations (Figure 2), free tobramycin failed to eradicate growth at DNA/F-actin concentrations of 125 to 1,000 mg/l. Higher concentrations of these inhibitory factors (500 to 1,000 mg/l) however, were required to hinder the liposomal tobramycin activity. On the other hand, bactericidal activity of liposomal polymyxin B co-incubated with DNA/F-actin remained the same as antibacterial activity was not impaired within 3 hr (data not shown). Under the same conditions discussed above, the effects of bacterial surface components LPS and LTA on the activity of antibiotics were investigated. Free tobramycin (Figure 3A) activity was increasingly inhibited at LPS/LTA concentrations of 1 to 1,000 mg/l. While the lower concentrations (1 to 10 mg/l) did not have any effect, higher concentrations (100 to 1,000 mg/l) of LPS/LTA were able to inactivate liposomal tobramycin. Polymyxin B formulations behaved the same as tobramycin in the presence of LPS/LTA, as indicated in Figure 3B.

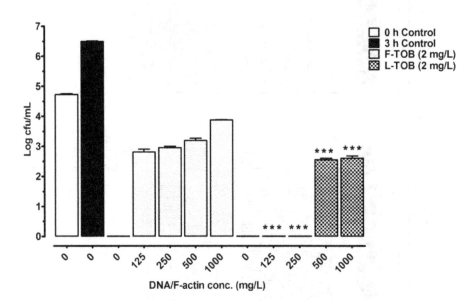

Figure 2. Bactericidal activity and inhibition of tobramycin by DNA and F-actin. Bactericidal concentrations of free tobramycin (F-TOB), and liposomal tobramycin (L-TOB) at 2 mg/l were incubated with *P. aeruginosa* (ATCC 27853), or in presence of DNA/F-actin (125 to 1,000 mg/l). Growth controls are represented at 0 hr, and 3 hr. Comparisons between free and liposomal tobramycin was made by ANOVA one-way post t-test, and P-values were considered significant when (***) p < 0.001.

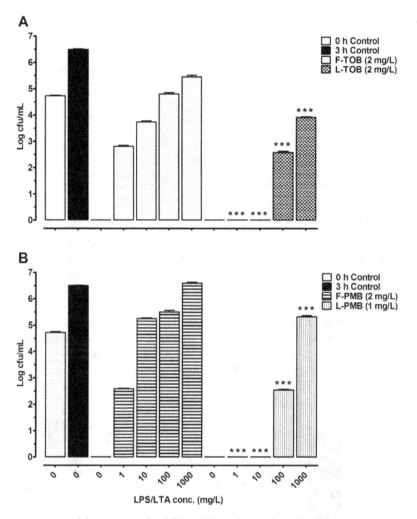

Figure 3. Bactericidal activity and inhibition of antibiotics by LPS and LTA. A) Bactericidal concentrations of free tobramycin (F-TOB) and liposomal tobramycin (L-TOB) were incubated in presence of LPS/LTA (1 to 1,000 mg/l). B) Bactericidal concentrations of free polymyxin B (F-PMB) and liposomal polymyxin B (L-PMB) were incubated in presence of LPS/LTA (1 to 1,000 mg/l). Growth controls are represented at 0 hr (empty bar), and 3 hr (dark bar). Comparisons between free and liposomal formulations were made by ANOVA one-way post t-test, and P-values were considered significant when (***) p<0.001.

Since negatively charged polyanions hindered bactericidal activity in a short period of time (3 hr exposure), their effect on the MBCs in an 18 hr period were also investigated (Table 2). MBC levels increased 16-fold for free tobramycin (16 mg/l) compared to 4-fold for its liposomal form (8 mg/l). Free (32 mg/l) and liposomal polymyxin B (16 mg/) were inhibited equally by the polyanions (64-fold increase in MBC).

Table 2. Minimum Bactericidal Concentrations.

Formulations	MBC (mg/L)	
	CAMH broth	Inhibitory factors
Free tobramycin	1	16
Lipo tobramycin	2	8
Free polymyxin B	0.5	32
Lipo polymyxin B	0.25	16

Bactericidal activity of free and liposomal formulations against susceptible *P. aeruginosa* ATCC 27853 strain was carried out inbroth aloneor inpresence of DNNF-actin /LPS/L TA at a final concentration of 1000 m /L.

Antibacterial Activity on CF Sputum

To test the efficacy of entrapped versus free antibiotics in the CF sputa, pooled sputum was diluted and incubated with increasing concentrations of tobramycin and polymyxin B for 18 hr. As shown in Figure 4, bacterial counts were reduced, but neither of the formulations eradicated endogenous bacteria present in the sputum. Liposomal tobramycin (128 mg/l; 5.3 ± 0.1 logs) and polymyxin B (8 mg/l; 3.8 ± 0.1 logs) displayed higher bactericidal activity than free tobramycin (512 mg/l; 5.4 ± 0.2 logs) and polymyxin B (32 mg/l; 3.9 ± 0.1 logs). The sputum itself did not seem to have any antibacterial activity against endogenous strains as bacterial counts were increased from 0 hr (5.2 ± 0.1 logs) to 18 hr (7.8 ± 0.1 logs).

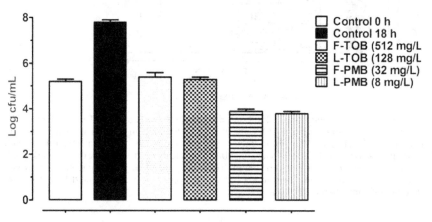

Figure 4. CF Sputum treatment with various antibiotic formulations. CFU counts were made after incubation of diluted CF sputum (1:10 w/v) in PBS with two-fold dilutions of free tobramycin at 512 mg/l (F-TOB), liposomal tobramycin at 128 mg/l (L-TOB), free polymyxin B at 32 mg/l (F-PMB), and liposomal polymyxin B at 8 mg/l (L-PMB). Growth controls are represented at 0 hr (empty bar), and 18 hr (dark bar).

ACKNOWLEDGMENTS

We would like to acknowledge Sharri-Lynne Zinger and the patients from the CF Clinic, and Beverly Harper from Memorial Hospital, Sudbury, Ontario, Canada.

AUTHORS' CONTRIBUTIONS

Conceived and designed the experiments: Misagh Alipour, Zacharias E. Suntres, Majed Halwani, Ali O. Azghani, and Abdelwahab Omri. Performed the experiments: Misagh Alipour and Majed Halwani. Analyzed the data: Misagh Alipour, Zacharias E. Suntres, Majed Halwani, Ali O. Azghani, and Abdelwahab Omri. Contributed reagents/materials/analysis tools: Misagh Alipour, Zacharias E. Suntres, Majed Halwani, Ali O. Azghani, and Abdelwahab Omri. Wrote the chapter: Misagh Alipour, Zacharias E. Suntres, Majed Halwani, and Abdelwahab Omri.

KEYWORDS

- **Cholesterol**
- **Cystic fibrosis**
- **1,2-Dimyristoyl-sn-glycero-3-phosphocholine**
- **Minimum bactericidal concentrations**
- ***Pseudomonas aeruginosa***

Chapter 7

Discovery of Novel Inhibitors of *Streptococcus pneumoniae*

Nan Li, Fei Wang, Siqiang Niu, Ju Cao, Kaifeng Wu, Youqiang Li, Nanlin Yin, Xuemei Zhang, Weiliang Zhu, and Yibing Yin

INTRODUCTION

Due to the widespread abusage of antibiotics, antibiotic-resistance in *Streptococcus pneumoniae* (*S. pneumoniae*) has been increasing quickly in recent years, and it is obviously urgent to develop new types of antibiotics. Two-component systems (TCSs) are the major signal transduction pathways in bacteria and have emerged as potential targets for antibacterial drugs. Among the 13 pairs of TCSs proteins presenting in *S. pneumoniae*, VicR/K is the unique one essential for bacterium growth, and block agents to which, if can be found, may be developed as effective antibiotics against *S. pneumoniae* infection.

Using a structure-based virtual screening (SBVS) method, 105 compounds were computationally identified as potential inhibitors of the histidine kinase (HK) VicK protein from the compound library SPECS. Six of them were then validated *in vitro* to be active in inhibiting the growth of *S. pneumoniae* without obvious cytotoxicity to Vero cell. In mouse sepsis models, these compounds are still able to decrease the mortality of the mice infected by *S. pneumoniae* and one compound even has significant therapeutic effect.

To our knowledge, these compounds are the first reported inhibitors of HK with antibacterial activity *in vitro* and *in vivo*, and are novel lead structures for developing new drugs to combat pneumococcal infection.

Streptococcus pneumoniae is a major risk factor with high morbidity and mortality world-widely, especially in the elderly and children. It is believed to be one of the four major infectious disease killers [1-5]. Meanwhile, an increasing number of bacterial strains with resistance are encountered in the clinic nowadays, among which antibiotic-resistant *S. pneumoniae* has caused many deaths due to antibiotics abusage in hospitals. Therefore, it is urgent to develop new types of antibiotics.

In prokaryotes, the two-component signaling systems, each pair of which are typically composed of HK and response regulator (RR), play important roles in drug-resistance, pathogenesis and bacterial growth [6-8]. The regulation of TCS on histidine phosphorylation in signal transduction distinct from that on serine/threonine and tyrosine phosphorylation in higher eukaryotes [9]. For some TCSs, both the HK and RR are essential for bacterial viability in several Gram-positive pathogens, including *Bacillus subtilis* (*B. subtilis*), *Enterococcus faecalis* and *Staphylococcus aureus*

(*S. aureus*) [10-13], and thus received attention as potential targets for antimicrobials [9, 14-17]. In *S. pneumoniae*, although at least 13 TCSs were identified, only TCS02 (also designated as VicR/K [18], MicA/B [19] or 492 hk/rr [20]) is essential for bacteria viability, which can be a potential target for antimicrobial intervention. To be detailed, in TCS02, only functional VicR appears to be essential for *S. pneumoniae* [21], without which *S. pneumoniae* can not grow or act as a pathogen [22]. However, the crystal structure of VicR is unsuitable for SBVS because the active site is too shallow to dock a small molecule [22, 23]. The reason that VicK does not seem to be essential for *S. pneumoniae* viability, was supposed to be that some currently unknown HKs also participate in the activation of VicR by phosphorylation [24, 25]. However, among these HKs, VicK it is best-known one with definite action on VicR. Moreover, recent researches showed a high-degree homology in the catalytic domain of these HKs [14-17]. Thus theoretically, selective inhibitors to VicK, a representative of HKs, can interrupt the phosphorylation of VicR and ultimately reduce the viability of *S. pneumoniae*.

The SBVS, an approach used widely in drug design and discovery, possesses many advantages, such as rapidness, economization, efficiency, and high-throughput. In the recent years, SBVS has attracted great attention in developing innovative antimicrobial agents. A case in point is the discovery of a lead-compound named diarylquinoline against *Mycobacterium tuberculosis* [26]. Our study here was designed to search the compound database for potential inhibitors targeting the VicK protein of *S. pneumoniae* by using *in silico* and experimental methods, which may provide much valuable information to develop new antibiotics against pneumococcal infection.

Sequence Analysis of the VicK TCS in *S. pneumoniae*

Domain analysis [http://smart.embl.de/smart/show_motifs.pl?ID=Q9S1J9] indicated that the VicK protein of *S. pneumoniae* contained one transmembrane segment and several domains: PAS, PAC, HisKA, and HATPase_c. Multi-alignment of the HATPase_c domain sequences showed that in most bacteria the sequences around the ATP binding site of VicK HKs are similar and have four conserved motifs: the N box, G1 box, F box, and G2 box [27]. This high homology of ATP binding domain of HKs in bacteria makes it reasonable to screen antibacterial agents by using this domain as a potential target [16].

Compared with VicK HATPase_c domain in *S. pneumoniae* (GenBank accession number: AAK75332.1), the most homologous sequence in the structural Protein Data Bank (PDB) was the similar domain of *Thermotoga maritime* (PDB entry: 2c2a) [28], a TCS molecule, with 33% sequence identity and 57% conservative replacements (Figure 1). This domain is the entire cytoplasmic portion of a sensor HK protein. The X-ray crystal structure of the domain of *Thermotoga maritima* was therefore used as a template for modeling the 3D structure of the VicK HATPase_c domain of *S. pneumoniae*.

```
VicK    -L--MENVTESKELERLKRIDRMKTEFIANISHELRTPLTAIKAYAETIYNSLGELDLSTLKEFLEVIIDQSNHLENLLNELLDFSRLERKSLQINREKV
2C2A    SGFISGLVAVLHDTTEQEKEERERRLFVSNVSHELRTPLTSVKSYLEALD--EGALCETVAPDFIKVSLDETNRMURNVTDLLHLSRIDNATSHLDVELI
         *:  ::  .  ::  :* :    *::*:*****#*****::*:* *::     * * :.  :*::* :*::*:: .::.:**.:**:: . : ::: * :


VicK    DLCDLVESAVNAIKEFASSHNVN--VLFESNVPCPVEAYIDPTRIRQVLLNLLNNGVKYSKKDAPDKYVKVILDEKDGGVLIIVEDNGIGIPDHAKDRIF
2C2A    NFTAFITFILNKFDKNKGQEKEKKYELVRDYPINSIWMEIDTDKNTQVVDNILNNAIKYSPDGG---KITVRMKTTEHDQNILSISDHGLGIPKQDLPRIF
         ::  ::   :*.:: ...: :   *,...   .:   **.  ::  **:  *:***:.:*** ...    1.*  .:. :::  :.*:*:**.:   ***


VicK    EQFYRVDSSLTYEVPGTGLGLAITKEIVELHGGRIWVESEVGKGSRFFVWIP--KDRAGEDNRQDN---
2C2A    DRFYRVDRARSRAQGGTGLGLSIAKEIIKQHKGFIVAKSEYGKGSTFIIVLPYDKDAVKEEVVEDEVED
         ::*****  : :   ******:*:***:: * * **.:** **** * : :*   **  .  *:  :*:
```

Figure 1. The sequence alignment of the HATPase_c domain of VicK in *S. pneumoniae* and 2c2a. The symbols below the alignment represent the similarity between two proteins. "*" denotes identical residues between two sequences, "means similar residues, ". means a bit different and blank means completely different. Schematic alignment diagram was made by the program ClustalX.

A 3D Model of the VicK HATPase_c Domain of *S. pneumoniae*

Based on the X-ray diffraction crystal structure of the homologous domain of the *Thermotoga maritima*, a 3D model for the VicK HATPase_c domain of *S. pneumoniae* was constructed. Figure 2A shows the final structure of this model that was checked and validated using structure analysis programs Prosa and Profile-3D [29]. This model of 3D structure contains five stranded β-sheets and four α-helices, which form a two-layered α/β sandwich structure. Figure 2B indicates that the model superposed well with the homologous domain of *Thermotoga maritima*, with a root-mean-square deviation (RMSD) of the Cα atoms being about 1.34 Å. The surface shape and general electrostatic feature of the HATPase_c domain of VicK were shown in Figure 2C. The ATP binding site consists of a relatively hydrophobic inner cavity and a larger hydrophilic outer cavity. Both cavities are connected by a gorge-like channel, and are consisted of highly conserved residues which can bind and fix the substrate. The inner part lack of polar amino acid residues can accommodate the adenosine, while the outer one rich in charged residues can bind the triphosphate.

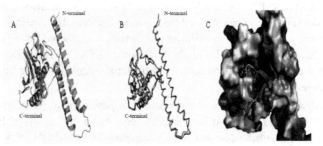

Figure 2. The modeled structure of the VicK HATPase_c domain of *S. pneumoniae*. (A) The solid ribbon representation of the structure model of the VicK HATPase_c domain. (B) Structure superposition of sketch of modeled VicK structure with the template. (C) Shape and surface features of the ATP-binding pocket of the VicK HATPase_c domain. The color denotes electrostatic potential of the protein surface. The red and blue color show negative and positive charged potential respectively, and the white surface means neutral potential of non-polar hydrophobic residues. The ATP-binding pocket is divided into "inner" and "outer" parts. The loop covered on the pocket is shown as tube for the sake of clearly demonstrating the hydrophobic inner part. The outer part of pocket is hydrophilic because of many polar residues in the entrance of the pocket, including the polar loop structure. All the pictures were generated by PyMol [http://www.pymol.org/].

Discovery of Potential Inhibitors of the *S. pneumoniae* VicK HK by Virtual Screening

The target site for high throughput virtual screening (HTVS) was the ATP-binding pocket of the VicK HATPase_c model of *S. pneumoniae*, which consisted of residues within a radius of 4 Å around the ATP site. In the primary screening, the database SPECS containing about 200,000 molecules was searched for potential binders using the program DOCK4.0 [30, 31]. Subsequently, structures ranked in the first 10,000 were re-scored by using the Autodock 3.05 program [32]. As a result, about 200 molecules were filtered out by these highly selective methods. Finally, we manually selected 105 molecules according to their molecular diversity, shape complementarities, and the potential to form hydrogen bonds and hydrophobic interactions in the binding pocket of the VicK HATPase_c domain.

Inhibition of the VicK' Protein ATPase Activity *In Vitro*

In order to confirm the interaction of the potential VicK inhibitors with their putative target protein, we expressed and purified His-tagged VicK' protein by using the pET28a plasmid in BL21(DE3) as shown in Figure 3A. The kinase activity of VicK' protein was measured by quantifying the amount ATP remained in solution after the enzymatic reaction (Figure 3B). These results indicated that the purified VicK' protein possessed the ATPase activity, which can hydrolyze ATP *in vitro*. Using the purified active VicK, we obtained 23 compounds from the 105 candidate inhibitors which could decrease the ATPase activity of VicK' protein by more than 50%, indicating these compounds may also be potential VicK inhibitors in *S. pneumoniae*.

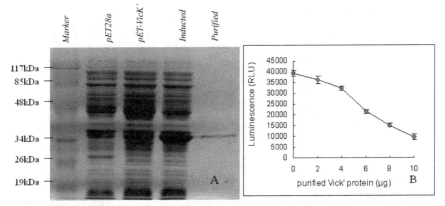

Figure 3. (A) SDS-PAGE analysis of VicK' purification (B) Identification of kinase activity of VicK' protein *in vitro*. Variant amounts of VicK' proteins were added into reaction systems containing a constant ATP concentration (5 µM). Each assay was performed in quadruplicate and repeated three times. Luminescent output is inversely correlated with the concentration of the kinase.

Antimicrobial Activities of Potential Vick' Inhibitor and Cytotoxicity of the Antimicrobial Compounds *In Vitro*

We investigated the bactericidal activity of these 23 compounds against *S. pneumoniae* using a standard minimal bactericidal concentration assay (MIC) (Table 1). Six compounds (Figure 4), each inhibiting the VicK' activity by more than 50% (52.8%, 54.8%, 51.6%, 61.9%, 71.1%, and 68.8%, respectively) (Figure 5), could obviously inhibit the growth of *S. pneumoniae*, with MIC values below 200 μM. Moreover, their MIC values were positively correlated with the corresponding IC50 (the concentration of inhibiting 50% VicK' protein autophosphorylation) values (r = 0.93), which indicates that the bactericidal effects of these chemicals were realized by disrupting the VicK/R TCS system in *S. pneumoniae*. Chemical structures of these six compounds are shown in Figure 4, which belong to three different classes of chemicals: one imidazole analogue, four furan derivatives and one derivative of thiophene (Figure 4).

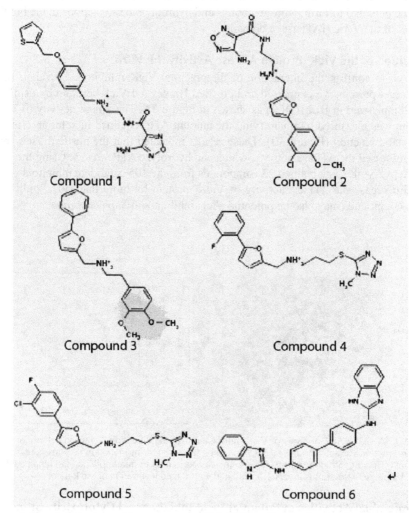

Figure 4. Chemical structures of the compounds with inhibitive effects on the growth of *S. pneumoniae*. These six inhibitors belong to three different classes of chemical structures: one imidazole analogue (compound 6), four furan derivatives (compound 2, 3, 4, and 5) and one derivative of thiophene (compound 1).

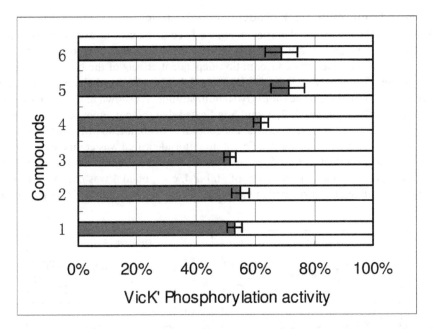

Figure 5. Inhibition ratio of VicK′ protein autophosphorylation by six lead compounds with antibacterial effects (from the 23 compounds). The inhibitory activities of the compounds for the ATPase activity of the VicK′ protein was measured using the Kinase-Glo™ Luminescent Kinase Assay. Briefly, purified VicK′ protein(6 µg/50 µl) was pre-incubated with compounds(final concentration, 200 µM) in a reaction buffer containing 40 mM Tris-HCl (pH 7.5), 20 mM $MgCl_2$ and 0.1 mg/ml BSA, at room temperature for 10 min. Then ATP (5 µM) was added for another incubation of 10 min at room temperature, and detected the rest amount of ATP.

Table 1. Biological effects of six potential inhibitors of the VicK histidine kinase.

Chemical inhibitor	MIC (µM)	MBC (µM)	CC50 (µM) on Vero cell	IC50 (µM) for VicK′ protein
Compound 1	100	>200	213	542.25
Compound 2	50	200	321.33	562.41
Compound 3	100	>200	274.22	502.63
Compound 4	200	>200	360	>1000
Compound 5	100	>200	516.17	598.11
Compound 6	0.28	25	392	32.60
Compound 7	0.02	2.0	undone	undone

A 3-(4, 5-dimethylthiazol-2-yl)-2, 5-diphenyl tetrazolium bromide (MTT) assay was carried out on Vero cell line to determine the CC50 (concentration that induces a 50% cytotoxicity effect) values of these compounds. As shown in Table 1, the CC50 values of all these six compounds were larger than 200 µM and than their respective MIC values, indicating low cytotoxicity effects on Vero cell. Collectively, these compounds inhibited bacterial growth with low toxic effects.

Time- and Concentration-dependent Growth Curve

While several compounds identified in our study could be used as excellent drug leads *in vitro*, the best and most valuable ways would be *in vivo* validation. The following results of the time- and concentration-dependent effects of the lead inhibitors on the growth of *S. pneumoniae* further illustrated their antibacterial characteristics, and would be an important guide for *in vivo* administration. As shown in Figure 6, the similar curves of compounds 1, 2, 3, and 5 indicated that these compounds have significant activity against *S. pneumoniae* at concentration of about 200 µM, and this activity could last at least 8 hr. The most efficient inhibitor identified was compound 6, which had bactericidal effect against *S. pneumoniae* even at concentration of as low as 0.2 µM. However, even at concentration of 400 µM, compound 4 was not likely to have bactericidal effect, but it seemed to have delayed the multiplication of *S. pneumoniae*.

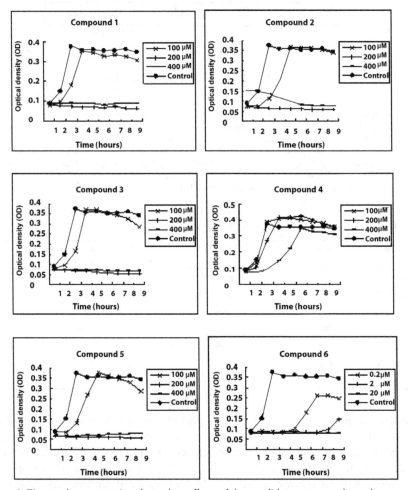

Figure 6. Time and concentration-dependent effects of the candidate compounds on the growth of *S. pneumoniae in vitro*.

Therapeutic Effects of the Lead Compounds in Mouse *S. pneumoniae* Infections

Mouse sepsis models by *S. pneumoniae* (ATCC7466) were successfully established by intraperitoneal injection of 100 μl *S. pneumoniae* (5×10^3 CFU/ml). Generally, these mice began to die within 24 hr and could not survive more than 48 hr unless they got appropriate therapeutic treatments. For facilitation of comparisons between the effects of these compounds and positive control (penicillin), the concentration of penicillin used in this study almost equaled to that of the lead compounds. To rule out the direct antibacterial effects that may compromise with the efficiency of this model, the lead compounds and penicillin were administrated through caudal vein. As shown in Figure 7, these compounds were able to decrease, though slightly, the mortality of the infected mice in the first 24 hr as compared to negative control (normal sodium, NS) ($p < 0.01$). Significant treatment effects were found among the groups ($p < 0.01$) by an overall comparison. Pairwise comparisons revealed that compounds 1–6 prolonged survival time in mouse sepsis models as compared to negative control (p < 0.01). However, compound 1, 2, 3, and 6 were less effective than positive control PNC ($p < 0.05$ or $p < 0.1$). Although, these compounds could not reverse the fatal pneumococcal infection with concentration used in this study, *in vivo* antibacterial activity of these six compounds suggested that it would be promising to develop lead-compound-based drugs against pneumococcal infection.

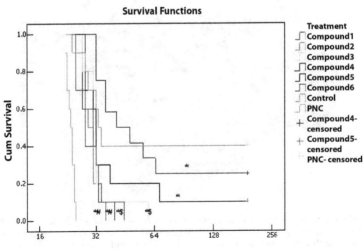

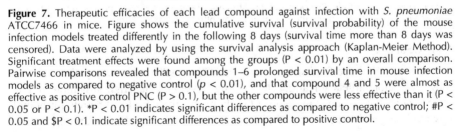

Figure 7. Therapeutic efficacies of each lead compound against infection with *S. pneumoniae* ATCC7466 in mice. Figure shows the cumulative survival (survival probability) of the mouse infection models treated differently in the following 8 days (survival time more than 8 days was censored). Data were analyzed by using the survival analysis approach (Kaplan-Meier Method). Significant treatment effects were found among the groups (P < 0.01) by an overall comparison. Pairwise comparisons revealed that compounds 1–6 prolonged survival time in mouse infection models as compared to negative control (p < 0.01), and that compound 4 and 5 were almost as effective as positive control PNC (P > 0.1), but the other compounds were less effective than it (P < 0.05 or P < 0.1). *P < 0.01 indicates significant differences as compared to negative control; #P < 0.05 and $P < 0.1 indicate significant differences as compared to positive control.

Molecular Modeling of VicK' Protein and Its Potential Inhibitors

In order to get insight into the mechanism of inhibition, further studies were carried out to verify the interaction modes between six compounds and the modeled structure of VicK' protein. Autodock 3.05 software was used for the docking simulation. The binding conformations of these inhibitors in the ATP-binding pocket of the VicK HATPase_c domain were shown in Figure 8. Although these structures are diverse, the binding models of six potential inhibitors are similar, especially in the inner part of the conserved domain. The surface of the binding pocket (Figure 2C) is divided into two parts, one is hydrophobic inner part composed of residues ILE146, ILE175, LEU180, ILE182, PHE238, and the other is the outer hydrophilic part consisted of residues ASN149, LYS152, TYR153, ARG196, ARG199. All six compounds bind in the pocket with rigid aromatic ring parts inserting into the inner part. In the large and flexible outer part, these compounds adopt different interactions. All of them have hydrogen bond acceptors in the binding outer part. They could form hydrogen networks with the polar residues to stabilize the substrate interactions. Their binding models resemble natural substrate ATP much.

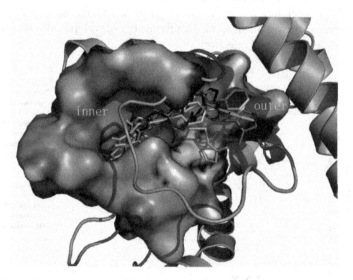

Figure 8. Three-dimensional structural binding modes of six potential inhibitors to VicK' protein derived from the docking simulations. The loop covered on the pocket was shown in tube. Six compounds were shown in stick with different colors. Their binding conformations showed similar interaction modes in the inner pocket. The binding diversity was restrained by small space and hydrophobic characteristic. By contrast, these structures bound in the outer pocket in various ways. This image was generated using the PyMol program [http://www.pymol.org/].

DISCUSSION

In bacteria, HKs have fundamental roles in TCS signal transduction pathways. Thus they are major targets for antibacterial drug development. High structural and sequence homology of this kinase gene family makes the HKs ideal targets for homology modeling and structure based virtual screening. The SBVS is an approach based on the three-dimensional structures of macromolecular to identify chemical entities binding to the targets and to elicit potential biological mechanisms with the advantages of speed, efficiency and high-throughput. The availability of the small molecular lead-compound library and the modeled 3D target structure makes it possible to use SBVS to screen out a limited number of promising candidates that can interrupt the TCS signal transduction by interacting with the HKs substrate of *S. pneumoniae*.

The HKs, as novel antibacterial targets, have attracted many attentions due to their essentiality in the viability of microbes and their deficiency in animals. The HKs are involved in the regulation of bacterial growth and virulence in many bacterial species. Previously, a HK named VicK has been used to screen lead compound inhibitors in *B. subtilis* and *S. epidermidis*. We here for the first time obtained 105 candidate chemical compounds directly aiming at *S. pneumoniae* VicK by screening 200,000 possible compounds in *silico*. Compounds that can bind to the purified target protein VicK' and compete with its substrate ATP were further verified by *in vitro* and *in vivo* antibacterial assays. Eventually, we obtained six compounds with antibacterial activity that may be used as novel drug leads.

Commonly, the response regulator YycF and the histidine kinase YycG are the only essential TCS for viability in *B. subtilis* and *S. aureus* [10, 12]. In *S. pneumoniae*, the VicR/K TCS regulates the expression of several critical genes, such as those encoding surface proteins and virulence factors [21, 33]. However, only the response regulator VicR was found to be essential [20, 34]. The signal transduction of VicK was possibly bypassed by other TCS HKs [35]. VicK has conserved ATP-dependent HATPase_c domains accounting for autophosphorylation. Even non-cognate HKs from other bacteria can phosphorylate the purified VicR from *S. pneumoniae* [18]. In a previous study [36], the MIC values of the lead compounds screened out by SBVS targeting the YycG of *S. epidermidis* were almost equal to the corresponding IC_{50} (for YycG) values, with a correlation coefficient of 0.959, which suggested that inhibition of 50% the YycG protein activity would interfere with the growth of *S. epidermidis*. If this case is true in *S. pneumoniae*, the result that the MIC values of the lead-compounds were far less than the corresponding IC50 values may be explained as bypass effects of these compounds on other HKs. In a word, these lead compounds are most likely having a "cross-inhibition" on other HKs in *S. pneumoniae*, which can enhance their antibacterial effects, although they were not verified in this study.

Although, the VicK protein in *S. pneumoniae* can be homologous to YycG in other gram-positive strains, such as *S. epidermidis*, *Enterococcus faecalis* and *S. aureus*, different strains generally have different characteristics of the HATPase_c domain structure of HKs. These characteristics will determine the binding specificity of the lead compounds screened out by SBVS. Moreover, a different template for homologous modeling and different parameters for SBVS were used, which can guarantee the specificity of the lead compounds binding to the VicK' discovered. What is more, 23 compounds can inhibit the purified VicK' protein activity by more than 50%, 6 of which displayed different degrees of antibacterial effects *in vitro* and *in vivo*. Regretfully, the *in vivo* activities of these compounds were not quite consistent with their corresponding *in vitro* activity, and some compounds displayed obvious cytotoxicity, which would challenge our future investigation. Moreover, it seems to be a paradox that compound 4 have less bactericidal effects in the time- and concentration-dependent antibacterial assays, but demonstrated significant therapeutic effects in mice infected by *S. pneumoniae*. However, due to the VicK' is not essential in *S. pneumoniae*, this chemical may have a possibility to interrupt the invasion and virulence rather than cause numerous death of the bacterium, which decreases the selection pressure and contributes to the maintenance of species diversity, thus reduces the emergence of drug-resistant strains. Anyway, the subtle mechanisms need our future work.

MATERIAL AND METHODS

Bacterial Strains, Media, and Reagents

S. pneumoniae (D39) ATCC7466 was purchased from the American Type Culture collection (ATCC, USA). *S. pneumoniae* D39 was grown in C + Y medium. Plasmids were transformed into *Escherichia coli* (*E. coli*) strains that were grown in Luria-Bertani (LB) broth. For selection of *E. coli* transformants, kanamycin (50 μg/ml, final concentration) was added to the growth medium.

All compounds screened out in our study were purchased from the SPECS Company in the Netherlands. Stock solutions of the compounds were prepared in Dimethyl Sulfoxide (DMSO). Other chemicals were purchased from Sigma.

Bioinformatics Analysis

Domain analysis was performed based on the SMART database. The complete genome sequences of the *S. pneumoniae* strain ATCC7466 were accessed from the National center for Biotechnology information (NCBI) genome database. For the homologous sequences with the VicK HATPase_c domain of *S. pneumoniae* ATCC7466, the PDB was searched by using the Blastp program. ClustalX was used to align the protein sequences.

3D Structure Modeling of the VicK HATPase_c Domain

The sequence of *S. pneumoniae* VicK was retrieved from GenBank (accession number: AAK75332.1). The Align123 module in Insight II was used in the pairwise sequence alignment. Using the secondary structure information of *Thermotoga maritima* (PDB entry: 2c2a), the sequence alignment was adjusted manually to obtain a fine alignment for 3D structure construction. The 3D model of the VicK HATPase_c domain was generated by using the MODELLER module in Insight II. Several structural analysis programs such as Prostat and Profile-3D were used to check the structure quality. The Prostat module of Insight II was used to analyze the properties of bonds, angles, and torsions. The profile-3D program was used to check the structure and sequence compatibility.

Structure-based Virtual Screening

The SBVS was performed as described previously [36], with modification. Briefly, the binding pocket of the VicK HATPase_c domain was used as a target for screening the SPECS database by using the docking approach. A primary screening was conducted by using the program DOCK4.0. Residues within a radius of 4 Å around the ATP-binding pocket of the VicK HATPase_c domain were used for constructing the grids for the docking screening. Subsequently, the 10,000 compounds with the highest score as obtained by DOCK search were selected for a second round docking by using the Autodock 3.05 program, followed by our own filter of drug likeness to eliminate the non-drug-able molecules. Finally, we manually selected 105 molecules according to their molecular diversity, shape complementarities, and potential to form hydrogen bonds in the binding pocket of the VicK HATPase_c domain.

Molecular Modeling of the Interaction between Inhibitors and the Target Protein

To determine the binding modes, Autodock 3.05 was used for automated docking analysis. The Lamarchian genetic algorithm (LGA) was applied to deal with the protein-inhibitor interactions. Some important parameters were set as follows: the initial number of individuals in population is 50; the elitism value is 1, which automatically survives into nest generation. The mutation rate is 0.03, which is a probability that a gene would undergo a random change. The crossover rate, the probability of proportional selection, is 0.80. Every compound was set to have 10 separated GA runs and finally 10 conformations would be generated. The conformations were clustered automatically and the conformation with minimum binding free energy in the cluster with minimum RMSD value was selected as the representative conformation of the inhibitor.

Cloning, Expression, and Purification of the VicK Protein

The *VicK* gene fragment containing the cytoplasmic signal domains (the HATPase_c and HisKA domain) of VicK (coding 200–449 aa) was amplified by PCR. The upstream and the downstream primers were 5'-CGGGATCCGAGCAGGAGA-AGGAAGAAC-3' and 5'-CGCTCGAGGTCTTCTACTTCATCCTCCCA-3' respectively. Subsequently, the fragment was digested with *EcoR* I and *Xho* I (TaKaRA, Japan) and ligated into the corresponding sites of pET28a to obtain a recombinant plasmid pET28/VicK'. After being transformed into *E. coli* strain BL21 (DE3), this recombinant plasmid was induced to express the protein of VicK' by 0.2 mM isopropyl-1-thio-β-D-galactopyranoside (IPTG) at 24°C for 20 hr. Cells were harvested and sonicated, and then the debris was removed by centrifugation. The fraction containing the cytoplasmic domain was isolated from the supernatant solution through a His-tagged column, with a purity of more than 95%, as assessed by gel electrophoresis and Coomassie Blue staining.

Inhibition Assay for the ATPase Activity

The inhibitory activity of the compounds for the ATPase activity of the VicK' protein was measured using the Kinase-Glo™ Luminescent Kinase Assay (Promega, Madison, USA). Briefly, 6 μg purified VicK' protein was pre-incubated with a series of dilutions of compounds in a reaction buffer containing 40 mM Tris-HCl (pH 7.5), 20 mM $MgCl_2$ and 0.1 mg/ml BSA, at room temperature for 10 min. Then 5 μM ATP was added for another incubation of 10 min at room temperature, and Kinase-Glo™

Reagent was added to detect the rest amount of ATP, as reflected by luminescence intensity (Lu). In parallel, the VicK' protein with no addition of compounds was used as control and ATP only was used as blank. The rate of inhibiting protein phosphorylation (R_p) by the compounds was calculated by the following equation: $R_p = (Lu_{compound} - Lu_{control})/(Lu_{blank} - Lu_{control}) \times 100\%$. IC_{50} (the concentration of inhibiting 50% VicK' protein autophosphorylation) was calculated by using the SPSS 11.0 software.

Minimal Inhibitory Concentration (MIC) and Minimal Bactericidal Concentration (MBC) Assays

The MIC assays for the antibacterial activities of the compounds were performed according to the broth micro-dilution (in 96-well plate) methods of the Clinical and Laboratory Standards Institute (CLSI) of America. The MBC was obtained by subculturing 200 µl from each negative (no visible bacterial growth) well in the MIC assay which were then plated onto Columbian blood plates. The plates were incubated at 37°C for 24 hr, and the MBC was defined as the lowest concentration of substance which produced subcultures growing no more than five colonies on each plate. Each assay was repeated at least three times.

Time- and Concentration-dependent Curve

Streptococcus pneumoniae strains ATCC7466 were grown at 37°C in C + Y medium till OD_{550} reaching 0.1. Then 200 µl of the suspending bacteria was extracted into the wells of a 96-well plate for incubation at 37°C with the additions of three different dilutions of the six compounds. Subsequently, the plate was detected by spectrophotometer per hour for drawing the time- and concentration-dependent curve. All samples were assayed in triplicate, and each assay was repeated at least three times.

In Vitro Cytotoxicity

Cytotoxicity of the antibacterial compounds on cultured Vero cell was measured by using the Cell Proliferation Kit I (MTT) (Sigma). Briefly, a series of dilution of the compounds were added into the medium, containing 1% of DMSO, to culture Vero cell. Cytotoxicity of the different concentration of chemicals was determined according to the kit protocol. Each assay was performed in quadruplicate and repeated three times. The results were converted to percentages of the control (cells only treated with 1% DMSO) and CC_{50} (concentrations that produce a 50% cytotoxicity effect on Vero cell) was calculated by using the SPSS 11.0 software.

In Vivo Assays

Male and female BALB/c mice, aged 6–8 weeks (approx. 18–20 g), were used to evaluate the *in vivo* effects of the compounds. Briefly, these mice were randomly assigned to eight groups (10–12 per group, half in each sex): six compound-treated groups, one negative control and one positive control. All the mice were administrated with 100 µl suspended *S. pneumoniae* strain ATCC7466 (5×10^3 CFU/ml in phosphate buffered saline) by intraperitoneal injection route. Compounds (1–6) were diluted to the concentration of MIC respectively (1.27 mg/kg/d, 0.65 mg/kg/d, 1.13 mg/kg/d, 2.32 mg/kg/d, 1.27 mg/kg/d, 0.014 mg/kg/d, respectively) with normal sodium and 200 µl was administered by vena caudalis route after infection. Two control groups were administered with 200 µl normal sodium (negative control) and penicillin (0.42 mg/kg/d, positive control) respectively by the same injection route. Treatments were continued 3 times a day for 3 consecutive days, and these levels of chemicals caused few toxic influences on normal mice. The results are expressed as cumulative survival rates over the following 8-day observation.

CONCLUSION

To summarize, we have successfully found out several promising lead compounds for further drug development in this study, which also can be used as inhibitors to explore the mechanism of autophosphorylation by VicK as well as other HKs. Important work in future would be validation of their antibacterial effects in different strains and structural modification for more effective derivatives with less *in vivo* toxicity, and investigation into whether they can bind to other ATP-dependent kinase is also necessary.

KEYWORDS

- **Autophosphorylation**
- **Cytoplasmic portion**
- **Cytotoxicity**
- ***Streptococcus pneumoniae***

AUTHORS' CONTRIBUTIONS

Xuemei Zhang and Yibing Yin conceived of the study and participated in its design and coordination. Nan Li, Fei Wang, and Weiliang Zhu carried out the modeling of VicK protein and structure-based virtual screening. Nan Li, Siqiang Niu, Youqiang Li, Kaifeng Wu, and Ju Cao participated in the biological experiments of the *in vivo* assays and the *in vitro* assays. Nan Li, Fei Wang, and Nanlin Yin participated in analyzed the data and produced figures. Nan Li, Fei Wang, Weiliang Zhu, Xuemei Zhang and Yibing Yin drafted the manuscript. All the authors have read and approved the final manuscript.

ACKNOWLEDGMENTS

This work was supported by the National Natural Science Foundation of China (No. 30671868, 20721003).

Chapter 8

Archaeosomes Made of *Halorubrum tebenquichense* Total Polar Lipids

Raul O. Gonzalez, Leticia H. Higa, Romina A. Cutrullis, Marcos Bilen, Irma Morelli, Diana I. Roncaglia, Ricardo S. Corral, Maria Jose Morilla, Patricia B. Petray, and Eder L. Romero

---

INTRODUCTION

Archaeosomes (ARC), vesicles prepared from total polar lipids (TPL) extracted from selected genera and species from the Archaea domain, elicit both antibody and cell-mediated immunity to the entrapped antigen, as well as efficient cross priming of exogenous antigens, evoking a profound memory response. Screening for unexplored Archaea genus as new sources of adjuvancy, here we report the presence of two new *Halorubrum tebenquichense* strains isolated from gray crystals (*GCs*) and black mud (*BM*) strata from a littoral Argentinean Patagonia salt flat. Cytotoxicity, intracellular transit, and immune response induced by two subcutaneous (sc) administrations (days 0 and 21) with bovine serum albumin (BSA) entrapped in ARC made of TPL either form *BM* (ARC-*BM*) and from *GC* (ARC-*GC*) at 2% w/w (BSA/lipids), to C3H/HeN mice (25 µg BSA, 1.3 mg of archaeal lipids per mouse) and boosted on day 180 with 25 µg of bare BSA, were determined.

The DNA G + C content (59.5 and 61.7% M *BM* and *GC*, respectively), 16S rDNA sequentiation, DNA–DNA hybridization, arbitrarily primed fingerprint assay, and biochemical data confirmed that *BM* and *GC* isolates were two non-previously described strains of *H. tebenquichense*. Both multilamellar ARC mean size were 564 ± 22 nm, with –50 mV zeta-potential, and were not cytotoxic on Vero cells up to 1 mg/ml and up to 0.1 mg/ml of lipids on J-774 macrophages (XTT method). The ARC inner aqueous content remained inside the phago-lysosomal system of J-774 cells beyond the first incubation hour at 37°C, as revealed by pyranine loaded in ARC. Upon sc immunization of C3H/HeN mice, BSA entrapped in ARC-*BM* or ARC-*GC* elicited a strong and sustained primary antibody response, as well as improved specific humoral immunity after boosting with the bare antigen. Both IgG1 and IgG2a enhanced antibody titers could be demonstrated in long-term (200 days) recall suggesting induction of a mixed Th1/Th2 response.

We herein report the finding of new *H. tebenquichense* non alkaliphilic strains in Argentinean Patagonia together with the adjuvant properties of ARC after sc administration in mice. Our results indicate that ARC prepared with TPL from these two strains could be successfully used as vaccine delivery vehicles.

In 1997, the pioneering work of Sprott showed that parenteral administration of nano-sized vesicles (ARC) prepared with TPL extracted from microorganisms of the

Archaea domain of life [1], produced a strong humoral response in mice [2]. Archaeal lipids exhibit radically different hydrocarbon backbones and polar head groups, as compared to polar lipids synthesized by organisms from Eukarya and Bacteria domains. Archaeal lipid backbones possess ether linkages and isoprenoid chains, mainly phytanyl and bysphythanediyl—archaeols and caldarchaeols—in sn-2,3 enantiomeric configuration, in contrast to the ester linkages, straight fatty acyl chains and sn-1,2 configuration of the glycerophospholipids from Eukarya and Bacteria domains [3]. Ether links are more resistant to acid hydrolysis than esters and the backbone/head group cross section of archaeal lipids is almost two folds higher than that of glycerophospholipids from Eukarya and Bacteria domains [4, 5]. The same as liposomes, ARC can be prepared by self association of archaeal lipids upon a small input of energy in aqueous media (thin film hydration). However, beyond those apparent similarities, there are remarkable structural differences: the ARC surface is highly entropic, possessing half the surface tension than that of liposomes [5, 6] and its permeability to protons and sodium cation is nearly one-third of that determined for liposomes; the inclusion of macrocyclic archaeols and caldarchaeols further impairs ARC permeability to water and small solutes [7, 8].

Those structural features make the ARC capable of establishing unique interactions with the biological environment, specifically eliciting adjuvancy to foreign proteins upon sc administration in preclinical models by strongly stimulating both the humoral as well as the cellular response, together with a sharp memory recall. Additionally, lipopolysaccharides are absent in Archaea [9] and, opposite to conventional immunomodulators that usually must be included into the liposomal structure [10, 11], no toxicity has been found after parenteral administration of ARC, even at high or multiple dosage [12, 13].

For strategic development of Third World countries, it is of crucial importance to count on vaccines of unproblematic storage-conservation, with high resistance to hydrolysis, oxidation and mechanical destruction, or easily reconstitutable upon lyophilization [14]. Adjuvants should preferably be biodegradable, non toxic, abundant, cheap, and available from sustainable sources. Because ARCs are suitable candidates to fulfill those requirements, it is relevant to survey the adjuvant properties of ARC made of TPL extracted from unexplored archaeal genera and species. In such context, we determined the cytotoxicity, intracellular transit, and adjuvant activity of ARC prepared with TPLs of two *H. tebenquichense* strains isolated from Argentine Patagonia, upon two sc BSA doses followed by a single boosting inoculation in C3H/HeN mice.

Strain Isolation, Growth, and Characterization

Halophilic archaea isolated from the upper *GC* and the deeper *BM* strata grown in enriched medium were characterized as disc-shaped (0.2 × 0.8 mm), motile Gram-negative microorganisms. Differentially, *GC* and *BM* colonies exhibited orange–red and reddish pigmentation, respectively. However, when grown in basal medium, a mixture of pleomorphic rods and disc-shaped motile archaebacteria displaying undefined gram staining was observed.

The colonies grew in 10–20% NaCl-containing media, without Mg^{+2} requirements, but optimum growth occurred at 20% NaCl, 40°C and pH 7.3–7.5. Remarkably, *GC* was able to grow at increased temperature (50°C). Both colonies were non alkaliphilic, aerobic, oxidase, and catalase positive. Only *GC* produced acid from a fructose source. Fructose, pyruvate, trehalose, and galactose, but not starch, were used as sole carbon and energy sources by both colonies (Table 1). Both *GC* and *BM* organisms were chloramphenicol resistant and experienced no lysis even in the total absence of sodium salt.

Table 1. Phenotypic characteristics of the two colonies isolated from Argentinean Patagonia and the *Halorubrum tebenquichense* strain ALT6-92 isolated from Atacama saltern.

Characteristics	BM	GC	ALT6-92[a]
Growth salt concentration (% w/v)			
10	+	+	-
15	+	+	+
20	+	+	nr
Growth at 50°c	-	+	+
Growth at pH I 0	-	-	+
Cabalase	+	+	+
Oxidase	+	+	+
Acid from			
Xylose	-	-	-
Fructuose	-	+	-
Glucosa	-	-	-
Utilization of			
Fructose	+	+	+
Piruvate	+	+	+
Starch	-	-	+
Trehalose	+	+	+
Galactose	+	+	+
Hydrolysis of starch	-	-	-
G+C content (%)	59.5	61.7	63.2
% DNA-DNA similarity [b]	88.4 (94.0)	94.4 (91.9)	

[a]Data obtained from Lizama et al., 2002
[b]Hybridization against H. tebenquichense. Values in parentheses are results of measurements in duplicate Symbols: + positive result; - negative result; +/- slight mark; nr not reported.

For each isolate, a nearly complete 16S rDNA gene region (1,300 bp) was sequenced (GenBank accession numbers GQ182977 and GQ182978) and compared with the same segment from other halophilic archaea. Phylogenetic analyses showed that both isolates belong to the Archaea domain and are related to the *Halorubrum* cluster, with the highest similarity to *Halorubrum tebenquichense* (98%). The G + C contents of the two isolates and the reference strain ALT-92T [15] were determined.

The *BM* and *GC* had a G + C content of 59.5 and 61.7 M%, respectively (Table 1), whereas ALT-92T presented 63.2 M%. On the basis of the *ad hoc* committee recommendations [16, 17] of a threshold value of 70% DNA–DNA similarity for the definition of bacterial species, the two strains formed an homogeneous cluster with a high degree of internal similarity (DNA–DNA similarity > 90.5%) and should be considered as members of the same *H. tebenquichense* species.

The random amplified polymorphic DNA (RAPD) or arbitrarily primed polymerase chain reaction (AP-PCR) DNA fingerprinting technique provides one of the most sensitive and efficient of current methods for distinguishing different strains of a species [18]. In order to corroborate such differences between the two isolates, AP-PCR fingerprint analysis was performed (Figure 1). The band pattern showed visible differences between *BM* and *GC* fingerprint (absence or presence of 2, 4, 6 and 7 bands), in concordance with biochemical tests and 16S rDNA sequence data. In sum, differences in fructose metabolism, utilization of starch and optimum growth conditions (Table 1), together with those revealed by the AP-PCR, suggested that *BM* and *GC* isolates were different not only from those described for *H. tebenquichense* ALT-92 T [15], but also between each other and could be classified as new *H. tebenquichense* strains.

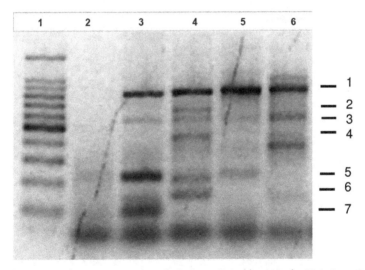

Figure 1. The AP-PCR finger print DNA analysis. Lane 1: Ladder 100 pb -PB-L, Lane 2: negative control, Lane 3: *H. tebenquichense*, Lane 4: *BM*, Lane 5: *GC*, Lane 6: *E. coli*.

Lipid Extraction and Characterization

The TPL, defined as the acetone insoluble portion of the total lipids, extracted from frozen cells obtained from a batch of 8 l, ranged between 90 and 120 mg for each isolate.

A high content of phosphate groups is reported in the literature for the *Halorubrum* genus. Hence, for TPL quantification two colorimetric methods were applied: one, for detecting total phosphate (Bötcher) and the other for detecting organic phosphate

(Stewart). Two calibration curves were prepared employing dry mass of TPL as standards that resulted linear between 10–40 µg for Bötcher and 20–400 µg for Stewart, with correlation coefficients exceeding 0.997, for each source of TPL.

On the other hand, calibration curves using inorganic phosphate (NaH_2PO_4) as standard were prepared with the aim of determining the phosphate percentage of archaeal lipids by the Bötcher method. A linear plot of dry mass of TPL versus µg inorganic phosphate was determined and the relative amount of phosphate in the extracted TPL was similar for both isolates, in the order of 6.25% w/w.

ESI-MS Polar Lipids Profile

The Electrospray ionization-mass spectrometry (ESI-MS) spectrum (negative ions) analyses of the TPL of *H. tebenquichense*, *BM*, and *GC* showed three main peaks at m/z 805, 899, and 1055.9 (Figure 2). In addition, *GC* displayed intense ion peak at m/z 1,521 that was less intense for *H. tebenquichense* and *BM* extracts. The peaks

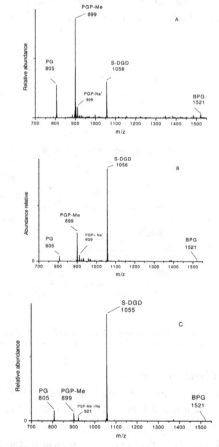

Figure 2. Electrospray ionization-mass spectrometry (ESI-MS) analysis of polar lipids from (A) *H. tebenquichense*, (B) *BM*, and (C) *GC* extract.

at m/z 731.5 (diagnostic of PA) and 886 (PGS) that were reported for *Halorubrum* sp. [19] could not be detected in any of our samples. Negative ion ESI-MS (Figure 2A, B, and C) identified major signals corresponding to PG m/z 805.7, PGP-Me m/z 899.5 (as monocharged peak), 449 (as bicharged peak) and S-DGD m/z 1055.7 (as monocharged peak). The diagnostic peak of archaeal cardiolipin (BPG) at m/z 760 (as bicharged peak) and 1,521 (monocharged peak) was detected only as a small signal in the three TPL extracts.

ARC Characterization

As revealed by transmission electron microscopy, the two ARC preparations were multilamellar, with a mean size of 564 ± 22 nm and zeta-potential near to -50 mV. The BSA incorporation did not modify size or zeta-potential, the protein/lipid ratio was 20 µg/mg and the encapsulation efficiency around 3–4%.

ARC Uptake by Cells and Cellular Toxicity

None of the ARC or HSPC:cholesterol liposomes significantly reduced the viability of non phagocytic cells (Vero cell line) upon 24 hr incubation (Figure 3A). On the other hand, 10 µg/ml ARC-*GC* was non cytotoxic but the higher concentrations reduced cell viability by 25%, while increased concentrations (up to 100 µg/ml) of ARC-*BM* and HSPC:cholesterol liposomes did not affect cultured macrophages (J-774 cell line) (Figure 3B).

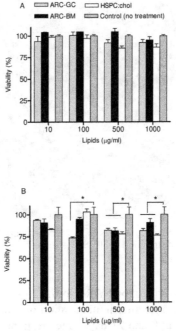

Figure 3. Viability of Vero (A) and J-774 cells (B) upon 24 hr incubation with ARC-*GC*, ARC-*BM* or HSPC:cholesterol liposomes, as function of concentration. Values represent the average of triplicates $\pm$ S.D. ANOVA, * $p < 0.05$.

As shown in Figure 4A, and independently of the TPL source, a minimal time period of 30 min was required for ARC uptake by phagocytic cells, as judged by the detection of intracellular fluorescence after incubation with both ARC-HPTS/DPX. In addition, J-774 macrophages incubated 45 min with both ARC-HPTS/DPX showed punctual fluorescence (as shown by arrows in Figure 4A) that resulted from the confinement of the HPTS/DPX into vesicular compartments in the cytoplasm. In case membrane fusion between ARC and the phagosome occurred, the ARC inner content should be released to the cytoplasm. Consequently, the HPTS would be dequenched from DPX and an homogeneous brightness filling the cytoplasm should be observed [20, 21]. However, the punctual emission upon excitation at 440 nm, indicating confinement of the pair HPTS/DPX in acidic compartments, remained unchanged. Hence, the ARC did not fuse/disrupt, staying inside the phagosomes for at least 60 min (Figure 4B).

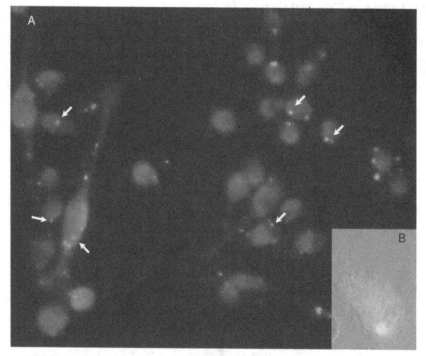

Figure 4. (A) Fluorescence microscopy images of J-774 cells upon 30 min incubation with ARC-*BM*-HPS/DPX. (B) Intracellular distribution of HPTS in green vesicles persistent 60 min after uptake.

Antibody Response

To evaluate the adjuvant activity of both ARC formulations, we tested the antigen-specific humoral immune response after immunization of mice with BSA-loaded ARC. Following sc inoculations at 0 and 21 days, mice responded by day 28 with similar anti-BSA antibody (total IgG) titers for both ARC-*BM*/*GC*-BSA (Figure 5A), and presented a strong enhancement (~2 log, $p < 0.01$) of antibody titers over those of BSA and BSA-Al groups. Immunization with either empty ARC-*BM/GC* failed to evoke any anti-BSA IgG response, whereas only one out of four mice receiving adjuvant-free BSA displayed detectable specific antibodies. Mean titers in ARC-*BM/GC*-BSA groups were sustained until at least day 60 and declined by day 135. In all cases, very little anti-BSA IgM reactivity was detected throughout the study (data not shown).

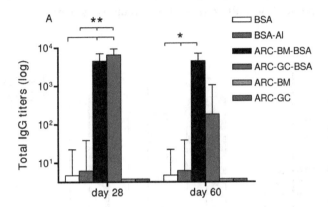

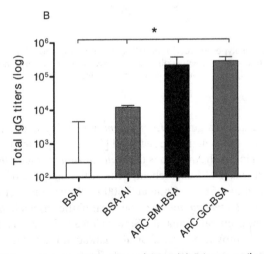

Figure 5. Humoral response to ARC-entrapped BSA. (A) Primary antibody response. Mice were immunized sc on days 0 and 21 with 25 µg of BSA alone, BSA adsorbed on Al_2O_3, BSA entrapped in TEB-*BM* and TEB-*GC* or empty TEB. Serum samples were collected on days 28 and 60, and ELISA assayed for anti-BSA antibody titers. (B) Long-term memory response to ARC-entrapped BSA. Mice were immunized sc on days 0 and 21 with 25 µg of BSA alone, BSA adsorbed on Al_2O_3 or BSA entrapped in ARC, and boosted on day 180 with 25 µg of BSA (without adjuvant). Serum samples were collected 3 weeks later and assayed for anti-BSA antibody titers. Values represent mean titers ± SEM. ANOVA, * $p < 0.05$, ** $p < 0.01$.

To investigate the ability of ARC to generate long-term memory immunity, a boost-dose with BSA alone was injected on day 180. Three weeks later, a significantly higher serum antibody response (>1 log, $p < 0.05$) was detected in ARC-*BM/GC*-BSA compared with the BSA group (Figure 5B).

We further examined the IgG isotype distribution in the sera of immunized mice on day 200. Both IgG1 and IgG2a increased antibody titers could be demonstrated in long-term responses from ARC-*BM/GC*-BSA groups (Figure 6).

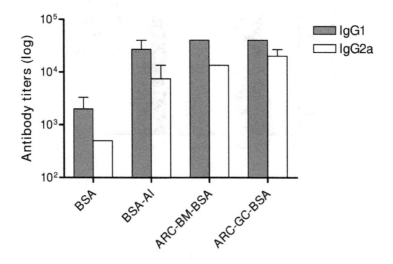

Figure 6. Anti-BSA antibody isotypes induced after ARC-based immunization. Mice were immunized as indicated in Figure 5. Serum samples were collected three weeks after boosting and ELISA analyzed for IgG subclasses (IgG1 and IgG2a). Values represent mean titers ± SEM.

DISCUSSION

The extreme halophilic archaea genus *Halorubrum* is worldwide distributed [22-28] either at high altitude as well as under the sea level [15, 25, 29]. The only *Halorubrum* species so far described in South America is the *Halorubrum tebenquichense* ALT6-92 strain isolated from the water of Lake Tebenquiche at the unique environment of the Atacama saltern, located in northern Chile at 2,300 m above sea level [15]. Here we report the finding of new *H. tebenquichense* non alkaliphilic strains in a salt flat from the littoral of the Argentinean Patagonia, the Salina Chica in Península de Valdés (42° 32' S, 63° 59' W), at the province of Chubut. Remarkably, the Salina Chica is separated from Lake Tebenquiche by the highest (nearly 7,000 m) and younger (70 million years) mountains in America, the Cordillera de los Andes, distant 2,500 km south–east away and placed 40 m below sea level in a zone of temperate weather.

The membranes of extremely halophilic archaea such as *H. tebenquichense* have several unique characteristics that vary little within specific genera [30]. Phospholipids with ethanolamine, inositol, and serine head groups are generally absent and specific phosphatidylglycerol (PG) phospholipids, sometimes including several sulfated gly-colipids, predominate. Extreme halophiles contain 50–80% archaetidylglycerol methylphosphate (PGP-Me), an archaeal analogue of PG methylphosphate [31] that contributes to membrane stability in hypersaline environments [32], archaetidylglycerol (PG), and also some strains have minor amounts of sulfated PG [33].

Analysis of ESI-MS spectra of polar lipid isolates from *H. tebenquichense, BM,* and *GC*, through comparison with the reference data [34, 35] suggested the main phospholipids present in the extracts were PG, phosphatidylglycerophosphate methyl ester (PGP-Me), and sulfated diglycosyl diphytanylglycerol diether (S-DGD). The prominent peak at 1055.9 corresponded to S-DGD and is representative of the *Halorubrum* genus [36]. The archaeal cardiolipin, bisphosphatidyl glycerol (BPG) was only found as traces in the three samples. While the complete lipid quantitation from each sample as well as the RMN identification of the S-DGD sulfonolipid will be accomplished in ongoing research.

Up to now, only ARC made of TPL extracted from *Methanobrevibacter smithii,* a methanogen archaea, are known to enhance the recruitment and activation of antigen presenting cells [37, 38], induce co-stimulatory molecules expression [38], elicit CD8+ cytotoxic responses even in the absence of CD4+ helper T-lymphocytes [39], and to evoke a profound memory response, those all important stimuli that other vesicular systems such as liposomes, niosomes, and ISCOMS are unable to trigger [40]. Additionally, although inflammation in innate responses can enable further adaptive responses [41], ARC made of TPL from *M. smithii* evoke immunologic memory in the lack of visible inflammation, activating dendritic cells in the absence of IL-12-dependent initial inflammatory response [42].

As regard to the extreme halophiles, ARC made of TPL extracted from *Halobacterium salinarum* was reported to induce a robust initial CTL and antibody response in mice [43]. Such titers, however, were not increased after boost administration and failed to induce a significant memory response. Only ARC made of TPL extracted from the hyperthermophile *Thermoplasma acidophilum* produced the most intense antibody memory responses to antigen challenge, together with a parallel induction of intense cell cycling in CD4+ T cells suggesting efficient maintenance of T-cell memory [43]. A further study showed that among ARC from nine archaeal strains tested, all triggered remarkable primary CTL responses but only *M. smithii* and *T. acidophilum*, both rich in caldarchaeols, evoked a strong recall at >50 weeks [44].

This is the first report of adjuvancy after parenteral administration of ARC-*BM/GC*-BSA. Both ARC elicited a strong and lasting primary antibody response, and remarkably in the absence of caldarchaeols in their lipid compositions, an enhanced memory humoral response after boosting with the bare antigen in C3H/HeN mice upon sc immunization. The ARC-*BM/GC*-BSA initially triggered higher antibody titers than BSA formulated in alum adjuvant. Additionally, IgG isotype analysis of immunized mice revealed that both BSA-specific IgG1 and IgG2a antibodies were raised by both ARC-*BM/GC*-BSA, suggesting induction of a mixed Th1/Th2 response, in agreement with a previous report [39]. Our preliminary finding of humoral immune memory response elicited by ARC from extreme halophiles, together with established data on primary CTL and memory responses from ARC made of TPL from methanogens and hyperthermophiles, suggests that the headgroups are of crucial importance in the induction of immune responses, as recently determined by employing synthetic glycoarchaeols [45, 46].

To sum up, even though our results for IgG isotyping in a single strain of mice (C3H/HeN) may not be extrapolated to the response in humans, it is promising to note that, in preclinical evaluation in animal model, ARC-*BM*/*GC*-BSA would appear as an efficient adjuvant delivery system to promote both humoral and, probably, cell-mediated immunity to the entrapped antigen.

MATERIALS AND METHODS

Sodium 3'-[1-(phenylamino-carbonyl)-3,4-tetrazolium]-bis-(4-methoxy-6-nitro) benzene sulfonic acid hydrate (XTT), BSA, antifoam 204 and cholesterol were provided by Sigma-Aldrich (Argentina). Hydrogenated phosphatidylcholine from soybean (HSPC) was obtained from Northern Lipids (Vancouver, Canada). The RPMI 1640 culture medium was purchased from Invitrogen Corporation. Endotoxin-free fetal bovine serum (FBS) was bought from Hyclone. The L-Glutamine, Trypsin, EDTA and penicillin/streptomycin were provided by PAA Laboratories GmbH (Austria). The fluorophore 8-hydroxypyrene-1,3,6-trisulfonic acid (HPTS) and the quencher p-xylene-bis-pyridinium bromide (DPX) were purchased from Molecular Probes (Eugene, OR, USA). *Halorubrum tebenquichense* strain ALT6-92 was purchased from Deutsche Sammlung von Mikroorganismen und Zellkulturen (DSMZ). Tris buffer and all the other analytical grade reagents were from Anedra (Argentina).

Culture Growth and Characterization

Soil samples were collected from Salina Chica, Península de Valdés, Chubut, Argentina. Samples were classified according to their strata source as upper *GCs* and deeper *BM*. Isolation of microorganisms from each strata was done by seeding an aliquot on basal growth medium [48] with 1.6% agar (solid basal medium) at 37°C. The resulting colonies were grown on liquid medium at 40°C, at 160 rpm with chloramphenicol (30 mg/l) and then seeded on solid basal medium or brain-heart broth supplemented with yeast extract, glucose or blood (enriched medium).

Optimal NaCl concentration, pH, temperature and Mg^{+2} requirements for growth were determined as described by Oren [49]. Gram staining as described by Dussault [50], cell shape and pigmentation examined by optical microscopy, were performed on colonies grown in basal medium and also in medium supplemented with blood or with brain-heart broth. Standard biochemical test methods were assessed for each colony.

Salt requirement for maintaining stability of the cell envelope was determined as described by Oren [49] by measuring the loss of turbidity of cell suspensions in medium of decreasing concentrations of NaCl (8, 6, 4, 2, and 0% w/v).

Biomass was generated in 8 l batch cultures in basal medium supplemented with yeast extract, glucose, and antifoam (20 μl/l). Cultures were monitored by absorbance at 660 nm and harvested in late stationary phase for storage as frozen cell pastes.

DNA Isolation, G + C Content

The DNA was isolated using a French pressure cell (Thermo Spectronic, Rochester, NY) and was purified by chromatography on hydroxyapatite as described [51]. The G + C content (M%) of the DNA was determined by the midpoint value of the thermal denaturation profile (Tm) with a Perkin Elmer spectrophotometer at 260 nm, programmed for temperature increases of 1.0°C/min [52]. Tm was determined by the graphic method described by Ferragut and Leclerc [53] and the G + C content was calculated using Owen and Hill's equation (1979).

PCR Amplification of the 16S rDNA Gene Coding Sequence and Sequencing

Purified genomic DNA was used for PCR amplification of the16S rDNA gene. The following two sets of primers were used on the basis of the highly conserved regions of halobacterial 16S rDNA sequences as described by Ihara [48]: f1 (5' ATTCCGGTT-GATCCTGC 3'), r1 (5' TTTAAGTTTCATCCTTG 3'), and f2 (5' AACCGGATTAG-ATACCC 3'), r2 (5' GTGATCCAGCCGCAGATTCC 3'). The PCR was performed with 35 cycles of 30 s at 94°C, 30 s at 37°C and 90 s at 72°C. The PCR products were analyzed by 1.5% agarose gel electrophoresis. The products were purified from the gel using S.N.A.P.™ Gel Purification Kit (Invitrogen) and sequenced directly by the dideoxy chain termination method [54].

Phylogenetic Analysis

The obtained sequences were compared with previously described 16S rDNA sequences of halophilic archaeas from the NCBI database. The sequences were aligned by using CLUSTAL X 1.83 [55, 56] and phylogenetic trees were constructed by the neighbor-joining method with Kimura two-parameter calculation in MEGA 3.1 software. The confidence levels for branching orders were evaluated by the bootstrap method [57].

DNA–DNA Hybridization

The DNA–DNA hybridization studies were performed as described by De Ley [58] under consideration of the modifications described by Huss [59] using a Cary® 100 Bio UV/VIS-spectrophotometer equipped with a Peltier-thermostatted 6 × 6 multicell changer and a temperature controller with *in-situ* temperature probe (Varian, Inc.).

DNA Fingerprint

The DNA from the isolates was used for AP-PCR fingerprint assay. Three small primers were used: T3GC (5' CCCAKTCGTGAWTCATGCT 3'), T3R, (5' TCCTCAYT-TAATNAMCATGCT 3'), and Rsh (5' ATCAAAAT 3'). The PCR was performed for 35 cycles, starting with 10 s of denaturation at 92°C followed by 60 s of annealing at 30°C and 90 s of elongation at 72°C. The reaction mixture contained 10 ng of genomic DNA, 3 mM $MgCl_2$, 1 μM of each primer, 0.2 μM of dNTPs, and 2 U of Taq Polymerase (Invitrogen) in a final volume of 20 μl. The PCR results were visualized in a 1.5% agarose gel electrophoresis and analyzed by Image Kodak Software.

Lipid Isolation

Lipids were extracted by the method of Bligh and Dyer as modified for extreme halophiles [60], from frozen and thawed biomass of each colony with $Cl_3CH:CH_3OH:H_2O$ (1:2:0.9; v:v), and the TPL fraction was collected by precipitation from cold acetone. Phospholipids of the TPL were quantified by Bötcher [61] and Stewart [62] methods.

ESI-MS Analysis

The ESI-MS analysis was performed using a Thermo Finnigan LCQ Ion Max mass spectrometer (Thermo Finnigan MAT, San Jose, CA, USA) equipped with a electrospray ionization source. Analyses were carried out in the loop injection mode with dried lipid extracts dissolved in chloroform-methanol (1:1 vol/vol). Samples (5 µl) injected via a 10 µl loop were transferred to an MS electrospray interface (ESI) at flow rate of 10 µl/min. Interface conditions were as follows: nebulizer gas (air) 12 l/min, curtain gas (nitrogen), 1.2 l/min needle voltage –5.0 kV (negative ions), mass range 50–2,000 amu.

ARC Preparation and Physicochemical Characterization

Two different types of ARC were prepared from TPL isolated from each colony: ARC-GC and ARC-*BM*. Briefly, 20 mg of TPL from $CHCl_3$: CH_3OH (9:1, v/v) solution was rotary evaporated at 40°C in round bottom flask until organic solvent elimination. The thin lipid film was flushed with N_2 and hydrated at 40°C with 1 ml of 10 mM Tris-HCl buffer plus 0.9% w/v NaCl, pH 7.4 (Tris buffer) (empty ARC) or with 1 ml of 10 mg/ml BSA solution in Tris buffer (ARC-CG/*BM*-BSA). The resultant suspensions were sonicated (20 min in a bath type sonicator 80 W, 40 KHz) and submitted to five cycles of freeze/thaw between –80 and 37°C. Free BSA was removed by centrifugation and washing with Tris buffer (10,000 × g for 20 min). Liposomes made of HSPC:cholesterol (1:1 M:M) were prepared in the same way.

The BSA/phospholipid weight ratio from each preparation was determined by phospholipid and protein quantitation. Phospholipids were quantified by Bötcher [61] whereas BSA content was measured by Bradford [63].

Mean ARC size was determined by dynamic light scattering with a 90 Plus Particle size analyzer (Brookhaven Instruments) and zeta-potential was determined with a Zetasizer 4 (Malvern). Electron microscopy images of ARC upon phosphotungstic acid negative staining were obtained with a TEM Jeon 1210, 120 Kv, equipped with EDS analyzer LINK QX 2000.

CYTOTOXICITY

Upon incubation with ARC, cell viability was measured as mitochondrial dehydrogenase activity employing a tetrazolium salt (XTT) on Vero cells and the murine macrophage-like cell line J-774 [64].

Cells maintained at 37°C with 5% CO_2, in RPMI 1640 medium supplemented with 10% heat-inactivated FBS, 2 mM glutamine, 100 UI/ml penicillin and 100 µg/ml streptomycin, were seeded at a density of 5 × 10^4 cells/well in 96-well flat bottom microplates. Culture medium of nearly confluent cell layers was replaced by 100 µl of medium containing 10, 100, 500, or 1,000 µg/ml of lipids. Upon 24 hr at 37°C, the cells were washed with 0.1 M phosphate buffered saline (PBS) pH 7.2, and incubated with 200 µg/ml of XTT for 4 hr. The XTT solution was removed and the extent of reduction of XTT to formazan within the cells was quantified by measuring the absorbance at 450 nm using an ELISA reader. Liposomes made of HSPC:cholesterol (1:1, M:M) were used as control.

Cell Uptake
Cell uptake and intracellular fate of ARC loaded with the fluorophore/quencher pair HPTS/DPX was followed upon incubation with J-774 macrophages by fluorescence microscopy. The ARC loaded with HPTS/DPX was prepared as stated before, except that the lipid films were hydrated with a solution of 35 mM HPTS and 50 mM DPX in Tris buffer. The J-774 cells grown to near confluence on rounded coverslips in 24-well plate were incubated with 10 µg/ml µg of ARC-*GC*/*BM*-HPTS/DPX at 37°C for 10, 20, 30, 45, and 60 min. After incubation, suspensions were removed, cells were washed, and coverslips were mounted on a fluorescence microscope. Cell-associated HPTS fluorescence was monitored using a Nikon Alphaphot 2 YS2 instrument.

Similarly, the intracellular fate of both fluorescent ARC upon 45 min-incubation with J-774 cells was followed by monitoring the HPTS fluorescent signal along 1 hr.

Immunization
Female 6–8-week-old C3H/HeN mice were obtained from University of Buenos Aires, Argentina, and maintained under standard conditions. The study was conducted following the Institutional Experimental Guidelines for Animal Studies. Groups of five animals were immunized sc on days 0 and 21 with 25 µg of BSA; 25 µg of BSA with added 100 µg of Al_2O_3 (BSA-Al, Alhydrogel®, Superfos Biosector, Vedbaek, Denmark) or 25 µg of BSA entrapped in any type of ARC (ARC-*GC*-BSA or ARC-*BM*-BSA; 1.3 mg of archaeal lipids each ARC). Control mice were injected with equivalent amount of empty TEB. All groups were boosted with 25 µg of adjuvant-free BSA on day 180.

Evaluation of Antibody Response
Blood was collected from the tail vein at various time-points after immunization, as specified in the figure legends, and sera were analyzed by ELISA for the presence of anti-BSA antibodies. Briefly, microtiter plates (Nunc, Roskilde, Denmark) were coated overnight at 4°C with 45 µg/ml BSA diluted in 0.1 M carbonate-bicarbonate buffer (pH 9.6) and then blocked for 1 hr at 37°C with PBS containing 0.2% Tween 20 (0.2% PBST) after washing with 0.05% PBST. Another wash as above described was followed by the addition of 100 µl of 3-fold dilutions of individual sera in 0.05% PBST. After 2 hr at 37°C and further washing, the plates were incubated for 1 hr at 37°C with horseradish peroxidase-conjugated goat anti-mouse IgG (Pierce, Rockford, IL) diluted 1:2000 in 0.025% PBST. For antibody isotyping, horseradish peroxidase-conjugated rat anti-mouse IgG1 or IgG2a revealing antisera (PharMingen, San Diego, CA), diluted 1:1,000, were used. The plates were further washed and the reactions were developed by adding the ABTS substrate (2, 2'-azino-bis (3-ethylbenzthiazoline-

6-sulphonic acid), Sigma Chemical Co., St. Louis, MO). Color was allowed to develop for approximately 10 min at room temperature in the dark. The optical density was measured at 405 nm using an ELISA reader (Multiskan Ex, Thermo Labsystems, Finland). Antibody titers are represented as end-point dilutions exhibiting an optical density of 0.3 units above background.

Statistical Analysis

Statistical analyses were carried out with the Prisma 4.0 Software (GraphPad, San Diego, CA, USA). Group means were evaluated by ANOVA with Tukey's analysis to compare individual groups. Values of $p < 0.05$ were considered significant.

CONCLUSION

Two sc immunizations with either ARC-*BM/GC*-BSA, plus a boost with BSA alone rendered a long-term humoral response stronger than that achieved with BSA formulated in alum.

Remarkably, our results were elicited in C3H/HeN mice, less prone to render potent humoral responses than BALB/c and C57BL/6 backgrounds [43]. Such preliminary results merit deeper insights on the search of CD8[+] CTL activity and the induction of this type of long-term memory upon sc immunization with ARC. As judged by the phagosomal traffic followed by the pair HPTS/DPX loaded in ARC, our results indicated that there was neither fusion nor ARC content delivery to the cytoplasm, for at least 60 min post ARC uptake. Hence, cytoplasmic delivery of hydrosoluble material loaded in ARC-*BM/GC* should not happen, could either take longer than 60 min (considering that ARC-*BM/GC* were multilamellar, with negative zeta-potential at physiological pH, both factors that impair or delay intermembrane fusion) or could occur through mechanisms other than the fusion mechanism recently reported for *Methanobrevibacter smithii* archaeosomes [47]. Finally, in spite of their TPL invariance, extreme halophilic archaea are source of glycolipid fractions that probably markedly influence the induction of primary responses and memory recall [30]. In view of that, the complete composition of ARC TPL and the mechanisms of recruitment, uptake, and intracytoplasmic traffic of this antigen-delivery system are currently being analyzed in parallel with the ARC ability to induce expression of co-stimulatory molecules on professional APC.

KEYWORDS

- **Archaeal lipids**
- **Archaeosomes**
- **Black mud**
- **Electrospray ionization-mass spectrometry**
- **Gray crystals**
- *Halorubrum tebenquichense*
- *Methanobrevibacter smithii*

AUTHORS' CONTRIBUTIONS

Raul O. Gonzalez performed the preparation and structural characterization of ARC (size and Z potential, electronic microscopy), as well as cytotoxicity, cell transit, and immunization schemes. Leticia H. Higa grew the archaea strains, isolated the TPL, and grew the J-774 and Vero cells. Also contributed to the cytotoxicity, cell transit, and immunization schemes together with Raul O. Gonzalez. Marcos Bilen performed the DNA sequencing, C + G content, and finger print analysis. Irma Morelli performed the biochemical tests. Romina A. Cutrullis performed ELISA measurements. Diana I. Roncaglia performed the ESI-MS spectra and analysis. Patricia B. Petray and Ricardo S. Corral designed and developed the immunization protocols, participated to the discussion of the results and manuscript preparation, and provided minor financial support. Maria Jose Morilla coordinated the performance experiments and analysis of experimental results. Eder L. Romero coordinated the experiments, wrote the manuscript, and provided main financial support. All authors read and approved the final manuscript.

ACKNOWLEDGMENTS

This work was financially supported by Secretaria de Investigaciones, Universidad Nacional de Quilmes and the Consejo Nacional de Investigaciones Científicas y Técnicas (CONICET, Argentina). Maria Jose Morilla, Patricia B. Petray, Ricardo S. Corral, and Eder L. Romero are members of the Researcher Career Programme from CONICET. Leticia H. Higa and Romina A. Cutrullis are fellows from CONICET.

Chapter 9

Soil Microbe *Dechloromonas aromatica* Str. RCB Metabolic Analysis

Kennan Kellaris Salinero, Keith Keller, William S. Feil, Helene Feil, Stephan Trong, Genevieve Di Bartolo, and Alla Lapidus

INTRODUCTION

Initial interest in *Dechloromonas aromatica* strain RCB arose from its ability to anaerobically degrade benzene. It is also able to reduce perchlorate and oxidize chlorobenzoate, toluene, and xylene, creating interest in using this organism for bioremediation. Little physiological data has been published for this microbe. It is considered to be a free-living organism.

The a priori prediction that the *D. aromatica* genome would contain previously characterized "central" enzymes to support anaerobic aromatic degradation of benzene proved to be false, suggesting the presence of novel anaerobic aromatic degradation pathways in this species. These missing pathways include the benzylsuccinate synthase (*Bss*ABC) genes (responsible for fumarate addition to toluene) and the central benzoyl-CoA pathway for monoaromatics. In depth analyses using existing TIGRfam, clusters of orthologs gene (COG), and InterPro models, and the creation of *de novo* hidden Markov models (HMM) models, indicate a highly complex lifestyle with a large number of environmental sensors and signaling pathways, including a relatively large number of GGDEF domain signal receptors and multiple quorum sensors. A number of proteins indicate interactions with an as yet unknown host, as indicated by the presence of predicted cell host remodeling enzymes, effector enzymes, hemolysin-like proteins, adhesins, nitrous oxide (NO) reductase, and both type III and type VI secretory complexes. Evidence of biofilm formation including a proposed exopolysaccharide complex and exosortase (epsH) are also present. Annotation described in this chapter also reveals evidence for several metabolic pathways that have yet to be observed experimentally, including a sulfur oxidation (*soxFCDYZAXB*) gene cluster, Calvin cycle enzymes, and proteins involved in nitrogen fixation in other species (including RubisCo, ribulose-phosphate 3-epimerase, and *nif* gene families, respectively).

Analysis of the *D. aromatica* genome indicates there is much to be learned regarding the metabolic capabilities, and life-style, for this microbial species. Examples of recent gene duplication events in signaling as well as dioxygenase clusters are present, indicating selective gene family expansion as a relatively recent event in *D. aromatica*'s evolutionary history. Gene families that constitute metabolic cycles presumed to create *D. aromatica*'s environmental "foot-print" indicate a high level of diversification between its predicted capabilities and those of its close relatives, *A. aromaticum* str EbN1 and *Azoarcus* BH72.

Dechloromonas aromatica strain RCB is a gram negative Betaproteobacterium found in soil environments [1]. Other members of the Betaproteobacteria class are found in environmental samples (such as soil and sludge) or are pathogens (such as *Ralstonia solanacearum* in plants and *Neisseria meningitidis* in humans) and in general the genus *Dechloromonas* has been found to be ubiquitous in the environment.

A facultative anaerobe, *D. aromatica* was initially isolated from Potomac River sludge contaminated with benzene, toluene, ethylbenzene, and xylene compounds (BTEX) based on its ability to anaerobically degrade chlorobenzoate [1]. This microbe is capable of aromatic hydrocarbon degradation and perchlorate reduction, and can oxidize Fe(II) and H_2S [2]. Although several members of the Rhodocyclales group of Betaproteobacteria are of interest to the scientific community due to their ability to anaerobically degrade derivatives of benzene, *D. aromatica* is the first pure culture capable of anaerobic degradation of the stable underivitized benzene molecule to be isolated. This, along with its ability to reduce perchlorate (a teratogenic contaminant introduced into the environment by man) and inquiry into its use in biocells [3] has led to interest in using this organism for bioremediation and energy production. Since the isolation of *D. aromatica*, other species of *Azoarcus* have been found to possess the ability to anaerobically degrade benzene, but have not been genomically sequenced [4].

The pathway for anaerobic benzene degradation has been partially deduced [5], but the enzymes responsible for this process have yet to be identified, and remain elusive even after the intensive annotation efforts described here-in. Conversely, central anaerobic pathways for aromatic compounds described in various other species were not found to be present in this genome [6].

MATERIALS AND METHODS

Sequencing

Three libraries (3, 8, and 30 kb) were generated by controlled shearing (Hydroshear, Genomic Solutions, Ann Arbor, MI) of spooled genomic DNA isolated from *D. aromatica* strain RCB and inserted into pUC18, pCUGIblu21, and pcc1Fos vectors, respectively. Clonal DNA was amplified using rolling circular amplification [http://www.jgi.doe.gov/webcite] and sequenced on ABI 3700 capillary DNA sequencers (Applied Biosystems, Foster City, CA) using BigDye technology (Perkin Elmer Corporation, Waltham, MA). Paired end-reads [7] were used to aid in assembly, and proved particularly useful in areas of repeats.

The Phrap algorithm [8, 9] was used for initial assembly. Finishing and manual curation was conducted on CONSED v14 software [10], supplemented with a suite of finishing analysis tools provided by the Joint Genome Institute. *In silico* cross-over errors were corrected by manual creation of fake reads to guide the assembly by forcing the consensus to follow the correct path.

Gaps were closed through a combination of primer walks on the gap-spanning clones from the 3 and 8 kb libraries (identified by paired-end analysis in the CONSED software) as well as sequencing of mapped, unique polymerase chain reaction (PCR) products from freshly prepared genomic DNA.

The final step required to create a finished single chromosomal sequence was to determine the number of tandem repeats for a 672 base DNA sequence of unknown length. This was done by creating the full tandem repeat insert from unique upstream and downstream primers using long-range PCR. We then determined the size of product (amplified DNA) between the unique sequences.

Protein Sequence Predictions/Open Reading Frames (ORFs)

Annotation done at Oak Ridge National Laboratory consisted of gene calls using CRITICA [11], glimmer [12], and Generation [http://compbio.ornl.gov/webcite]. Annotation at the Virtual Institute for Microbial Stress and Survival (VIMSS) [http://www.microbesonline.org/webcite] used bidirectional best hits as well as recruitment to TIGRfam HMMs, as described in Alm et al. [13]. Briefly, protein coding predictions derived from NCBI, or identified using CRITICA, with supplemental input from Glimmer, were analyzed for domain identities using the models deposited in the Inter-Pro, UniProt, PRODOM, Pfam, PRINTS, SMART, PIR SuperFamily, SUPERFAMI-LY, and TIGRfam databases [13]. Orthologs were identified using bidirectional unique best hits with greater than 75% coverage. The RPS-BLAST against the NCBI COGs in the CDD database were used to assign proteins to COG models when the best hit E-value was <1e-5 and coverage was >60%.

Manual Curation

Each and every predicted protein in the VIMSS database [http://www.microbesonline.org/webcite] [13] was assessed to compare insights obtained from recruitment to models from several databases (TIGRfams, COGs, EC, and InterPro). Assignments that offered the most definitive functional assignment were captured in an excel spreadsheet with data entries for all proteins predicted in the VIMSS database. Extensive manual curation of the predicted protein set was carried out using a combination of tools including the VIMSS analysis tools, creation, and assessment of HMMs, and phylogenomic analysis, as described. Changes in gene functional predictions and naming were captured in the excel spreadsheet, and predictions with strong phylogenetic evidence of function posted using the interactive VIMSS web-based annotation interface.

Phylogenomic Analysis: Flower Power, SCI-PHY, and HMM Scoring

The HMMs were generated for a large subset of proteins of interest, as detailed, to predict functional classification with the highest confidence measures currently available. The HMMs allowed recruitment of proteins to phylogenetic tree alignments that most closely reflect evolutionary relatedness across species. The proteins were assembled within clades of proteins that are aligned along their full length (no missing functional domains), and that allow high confidence of shared function in each species.

Gene Family Expansion

A clustered set of paralogs was used to search for recent gene duplication events. After an initial assessment of the VIMSS gene information/homolog data, candidate proteins were used as seed sequences for Flower Power and internal tree-viewing tools or SCI-PHY analyses. These two approaches employed neighbor-joining trees using the Scoredist correction setting in the Belvu alignment editor, or the SCI-PHY utility and tree viewer. In either case resulting phylogenomic tree builds were reviewed, and contiguous protein alignments of two or more proteins from *D. aromatica* were considered to be candidates for a gene duplication event, either in the *D. aromatica* genome or in a predecessor species.

Resequencing to Verify Absence of Plasmid Structure

After finishing the *D. aromatica* genome, analysis of the annotated gene set revealed the notable absence of several anaerobic aromatic degradation pathways that were expected to be present, due to their presence in *A. aromaticum* EbN1 (an evolutionary near-neighbor, as determined by 16sRNA phylogeny). Because many catabolic pathways are encoded on plasmid DNA, we felt it was important to preclude this possibility. We re-isolated DNA from a clonal preparation of *D. aromatica* that experimentally supported anaerobic benzene degradation, using three different plasmid purification protocols, each based on different physical parameters. All three generated a single band of DNA. The protocol that generated the highest yield of DNA was used to create a complete, new library of 2 kb inserts, and the library was submitted to sequence analysis using the protocols previously cited.

DISCUSSION

Discussion of results and analyses concerning aromatic degradation, various predicted metabolic cycles, secretion, signaling, quorum-sensing, and gene family expansion are included in the relevant sections.

RESULTS

Overview of Gene and Protein Features

The finished sequence for *D. aromatica* reveals a single circular, closed chromosome of 4,501,104 nucleotides created from 130,636 screened reads, with an average G + C content of 60% and an extremely high level of sequence coverage (average depth of 24 reads/base). Specific probing for plasmids confirmed no plasmid structure was present in the clonal species sequenced, which supports anaerobic benzene degradation. It is noted however that the presence of two Tra clusters (putative conjugal transfer genes; VIMSS582582-582597 and VIMSS582865-582880), as well as plasmid partitioning proteins, indicates this microbial species is likely to be transformationally competent and thus likely to be able to support plasmid DNA structures.

The VIMSS, [http://www.microbesonline.org] and the Joint Genome Institute [http://genome.jgi-psf.org/finished_microbes/decar/decar.home.html] report 4,170 and 4,204 protein coding genes, respectively. Cross-database comparisons were done to assure the highest probability of capturing candidate orfs for analysis. The majority of proteins are shared between data sets. Variations in N-termini start sites were noted, both between JGI and VIMSS datasets and between initial and later annotation runs (approximately 200 N-termini differences between four runs of orf predictions were noted for the initial two annotation runs, Joint Genome Institute's, done at Oak Ridge National Laboratories—ORNL, and VIMSS).

The most definitive functional classification, TIGRfams, initially defined approximately 10% of the proteins in this genome; as of this writing, 33% of predicted proteins in the *D. aromatica* genome are covered by TIGRfams, leaving 2,802 genes with no TIGRfam classification. Many proteins in the current and initial non-covered sets were investigated further using K. Sjölander's HMM building protocols (many of which are available at [http://phylogenomics.berkeley.edu]), to supplement TIGRfams. The COG assignments were used for classification in the families of signaling proteins, but specific function predictions for these proteins also required further analyses. The metabolic and signaling pathways are discussed below, and the identity of orthologs within these pathways are based on analysis of phylogenomic profiles of clusters obtained by HMM analysis, with comparison to proteins having experimentally defined function.

Anaerobic Aromatic Degradation—Absence of Known Enzymes Indicates Novel Pathways

One of the more striking findings is the absence of known key enzymes for monoaromatic degradation under anaerobic conditions. One of the primary metabolic capabilities of interest for this microbe is anaerobic degradation of benzene. Fumarate addition to toluene via BssABCD is recognized as the common mechanism for anaerobic degradation by a phylogenomically diverse population of microbes [14-16] and has been called "the paradigm of anaerobic hydrocarbon oxidation" [17]. Benzoyl-CoA is likewise considered a central intermediate in anaerobic degradation, and is further catabolized via benzoyl-CoA reductase (BcrAB) [17]. Populated KEGG maps in the IMG and VIMSS databases, based on BLAST analyses, indicate the presence of some of the enzymes previously characterized as belonging to the Bss pathway in *D. aromatica*, yet more careful analysis shows the candidate enzymes to be members of a general family, rather than true orthologs of the enzymes in question. The majority of catabolic enzymes of interest for *D. aromatica* are not covered by TIGRfams or COGs families. For this reason Flower Power clustering, SCI-PHY subfamily clade analysis, and HMM scoring were used to ascertain the presence or absence of proteins of interest. The most reliable prediction-of-function approaches for genomically sequenced protein orfs are obtained using the more computationally intensive HMM modeling and scoring utilities. This allows the protein in question to be assessed by phylogenetic alignment to protein families or sub-families with experimentally known function, providing much more accurate predictions [18, 19].

To explore the apparent lack of anaerobic aromatic degradation pathways expected to be present in this genome, all characterized anaerobic aromatic degradation pathways from *A. aromaticum* EbN1 [20] were defined by HMMs to establish presence or absence of proteins in both the *D. aromatica* and *Azoarcus* BH72 genomes (these three genomes comprise nearest-neighbor species in currently sequenced species. In *A. aromaticum* EbN1, 10 major catabolic pathways have been found for anaerobic aromatic degradation, and nine of the 10 converge on benzoyl-CoA [21]. A key catalytic enzyme or subunit for each enzymatic step was used as a seed sequence to recruit proteins from a non-redundant set of Genbank proteins for phylogenetic analysis. The BssABC, present in *A. aromaticum* EbN1 [20, 22] as well as *Thauera aromatica* [6], and *Geobacter metallireducens* [23], is not present in either the *D. aromatica* or *Azoarcus* BH72 genomes (see Table 1). The BcrAB and BssABC, previously denoted as "central" to anaerobic catabolism of aromatics, are likewise absent. The set of recruited proteins for both BssABC and BcrAB indicate they are not as universally present as has been suggested. *Dechloromonas aromatica* does encode a protein in the pyruvate formate lyase family, but further analysis shows that it is more closely related to the *E. coli* homolog of this protein (which is not involved in aromatic catabolism) than to BssA. Anaerobic reduction of ethylbenzene is carried out by ethylbenzene dehydrogenase (EbdABCD1, 2) in *A. aromaticum*. This complex belongs to the membrane-bound nitrate reductase (NarDKGHJI) family. In *D. aromatica*, this complex of proteins is only present as the enzymatically characterized perchlorate reductase (PcrABCD; [24]) which utilizes perchlorate, rather than nitrate, as the electron acceptor. The EbdABCD proteins in *A. aromaticum* (VIMSS814904-814907 and VIMSS816928-816931) occur in operons that include (S)-1-phenylethanol dehydrogenases (Ped; VIMSS814903 and 816927) [25], both of which are absent from *D. aromatica*, as is the acetophenone carboxylase that catalyzes ATP-dependent carboxylation of acetophonenone produced by Ped.

Table 1. Anaerobic aromatic degradation enzymes in near-neighbor *Aromatoleum aromaticum* EbN1.

Proteins involved in the anaerobic aromatic pathways in Aromatoleum aromaticum str. EbN I	A. aromaticum EbN I - representative protein used for HMM models	Azoarcus BH72 ortholog	D. aromatica RCB ortholog
I) phenylalanine			
Pat	VIMSS813888:pat (COG 1448; EC 2.6. 1.57)	-	-
Pdc	VIMSS817385:pdc (COG3961)	-	-
Pdh	VIMSS816687:pdh (COG I 0 12)	-	-
IorAB	VIMSS813644:iorA (COG4321)	-	+
2) phenylacetate			
Pad BCD	VIMSS816693:padB	-	-
PadEFGHI	VIMSS816700:padl	-	-
Padj	VIMSS81670 I :padj	-	-
3) benzyl alcohol/benzaldehyde			
Adh	VIMSS815388:adh (COG I 062)	-	-
Ald	VIMSS816847:ald (COG 1012; ECI .2.1.28)	+	-
4) p-eresol			
PehCF	VIMSS813733:pchC (EC: 1.17.99. 1)	-	-
PehA	VIMSS815385:pchC	-	-
	VIMSS813734:pchF (EC1.1.3.38)	-	-
	VIMSS815387:pchF		-

Table 1. *(Continued)*

Proteins involved in the anaerobic aromatic pathways in Aromatoleum aromaticum str. EbN I	A. aromaticum EbN I - representative protein used for HMM models	Azoarcus BH72 ortholog	D. aromatica RCB ortholog
	VIMSS815384:pchA (COG I 0 12)	-	-
5) phenol			
PpsABC	VIMSS816923:ppsA phenylphosphate synthase	-	-
PpeABCD	VIMSS815367:ppcA	-	-
6) 4-hydroxybenzoate			
PcaK	VIMSS816471 :peaK (COG2271)	-	-
HbcL	VIMSS816681 :hbeLI 4-hydroxybenzoate CoA ligase		-
HcrCBA	VIMSS815644:hcrB	-	-
	VIMSS815645:hcrA	-	-
7) toluene			
BssDCABEFGH	VIMSS814633:bssA	-	-
BbsABCDEFGH(IJ)	VIMSS814644:bbsH	-	-
	VIMSS814645:bbsG	-	-
	VIMSS814647:bbsF	-	-
	VIMSS814649:bbsD	-	-
	VIMSS814651 :bbsB	-	-
8) ethylbenzene			
EbdABC	VIMSS814907:cbdA	-	+ (PerA)
Ped	VIMSS814906:ebdB	-	+ (PerB)
	VIMSS814905:ebdC	-	+(PerC)
	VIMSS814904:ebdD	-	+ (PerD)
	VIMSS814903:ped	-	-
9) benzoate			
BenK	VIMSS816652:benK	-	-
BelA	VIMSS8151 52: belA	+	-
BerCBAD	VIMSS813961 :bcrB	-	-
Deh Had Oah	VIMSS813959:berA	-	-

For all pathways except the ubiquitous phenylacetic acid catabolic cluster, which is involved in the aerobic degradation of phenylalanine, and the PpcAB phenylphosphate carboxylase enzymes involved in phenol degradation via 4-hydroxybenzoate, all key anaerobic aromatic degradation proteins present in *A. aromaticum* EbN1 are missing from the *D. aromatica* genome (Table 1), and the majority are also not present in *Azoarcus* BH72. The lack of overlap for genes encoding anaerobic aromatic enzymes between these two species was completely unexpected, as both *A. aromaticum* EbN1 and *D. aromatica* are metabolically diverse degraders of aromatic compounds. In general *Azoarcus* BH72 appears to share many families of proteins with *D. aromatica* that are not present in *A. aromaticum* EbN1 (e.g., signaling proteins, noted below).

Anaerobic degradation of benzene occurs at relatively sluggish reaction rates, indicating that the pathways incumbent in *D. aromatica* for aromatic degradation under anaerobic conditions might serve in a detoxification role. Another intriguing possibility is that oxidation is dependent on intracellularly produced oxygen, which is likely to be a rate-limiting step. *Alicycliphilus denitrificans* strain BC couples benzene degradation under anoxic conditions with chlorate reduction, utilizing the oxygen produced by chlorite dismutase in conjunction with a monooxygenase and subsequent catechol degradation for benzene catabolism [26]. A similar mechanism may account for anaerobic benzene oxidation coupled to perchlorate and chlorate reduction in *D. aromatica*. However, anaerobic benzene degradation coupled with nitrate reduction is also utilized by this organism, and remains enigmatic [5].

The extremely high divergence of encoded protein families in this functional grouping differs from the general population of central metabolic and housekeeping genes: *Azoarcus* BH72, *Azoarcus aromaticum* EbN1 and *D. aromatica* are evolutionarily near-neighbors within currently sequenced genomes, as defined both by the high level of protein similarity within housekeeping genes (defined by the COG J family of proteins), and 16sRNA sequence. *Azoarcus* BH72 and *A. aromaticum* EbN1 display the highest percent similarity between housekeeping proteins within this triad, with 138 of the 156 COG J proteins in *A. aromaticum* EbN1 displaying highest similarity to their BH72 counterparts. On average these two genomes display 83.5% amino acid identity across shared COG J proteins. *Dechloromonas aromatica* is an outlier in the triad, with higher similarity to *Azoarcus* BH72 than *A. aromaticum* EbN1 (43 of *D. aromatica's* 169 COG J proteins are most homologous to *A. aromaticum* EbN1 orthologs with an average 71% identity, and 67 are most homologous to *Azoarcus* BH72 with an average 72% identity).

Comparative genomics have previously established that large amounts of DNA present in one species can be absent even from a different strain within the same species [27]. In addition, the underestimation of the diversity of aromatic catabolic pathways (both aerobic and anaerobic) has been noted previously [28], and a high level of enzymatic diversity has been seen for pathways that have the same starting and end products, including anaerobic benzoate oxidation [29].

Aerobic Aromatic Degradation

Dechloromonas aromatica encodes several aerobic pathways for aromatic degradation, including six groups of oxygenase clusters that each share a high degree of sequence similarity to the phenylpropionate and phenol degradation (HPP and *Mhp*) pathways in *Comamonas* species [30, 31]. The *mhp* genes of *E. coli* and *Comamonas* are involved in catechol and protocatechuate pathways for aromatic degradation via hydroxylation, oxidation, and subsequent ring cleavage of the dioxygenated species. Only one of the clusters in *D. aromatica* encodes an *mhp*A-like gene; it begins with VIMSS584143 *Mhp*C, and is composed of orthologs of MhpABCDEF&R, and is in the same overall order and orientation as the *Comamonas* cluster as well as the *E. coli* *mhp* gene families [32] (see Figure 1, cluster 3). These pathways are also phylogenom-ically related to the biphenyl/polychlorinated biphenyl (Bhp) degradation pathways in *Pseudomonad* species [32]. For *Comamonas testosteroni*, this pathway is thought to be associated with lignin degradation [31]. Hydroxyphenyl propionate (HPP), an alkanoic acid of phenol, is the substrate for *Mhp*, and is also produced by animals in

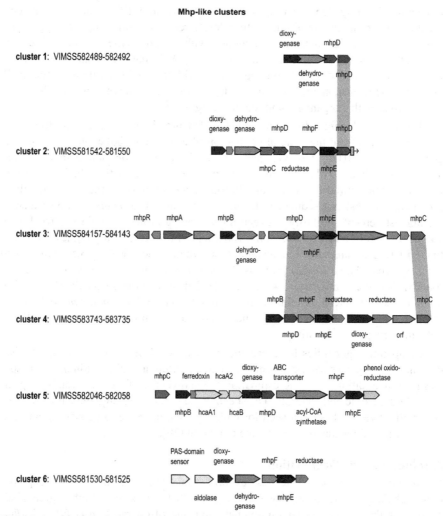

Figure 1. Aerobic degradation of aromatic compounds: multiple *Mhp*-like dioxygenase clusters. Each of the six *mhp*-like gene clusters in the *D. aromatica* genome is depicted. Recent gene duplications between individual proteins are shown by a purple connector between duplicates. Naming convention was chosen for simplicity and consistency, and names all proteins paralogous to a given *Mhp* protein with the *Mhp* name (MhpABCDEF or R), but does not imply enzymatic specificity for the substrates listed here-in, though the general enzymatic reaction is highly likely to be conserved. The *Mhp*: meta cleavage of HPP, (hydroxyphenyl)propionate. The *Mhp*A, 3HPP hydroxylase; *Mhp*B, DHPP 1,2-dioxygenase; *Mhp*C, 2-hydroxy-6-ketonona-2,4-dienedioate hydrolase; *Mhp*D, 2, deto-4-pentenoate hydratase; *Mhp*E, 4-hydroxy-2-ketovalerate aldolas; *Mhp*F, acetaldehyde dehydrogenase.

the digestive breakdown of polyphenols found in seed components [33]. Each gene cluster appears to represent a multi-component pathway, and is made up of five or more of various combinations of dioxygenase, hydroxylase, aldolase, dehydrogenase, hydratase, decarboxylase, and thioesterase enzymes.

The single predicted *Mhp*A protein in *D. aromatica* (VIMSS584155), which is predicted to support an initial hydroxylation of a substituted phenol substrate, shares 64.4% identity to *Rhodococcus* OhpB 3-(2-hydroxyphenyl) propionate monooxygenase (GI:8926385) versus 26.4% for *Comamonas testosteroni* (GI:5689247), yet the remainder of the *ohp* genes in the *Rodococcus ohp* clade do not share synteny with the *D. aromatica mhp* gene cluster.

Other Aromatic Oxygenases

Two chromosomally adjacent monooxygenase clusters, syntenic to genes found in *Burkholderia* and *Ralstonia* sp., indicate that *D. aromatica* might have broad substrate hydroxylases that support the degradation of toluene, vinyl chlorides, and TCE (Figure 2 and Table 2), and are thus candidates for benzene-activating enzymes in the presence of oxygen.

One monooxygenase gene cluster, composed of VIMSS581514 to 581519 ("tbc2 homologs", Figure 2), is orthologs to the tbuA1UBVA2C/tmoAECDBF/touABCDEF/ phlKLMNOP and tbc2ABCDEF gene families (from *P. stutzeri*, *R. pickettii*, and *Burkholderia* JS150). This gene cluster includes a transport protein that is orthologs to TbuX/TodX/XylN (VIMSS581520). Specificity for the initial monooxygenase is not established, but phylogenetic analysis places VIMSS581514 monooxygenase with near-neighbors TbhA [34], reported as a toluene and aliphatic carbohydrate monooxygenase (76.5% sequence identity), and BmoA [35], a benzene monooxygenase of low regiospecificity (79.6% sequence identity). The high level of similarity to the *D. aromatica* protein is notable. The region is also highly syntenic with, and homologous to, the tmoAECDBF (AY552601) gene cluster responsible for *P. mendocina's* ability to utilize toluene as a sole carbon and energy source [36].

Just downstream on the chromosome is a *phc/dmp/phh/phe/aph*-like cluster of genes, composed of the genes VIMSS812947 and VIMSS581535 to 581540 ("tbc1 homologs", Figure 2). Overall, chromosomal organization is somewhat different for *D. aromatica* as compared to *Ralstonia* and *Burkholderia*. *Dechloromonas aromatica* has a 14 gene insert that encodes members of the *mhp*-like family of aromatic oxygenases between the tandem tbc 1 and 2-like oxygenase clusters (see Table 2), with an inversion of the second region compared to *R. eutropha* and *Burkholderia*. Clade analysis indicates a broad substrate phenol degradation pathway in this cluster, with high sequence identity to the *TOM* gene cluster of *Bradyrhizobium*, which has the ability to oxidize dichloroethylene, vinyl chlorides, and TCE [37, 38]. The VIMSS581522 response regulator gene that occurs between the two identified monooxygenase gene clusters shares 50.3% identity to the *Thaurea aromatica tutB* gene and 48.2% to the *Pseudomonas* sp. Y2 styrene response regulator (occupying the same clade in phylogenetic analysis). The VIMSS581522 is likely to be involved in the chemotactic response in conjunction with VIMSS581521 (histidine kinase) and VIMSS581523 (methyl accepting chemotaxis protein), which would confer the ability to display a chemotactic response to aromatic compounds.

Monooxygenase clusters

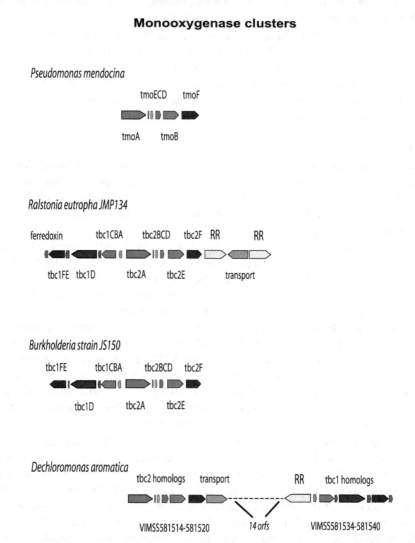

Figure 2. Catabolic oxygenases of aromatic compounds: Synteny between *D. aromatica, P. mendocina, Burkholderia,* and *R. eutropha.* Orthologs gene clusters for *P. mendocina, R. eutropha* JMP134, *Burkholderia* JS150, and *D. aromatica* are shown. *Dechloromonas aromatica* possesses two oxygenase gene clusters that are syntenic to the tbc1 and 2 catabolic gene clusters of Burkholderia JS150, but with an inversion and insertion in the chromosome. Also shown are the tmo (toluene mono-oxygenase) toluene degradative cluster of *P. mendocina* and the tbc1 and tbc2-like (tcb: toluene, chlorobenzene, and benzene utilization) gene cluster of *R. eutropha* (VIMSS896207-896222, Burkholderia protein names were used for consistency). The first seven orfs (encoding a tbc1-like cluster) of *R. eutropha* JMP134 are orthologs to the PoxABCDEFG (phenol hydroxylase) and P0123456 genes of *Ralstonia* sp. E2 and *R. eutropha* H16, respectively. Orthologs can be identified as having the same size and color scheme.

Table 2. Aromatic degradation in *D. aromatica*: mono- and di-oxygenases.

VIMSS id	Orthologs	Putative function	Size aas
581514	TbuA I /TmoA/TouA/PhiK/Tbc2A	methane/phenol/toluene hydroxylase	501
581515	TbuU/T moE/T ouB/PhiL/Tbc2B	toluene-4-monooxygenase	88
581516	TbuB/T moC/T ouC/PhiM/Tbc2C	ferredoxin subunit of ring-hydroxylating dioxygenase	111
581517	TbuV/TmoD/TouD/PhiN/Tbc2C	monooxygenase	146
581518	TbuA2/T moB/T ouE/PhiO/Tbc2E	hydroxylase	328
581519	TbuC/TmoF/TouF/PhiP/Tbc2F	flavodoxin reductase	338
581520	TbuX/TodX/XyiN	membrane protein; transport	464
581521	histidine kinase	signal transduction	963
581522	NarI	cheY like protein	208
581523	methyl-accepting chemotaxis protein	chemotaxis sensory transducer, membrane bound	532
581524	4-oxalocrotonate tautomerase	tautomerase	144
581525	oxidoreductase	oxidoreductase/dehydrogenase	254
581526	MhpE	4-hydroxy-2-oxovalerate aldolase	354
581527	MhpF	EC 1.2.1.1 0 Acetaldehyde dehydrogenase (acetylating)	305
581528	2-hydroxymuconic semialdehyde dehydrogenase	NAD+-dependent dehydrogenase (EC 1.2.1.60)	489
581529	ring-cleaving extradiol dioxygenase	catechol 2,3 dioxygenase (1.13.1 1.2)	311
581530	aldolase	4-hydroxyphenylacetic acid catabolism pathway	266
581531	S box domain	signal transduction	143
584293	orf	unknown	63
581532	orf	unknown	80
584294	EAL domain containing protein (obsolete in current YIMSS database)	diguanylate phosphodiesterase; signaling	65
581533	transcriptional regulator	LysR-type	300
581534	response regulator, tbuT family	activator of aromatic catabolism	558
812947	PhcK/DmpK/PhhK/PheA I /Tcb IA/AphK	monooxygenase	89
581535	PhcL/DmpL/PhhL/PheA2/Tcb I B/Aphl	hydroxylase	329
581536	PhcM/DmpM/PhhM/PheA3/T cb I C/AphM	monooxygenase	89
581537	PhcN/DmpN/PhhN/PheA4/T cb I D	aromatic hydroxylase	517
581538	PhcO/DmpO/Phh0/PheA5/T cb I E/AphO	aromatic hydroxylase	118
581539	PhcP/Dmp/PhhP/PheA6/Tcb I F/AphQ	hydroxylase reductase	353
581540	ferredoxin	2Fe-2S ferredoxin, iron-sulfur binding site	112
581541	transcriptional regulator	IPR000524: Bacterial regulatory protein GntR, HTH	235
581542	ring-cleaving extradiol dioxygenase	catechol 2,3 dioxygenase (EC 1. 13.1 1.2)	308
581543	orf	unknown	142
581544	2-hydroxymuconic semialdehyde dehydrogenase	NAD+-dependent dehydrogenase (EC 1.2.1.60)	484

Table 2. *(Continued)*

VIMSS id	Orthologs	Putative function	Size aas
581545	MhpC	2-hydroxy-6-ketonona-2,4-dienedioic acid hydrolase	274
581546/3337834	MhpD	2-keto-4-pentenoate hydratase	260
581547	oxidoreductase	3-oxoacyl-[acyl-carrier-protein] reductase (EC 1.1.1.1 00)	264
581548	MhpF	acetaldehyde dehydrogenase (acetylating; EC 1.2.1.1 0)	304
581549	MhpE	4-hydroxy-2-oxovalerate aldolase	343
581550	hydratase/decarboxylase	4-oxalocrotonate decarboxylase	262
581551	tautomerase	4-oxalocrotonate tautomerase	63

Overall, several mono- and di-oxygenases were found in the genome, indicating *D. aromatica* has diverse abilities in the aerobic oxidation of heterocyclic compounds.

There are several gene clusters indicative of benzoate transport and catabolism. All recognized pathways are aerobic. The benzoate dioxygenase cluster BenABCDR is encoded in VIMSS582483-582487, and is very similar to (and clades with) the xylene degradation (*xy*/XYZ) cluster of Pseudomonas.

There is also an *hca*A oxygenase gene cluster, embedded in one of the *mhp* clusters (see cluster 5, Figure 1). Specificity of the large subunit of the dioxygenase (VIMSS582049) appears to be most likely for a bicyclic aromatic compound, as it shows highest identity to dibenzothiophene and naphthalene dioxygenases.

Dechloromonas Aromatica's Sensitivity to the Environment

Cell Signaling

Dechloromonas aromatica has a large number of genes involved in signaling pathways, with 314 predicted signaling proteins categorized in COG T (signal transduction mechanisms) and a total of 395 proteins (nearly 10% of the genome) either recruited to COG T or possessing annotated signal transduction domains. Signaling appears to be an area that has undergone recent gene expansion, as nine recent gene duplication events in this functional group are predicted by phylogenetic analysis, as described in a later section (shown in Table 3).

Complex lifestyles are implicated in large genomes with diverse signaling capability, and in general genomes with a very large number of annotated orfs have high numbers of predicted signal transducing proteins, as shown in Figure 3, though some species, such as *Rhodococcus* RHA1 and *Psychroflexus* torques are notable exceptions to this trend. However, assessment of COG T population size relative to other genomes with a similar number of predicted orfs (Figure 3) indicates that *D. aromatica* is one of a handful of species that have a large relative number of signaling proteins versus similarly sized genomes. Other organisms displaying this characteristic include *Magnetospirillum magnetotacticum* MS-1, *Stigmatella aurantiaca*, *Myxococcus Xanthus* DK1622, *Magnetospirillum magneticum* AMB-1, *Oceanospirillum* sp. MED92, and *Desulfuromonas acetoxidans*. Within the Betaproteobacteria, *Chromobacterium violaceum*, and *Thiobacillus denitrificans* have a relatively large number of signaling cascade genes, but still have far fewer than found in *D. aromatica*, with 262 predicted COG T proteins (6% of the genome) and 137 COG T proteins (4.8% of the genome), respectively. Histidine kinase encoding proteins are particularly well-represented, with only *Stigmatella aurantiaca* DW4/3-1, *Magnetococcus* sp. MC-1, *Myxococcus xanthus* DK1622, and *Nostoc punctiforme* reported as having more. The 68 annotated histidine kinases include a large number of nitrate/nitrogen responsive elements. Furthermore, the presence of 47 putative histidine kinases predicted to contain two transmembrane (TM) domains, likely to encode membrane-bound sensors (see Figure 4), suggests that *D. aromatica* is likely to be highly sensitive to environmental signals. Nearly half (48%) of the predicted histidine kinases are contiguous to a putative response regulator on the chromosomal DNA, indicating they likely constitute functionally expressed kinase/response regulator pairs. This is atypically high for contiguous placement on the chromosome [39].

Table 3. Candidates for gene expansion in the *D. aromatica* genome.

Protein/protein family function	Number of duplicates	Number of triplicates
Transport (membrane)	12	
Signal transduction or regulatory- includes:	9	
FlhD homolog	(1)	
FlhC homolog	(1)	
Nitrogen regulatory protein PII homolog	(1)	
Hydrolase/transhydrogenase or hydratase	4	1
Cytochromes	3	2
Mhp family	2	2
Phospholipase/phosphohydrolase	2	1
Phasin	1	
Dioxygenase	1	
NapH homolog	1	
NosZ homolog	1	
Unknown function	7	

A relatively high level of diguanylate cyclase (GGDEF domain [40-42]) signaling capability is implied in *D. aromatica* by the presence of 57 proteins encoding a GGDEF domain (Interpro IPR000160) and an additional 10 with a GGDEF response regulator (COG1639) [40]. The *E. coli*, for comparison, encodes 19. This gene family also appears to have undergone recent expansion in this microbe's evolutionary history. Microbes having a large number of proteins or even a diverse array of COG T elements do not a priori encode a large number of GGDEF elements, as *Stigmatella aurantiaca*, *Myxococcus*, *Xanthus* DK1622, and *Burkholderia pseudomallei*, by contrast, have very large genomes with extensive COG T populations, yet each have 20 or fewer proteins identified as having GGDEF domains, and *Prochloroccus* spp. appear to have none. Conversely, *Oceanospirillum* has a relatively small genome, yet has 112 proteins identified as likely GGDEF domain/IPR000160 proteins. The GGDEF/EAL domain response regulators have been implicated in root colonization in *Pseudomonas putida* (Matilla et al., 2007); in *E. coli* the GGDEF domain-containing YddV protein upregulates the transcription of a number of cell wall modification enzymes [42], and in point of fact, *D. aromatica's* VIMSS581804, a GGDEF domain containing homolog of the YddV *E. coli* protein, occurs upstream of a cluster of 16 cell wall division proteins (encoded by VIMSS581805-581820).

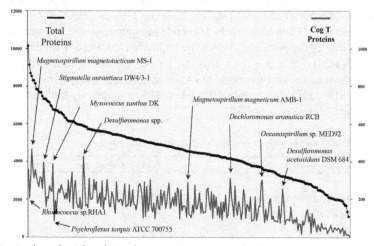

Figure 3. Number of predicted signaling proteins versus total protein count. Microbial genomes, displaying total number of predicted open reading frames (orfs, left axis) and total number of predicted signaling proteins (defined as COG T, right axis). Microbes displaying a high number of signaling orfs relative to total predicted proteins are labeled (above COG T line), as well as two large-sized genomes having a relatively low number of annotated COG T proteins (labeled below COG T line).

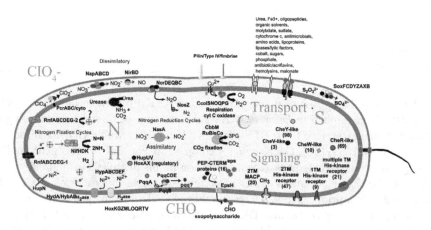

Figure 4. Overview of predicted metabolic cycles, membrane transporters, and signaling proteins in *D. aromatica*. Various metabolic cycles, secretory apparatus, and signaling cascades predicted in the annotation process are depicted. TM: transmembrane. Gene names are discussed in the relevant sections of this chapter. Areas of the cell depicting nitrogen, hydrogen, carbon, and sulfur cycles are indicated by "N", "H", "C", and "S".

Cellular Interactions with Community/Environment—Secretion

Type I Secretion

Fifteen transport clusters include a TolC-like outer membrane component, and recent gene family expansion is noted within several families of ABC transporters for this genome. The TolC was originally identified in *E. coli* as the channel that exports hemolysin [43], and hemolysin-like proteins are encoded in this genome. Two groups of ABC transporters occur as a cluster of five transport genes; these five-component transporters have been implicated in the uptake of external macromolecules [44].

The presence of putative lytic factors, lipases, proteases, antimicrobials, invasins, hemolysins, RTXs, and colicins near potential type I transport systems indicate that these might be effector molecules used by *D. aromatica* for interactions with host cells (e.g., for cell wall remodeling). Iron acquisition is likely to be supported by a putative FeoAB protein cluster (VIMSS583997, 583998), as well as several siderophore-like receptors and a putative FhuE protein (outer membrane receptor for ferric iron uptake; VIMSS583312). Other effector-type proteins, likely to be involved in cell/host interactions (and which in some species have a role in pathogenicity [45]), are present in this genome. Adhesins, haemagglutinins, and oxidative stress neutralizers are relatively abundant in *D. aromatica*. A number of transporters occur near the six putative soluble lytic murein transglycosylases, indicating possible cell wall remodeling capabilities for host colonization in conjunction with the potential effector molecules noted above. Homologs of these transporters were shown to support invasin-type functions in other microbes [45]. Interaction with a host is further implicated by: VIMSS581582, encoding a potential cell wall-associated hydrolase, VIMSS581622, encoding a predicted ATPase, and VIMSS3337824/formerly 581623, encoding a putative membrane-bound lytic transglycosylase.

Eleven tandem copies of a 672 nucleotide insert comprise a region of the chromosome that challenged the correct assembly of the genome, and finishing this region was the final step for the sequencing phase of this project (see Materials and Methods). Unexpectedly, analysis of this region revealed a potential orf encoding a very large protein that has been variously predicted at 4,854, 2,519, or 2,491 amino acids in size during sequential automated protein prediction analyses (VIMSS3337779/ formerly 582095). This putative protein, even in its smallest configuration, contains a hemolysin-type calcium-binding region, a cadherin-like domain, and several RTX domains, which have been associated with adhesion and virulence. Internal repeats of up to 100 residues with multiple copies have also been found in proteins from *Vibrio*, *Colwellia*, Bradyrhizobium, and *Shewanella* spp. (termed "VCBS" proteins as defined by TIGRfam1965).

Other potential effector proteins include: three hemolysin-like proteins adjacent to type I transporters, eight proteins with a predicted hemolysin-related function, including VIMSS583067, a hemolysin activation/secretion protein, VIMSS580979, hemolysin A, VIMSS583372, phospholipase/hemolysin, VIMSS581868, a homolog of hemolysin III, predicted by TIGRfam1065 to have cytolytic capability, VIMSS582079, a transport/hemolysin, and VIMSS581408, a general hemolysin. Five predicted proteins have possible LysM/invasin domains, including: VIMSS580547,

581221, 581781, 582766, and 583769. One gene, VIMSS583068, encodes a putative 2,079 amino acid filamentous haemagglutinin, as well as a hasA-like domain, making it a candidate for hasA-like function (hasA is a hemophore that captures heme for iron acquisition [46]).

Type II Secretion

Besides the constitutive Sec and Tat pathways, *D. aromatica* has several candidates for dedicated export secretons of unknown function, with 3–4 putative orthologs of PulDEFG interspersed with a lytic transglycosylase and a hemolysin (VIMSS582071-582085). The region from VIMSS581889 to VIMSS581897 includes pulDEFG type subunits and an exeA ATPase like protein. It is bracketed by signaling components comprised of a histidine kinase, adenylate cyclase, and a protein bearing similarity to the nitrogen response regulator glnG (VIMSS581898), which has been shown to be involved in NH_3 assimilation in other species [47].

In addition, there is a nine-gene cluster that encodes several proteins related to toluene resistance (VIMSS581899-581906).

A pilus-like gene cluster (which can also be classified as type IV secretion) occurs in VIMSS580547-580553, encoding a putative lytic transglycosylase, ABC permease, cation transporter, pilin peptidase, pilin ATPase, and PulF-type protein. This assembly resembles other pilin assemblies associated with attachment to a substrate, such as the pilus structure responsible for chitin/host colonization in *Vibrio cholerae* [48].

Another large pilus-like cluster (VIMSS584160-584173) occurs in close proximity to the *mhp*CEFDBAR oxygenase genes (see e.g., VIMSS584157, *mhp*R).

Type III Secretion

Dechloromonas aromatica has been shown to be chemotactic under various circumstances. The flagellar proteins (FliAEFGHIJKLMNOPQR, FlaABCDEFGHIJK) are followed by an additional cluster of 15 chemotaxis/signal transduction genes (VIMSS580462-580476), and homologs of FlhC and D regulatory elements required for the expression of flagellar proteins (VIMSS582640 and 582641) [49], identified by phylogenetic clustering, are also present. Since *D. aromatica* has a flagellum and displays chemotactic behavior, it is likely that the flagellar gene cluster is solely related to locomotion, though type III secretion systems can also encode dedicated protein translocation machineries that deliver bacterial pathogenicity proteins directly to the cytosol of eukaryotic host cells [50].

Type IV Secretion

There are two copies of a 21 gene cluster that includes 10 putative conjugal transfer (Tra) sex-pilus type genes in the *D. aromatica* genome (VIMSS582582-582601 and VIMSS582864-582884), indicating a type IV secretion structure that is related to non-pathogenic cell–cell interactions [51].

Type VI Secretion

A large cluster of transport proteins that is related to the virulence associated genetic locus HIS-1 of *Pseudomonas aeruginosa* and the VAS genes of *V. cholerae* [52, 53] includes homologs of hcp1, IcmF, and clpV (as VIMSS583005, 582995, and 583009, respectively, in *D. aromatica*). This IcmF-associated (IAHP) cluster has been associated with mediation of host interactions, via export of effector proteins that lack signal sequences [53]. Further evidence for type VI secretion is found in the presence of three proteins containing a Vgr secretion motif modeled by TIGRfam3361, which is found only in genomes having type VI secretory apparatus. Though most bacteria that contain IcmF clusters are pathogenic agents that associate with eukaryotic cell hosts [54], it has been reported that the host interactions supported by this cluster are not restricted to pathogens [55].

The type IV pili systems might be involved in biofilm development, as interactions with biofilm surfaces are affected by force-generating motility structures, including type IV pili and flagella [56]. Quorum sensing is a deciding input for biofilm formation, and the presence of an eps synthetic cluster lends further support for biofilm formation. Further, derivatives of NO, which is an evident substrate for *D. aromatica*, are a key signal for biofilm formation versus cell dispersion in the microbe *P. aeruginosa* [57].

Cellular Interactions with Community—Quorum Sensing

Quorum sensing uses specific membrane-bound receptors to detect autoinducers released into the environment. It is involved in both intra- and inter-species density detection [58, 59]. Cell density has been shown to regulate a number of cellular responses, including bioluminescence, swarming, expression of virulence factors, secretion, and motility (as reviewed in Withers et al., 2001 [60]).

Dechloromonas aromatica encodes six histidine kinase receptor proteins that are similar to the quorum sensing protein QseC of *E. coli* (VIMSS580745, 582451, 582897, 583274, 3337577, (formerly 583538), and 583893), five of which co-occur on the chromosome with homologs of the CheY like QseB regulator, and two of which appear to be the product of a recent duplication event (VIMSS583893 and 3337577). Of the six QseC homologs, phylogenetic analysis indicates VIMSS582451 is most similar to QseC from *E. coli*, where the QseBC complex regulates motility via the FlhCD master flagellar regulators (VIMSS582640 and 582641). The presence of several qseC/B gene pairs indicates the possibility of specific responses that are dependent on different sensing strategies. In other species, expression of ABC exporters is regulated by quorum sensing systems [46]; gene family expansion is indicated in the ABC export gene pool as well as the qseC/B sensors in *D. aromatica*.

N-acyl-homoserine lactone is the autoinducer typical for gram negative bacteria [61], yet *D. aromatica* lacks any recognizable AHL synthesis genes. *Ralstonia* Betaproteobacteria likewise encode several proteins in the qseC gene family and display a diversity of candidate cell density signaling compounds other than AHL [62]. The utility of having a diverse array of quorum sensing proteins remains to be determined, but appears likely to be associated with a complex, and possibly symbiotic, lifestyle for *D. aromatica*.

Cellular Interactions with the Environment—Stress

Carbon Storage

Poly-hydroxyalkanoates (PHAs) store carbon energy, are synthesized from the catabolism of lipids, and constitute up to almost 90% of the dry weight of the Betaproteobacteria species *Comamonas testosteroni* [63]. These lipid-like carbon/energy storage polymers are found in granular inclusions. The PhaR candidate VIMSS583509 is likely to be a regulatory protein for PHA synthesis, and is found near other proteins associated with PHA granule biosynthesis and utilization in *D. aromatica* (VIMSS583511-583513).

Phasins are relatively small proteins (180–200 aas) that have been shown to associate with PHA inclusions [64]. There are six copies of phasin-type proteins, with indications of recent gene duplication for three of the phasin-type proteins (VIMSS581881, 582264, and 3337571 (formerly 583582)). There are also three homologs of the active subunit poly-B-hydroxybutyrate polymerase (PhaC orthologs) and two PHA reductase candidates present in a direct repeat, which is also found in *Legionella pneumophila*. Interestingly, one PhaC-like protein, VIMSS583511, is 70% identical to NodG of *Azospirillum brasilense*, a nodulation protein [65]. No PhaA-like ketothiolase ortholog is present. The presence of an amplified gene pool for carbon storage granules in *D. aromatica* may confer the ability to survive under low nutrient conditions, and poly-3-hydroxybutyrate accumulation has recently been observed in *A. aromaticum* EbN1 cultures displaying reduced growth [66].

Phosphate

Inorganic polyphosphate storage appears likely, as both polyphosphate kinase (Ppk, VIMSS582444) and exopolyphosphatase (Ppx, VIMSS583870) are present. These genes are similar to those encoded in *Pseudomonas aeruginosa*, in that they are in disparate regions of the chromosome [67]. Polyphosphate has been implicated in stress response due to low nutrients in the environment [68], and also in DNA uptake [69].

Phosphate transport appears to be encoded in a large cluster of genes (VIMSS581746-581752), and response to phosphate starvation is likely supported by the PhoH homolog VIMSS583854.

Biofilm Formation

There is a large cluster of eps export associated genes, including a proposed exosortase (epsH, VIMSS582792). Presence of the eps family proteins (VIMSS582786, VIMSS582790-582801) indicates capsular eps production, associated with either host cell interactions (including root colonization [70]) or biofilm production in soil sediments [71]. *Dechloromonas aromatica* is one of a small number of species (19 out of 280 genomes assessed by Haft et al. [71]) that also encodes the PEP-CTERM export system. The PEP-CTERM signal, present in 16 proteins in this genome, is proposed to be exported via a potential exportase, represented in this genome by epsH (VIMSS582792). Additionally, the presence of proteins encoding this putative exportase is seen only in genomes also encoding the eps genes.

Metabolic Cycles

Nitrogen

Dechloromonas aromatica closely reflects several metabolic pathways of *R. capsulatus*, which is present in the rhizosphere, and its assimilatory nitrate/nitrite reductase cluster is highly similar to the *R. capsulatus* cluster [72]. Encoded nitrate response elements also indicate a possible plant association for this microbe, as nitrate can act as a terminal electron acceptor in the oxygen-limited rhizosphere. Alternatively, NO reduction can indicate the ability to respond to anti-microbial NO production by a host (used by the host to mitigate infection [73]). Several gene families are present that indicate interactions with a eukaryotic host species, including response elements that potentially neutralize host defense molecules, in particular nitric oxide (NO) and other nitrogenous species.

Nitrate is imported into the cytosol by NasDEF in *Klebsiella pneumoniae* [74] and expression of nitrate and nitrite reductases is regulated by the nasT protein in *Azotobacter vinelandii* [75]. A homologous set of these genes are encoded by the cluster VIMSS580377-580380 (NasDEFT), and a homolog of *nar*K is immediately downstream at VIMSS580384, and is likely involved in nitrite extrusion. Upstream, a putative nasA/nirBDC cluster (assimilatory nitrate and nitrite reduction) is encoded near the *nar*XL-like nitrate response element. The VIMSS580393 encodes a nitrate reductase that is homologous to the NasA cytosolic nitrate reductase of *Klebsiella pneumoniae* [76]. Community studies have correlated the presence of NasA-encoding bacteria with the ability to use nitrate as the sole source of nitrogen [77]. The large and small subunits of nitrite reductase (VIMSS580391 nirB and VIMSS580390 nirD) are immediately adjacent to a transporter with a putative nitrite transport function (VIMSS580389 NirC-like protein). The NirB orf is also highly homologous to both NasB (nitrite reductase) and NasC (NADH reductase which passes electrons to NasA) of *Klebsiella pneumoniae*. The HMMs created from alignments seeded by the NasB and NasC genes scored at $3.2e^{-193}$ and $4.0e^{-159}$, respectively, to the VIMSS580391 NirB protein. *Dechloromonas aromatica* is similar to *Methylococcus capsulatus*, *Ralstonia solanacearum*, *Polaromonas*, and *Rhodoferax ferrireducens* for *nas*A, *nir*B, and *nir*D gene clusters. However, the presence of the putative transporter *nir*C (VIMSS580389) shares unique similarity to the *E. coli* and *Salmonella nir*BCD clusters.

Putative periplasmic, dissimilatory nitrate reduction, which is a candidate for denitrification capability [78], is encoded by the *nap*DABC genes (VIMSS3337807/581796-581799). A probable cytochrome c', implicated in NO binding as protection against potentially toxic excess NO generated during nitrite reduction [79], is encoded by VIMSS582015. Although most denitrifiers are free living, plant-associated denitrifiers do exist [80]. There is no dissimilatory nitrate reductive complex *nar*GHIJ, but rather, *Nar*G and *Nar*H-like proteins are found in the evolutionarily-related perchlorate reductase alpha and beta subunits [24]. These proteins are present in the pcrABCDcld cluster, VIMSS582649-582652 and VIMSS584327, as previously reported for *Dechloromonas* species [81].

Ammonia incorporation appears to be metabolically feasible via a putative glu-ammonia ligase (VIMSS581081), an enzyme that incorporates free ammonia into the cell via ligation to a glutamic acid. An ammonium transporter and cognate regulator are likely encoded in the Amt and GlnK-like proteins VIMSS581101 and 581102.

Urea catabolism as a further source of nitrogen is suggested by two different urea degradation enzyme clusters. The first co-occurs with a urea ABC-transport system, just upstream of a putative nickel-dependent urea amidohydrolase (urease) enzyme cluster (VIMSS583666, 583671-583674, and VIMSS583677-583683; see Table 4). The second pathway is suggested by a cluster of urea carboxylase/allophanate hydrolase enzymes (VIMSS581083-581085, described by TIGRfams 1891, 2712, 2713, 3424, and 3425), which comprise four proteins involved in urea degradation to ammonia and carbon dioxide (CO_2) in other species, as well as an amidohydrolase [82].

Table 4. Putative nitrogen fixation gene cluster in *D. aromatica*.

VIMSS id	Ortholog	Size, aas
583652	FldA, flavodoxin typical for nitrogen fixation	186
583653	hypothetical protein	86
583654	NafY -I, nitrogenase accessory factor Y	247
583655	NifB, nitrogenase cofactor biosynthesis protein	500
583656	4Fe-4S ferredoxin	92
583657	nitrogenase-associated protein	159
583658	flavodoxin	423
583659	ferredoxin, nitric oxide synthase	95
583660	2Fe-2S ferredoxin	120
583661	NifQ	190
583662	DraG	326
583663	histidine kinase	1131
583664	Che-Y like receiver	308
583666	UrtA urea transport	420
583667/ 3337562	CynS cyanate lyase	147
583668	S-box sensor, similar to oxygen sensor arcB	794
583669	ABC transporter	393
3337561	Protein of unknown function involved in nitrogen fixation	72
583671	UrtB urea transport	525
583672	UrtC urea transport	371
583673	UrtD urea transport	278
583674	UrtE urea transport	230
583677	UreH urease accessory protein	288
583678	Urea amidohydrolase gamma	100
583679	Urea amidohydrolase beta	101
583680	Urea amidohydrolase alpha/UreC urease accessory protein	569
583681	UreE urease accessory protein	175
583682	UreF urease accessory protein	228
583683	UreG urease accessory protein	201
583685	nitroreductase	558

Table 4. *(Continued)*

VIMSS id	Ortholog	Size, aas
583686	ferredoxin, subunit of nitrite reductase	122
583691	DraT	328
583692	NifH nitrogenase iron protein (EC 1.18.6. 1)	296
583693	NifD nitrogenase molybdenum-iron protein alpha chain (EC 1.18.6.1)	490
583694	NifK nitrogenase molybdenum-iron protein beta chain (EC 1. 18.6.1)	522
583695/ 3337559	NifT	80
3337558	ferredoxin	63
583696	NafY -2 nitrogenase accessory factor Y	243
583710/ 3337556	NifW nitrogen fixation protein	113
3337555	NifZ	151
583711/ 3337554	NifM	271

Nitric Oxide (NO) Reductase

The chromosomal region around *D. aromatica*'s two *nos*Z homologs is notably different from near-neighbors *A. aromaticum* EbN1 and *Ralstonia solanacearum* which encode a *nos*RZDFYL cluster. *Dechloromonas aromatica*'s *nos*RZDFYL operon lacks the *nos*RFYL genes, and displays other notable differences with most nitrate reducing microbes. In *D. aromatica*, two identical *nos*Z reductase-like genes (annotated as *nos*Z1 and *nos*Z2, VIMSS583543 and VIMSS583547) are adjacent to two cytochrome c553s, a ferredoxin, and a transport accessory protein, and are uniquely embedded within a histidine kinase/response regulator cluster and include *nos*D and a *nap*GH-like pair that potentially couples quinone oxidation to cytochrome c reduction. This indicates the NO response might be involved in cell signaling and as a possible general detoxification mechanism for NO.

The Epsilonproteobacteria *Wolinella succinogenes* is quite similar to *D. aromatica* for NO reductase genes (both have two *nos*Z genes, a *nos*D gene and a *nap*-GH pair in the same order and orientation [83]), but the *W. succinogenes* genome lacks the embedded signaling protein cluster. Further, NO reductase homologs Nor-DQEBC (VIMSS582097, 582100-582103), along with the cytochrome c' protein (VIMSS582015), which has been shown to bind NO prior to its reduction [79], are all present, and potentially act in detoxification roles. It has been shown that formation of anaerobic biofilms of *P. aeruginosa* (which cause chronic lung infections in cystic fibrosis) require NO reductase when quorum has been reached [84], so a role in signaling and complex cell behavior is possible.

Wolinella succinogenes shares other genome features with *D. aromatica*. It encodes only 2,042 orfs, yet has a large number of signaling proteins, histidine kinases, and GGDEF proteins relative to its genome size. It also encodes *nif* genes, several genes similar to virulence factors, and similarity in the NO enzyme cluster noted above. *Wolinella succinogenes* is evolutionarily related to two pathogenic species

(*Helicobacter pylori* and *Campylobacter jejuni*), and displays eukaryotic host interactions, yet is not known to be pathogenic [85]. The distinction between effector molecules causing a pathogenic interaction and a symbiotic one is unclear.

Nitrogen Fixation

Nitrogen fixation capability in *D. aromatica* is indicated by a complex of *nif*-like genes (see Table 4), that include putative nitrogenase alpha (NifD, VIMSS583693) and beta (NifK, VIMSS583694) subunits of the molybdenum-iron protein, an ATP-binding iron-sulfur protein (NifH, VIMSS583692), and the regulatory protein NifL (VIMSS583623), that share significant sequence similarity and synteny to the free-living soil microbe *Azotobacter vinelandii*. *Dechloromonas aromatica* further encodes a complex that is likely to transport electrons to the nitrogenase, by using a six subunit rnfABCDGE-like cluster (VIMSS583616-583619, 583621, and 583622) that is phylogenomically related to the *Rhodobacter capsulatus* complex used for nitrogen fixation [86]. There is a second *rnf*-like NADH oxidoreductase complex composed of VIMSS583911-583916, of unknown involvement (see Figure 4). *Aromatoleum aromaticum* EbN1 and *Azoarcus* BH72 each encode two *rnf*-like clusters as well.

Embedded in the putative nitrogen fixation cluster are two gene families involved in urea metabolism (Table 4). This includes the urea transport proteins (UrtABCDE) and urea hydrolase enzyme family (Ure protein family).

Hydrogenases Associated with Nitrogen Fixation

Uptake hydrogenase is involved in the nitrogen fixation cycle in root nodule symbionts where it is thought to increase efficiency via oxidation of the co-produced hydrogen (H2) [87]. *Dechloromonas aromatica* encodes a cluster of 13 predicted orfs (Hydrogenase-1 cluster, VIMSS581358-581370; Table 5) that includes a hydrogenase cluster syntenic to the *hox*KGZMLOQR(T)V genes found in *Azotobacter vinelandii*, which reversibly oxidize H_2 in that organism [88]. This cluster is followed by a second hydrogenase (Hydrogenase-2 cluster, VIMSS581373-581383). The hydrogenase assembly proteins, *hyp*ABF and CDE are included (VIMSS581368-581370 and 581380-581381, and VIMSS3337851 (formerly 581382)) as well as proteins related to the hydrogen uptake (*hup*) genes of various rhizobial microbes [87]. The second region, with the *hyp* and *hyd*-like clusters, lacks overall synteny to any one genome currently sequenced. It does, however, display regions of genes that share synteny with *Rhodoferax ferrireducens*, which displays the highest percent identity across the cluster, both in terms of synteny and protein identity.

The VIMSS581384 encodes a homolog of the HoxA hydrogenase transcriptional regulator, which has been shown to be expressed only during symbiosis in some species [89]. Regulation is indicated by homologs of NtrX (VIMSS581123) and NtrY (VIMSS581124); the NtrXY pathway comprises a two-component signaling system involved in the regulation of nitrogen fixation in *Azorhizobium caulinodans* ORS571 [90].

Table 5. Hydrogenase clusters associated with nitrogen fixation.

VIMSS id	Orthologs	Putative function	Size, aas
581358	HoxK/HyaNHupS	hydrogenase- I small subunit	363
581359	HoxG/HyaB/HupL	hydrogenase- I, nickel-dependent, large subunit	598
581360	HoxZ/HyaC/HupC	Ni/Fe-hydrogenase I b-type cytochrome subunit	234
581361	HoxM/HyaD/HupD	hydrogenase expression/formation protein	204
581362	HoxUHypC/HupF	hydrogenase assembly chaperone	100
581363	HoxO/HyaE!HupG	hydrogenase- I expression	152
581364	HoxQ/HyaF/HupH	nickel incorporation into hydrogenase- I proteins	287
581365	HoxR/Hupl	rubredoxin-type Fe(Cys)4 protein	66
581366	Hupj/(similar to HoxT)	hydrogenase accessory protein	156
581367	HoxV/HupV	membrane-bound hydrogenase accessory protein	308
581368	HypA	hydrogenase nickel insertion protein	113
581369	HypB	hydrogenase accessory factor Ni(2+)-binding GTPase	352
581370	HypF	hydrogenase maturation protein	763
581371	ABC protein	periplasmic component, ABC transporter	260
581372	GGDEF domain	signal transduction, GGDEF	523
581373	HybO	hydrogenase-2 small subunit	394
581374	HybA	Fe-5-cluster-containing hydrogenase component	351
581375	HybB	cytochrome Ni/Fe component of hydrogenase-2	386
581376	HybC/HynA	hydrogenase-2 large subunit	570
581377	HybD/HynC	Ni, Fe-hydrogenase maturation factor	159
581378	HupF/HypC	hydrogenase assembly chaperone	96
581379	HybE/Hupj	hydrogenase accessory protein	183
581380	HypC	hydrogenase maturation protein	81
581381	HypD	hydrogenase maturation protein	374
581382/ 3337851	HypE	hydrogenase maturation protein	330
581383	HoxX/HypX	formation of active hydrogenase	558
581384	HoxA	response regulator with CheY domain (signal transduction)	495
581385	HoxB/HupU	regulatory [NiFe] Hydrogenase small subunit (sensor)	333
581386	HoxC/HupV	regulatory [NiFe] Hydrogenase large subunit (sensor)	472
581397	HupT	histidine kinase with PAS domain sensor	448
581398	HoxN/HupN/NixA	nickel transporter	269

Carbon Fixation via the Calvin–Benson–Bassham Cycle

The genes indicative of carbon fixation, using the Calvin cycle, are present in the *D. aromatica* genome. This includes Ribulose 1,5-bisphosphate carboxylase (RuBisCo, VIMSS581681), phosphoribulokinase (cbbP/PrkB, VIMSS581690), and a fructose bisphosphate (fba, VIMSS581693) of the Calvin cycle sub-type. The RuBisCo *cbbM* gene is of the fairly rare type II form. *Dechloromonas aromatica* cbbM displays a surprisingly high 77% amino acid identity to cbbM found in the deep-sea tube worm Riftia pachyptila symbiont [91]. In a recent study of aquatic sediments, Rhodoferax fermentans, Rhodospirillum fulvum and R. rubrum were also found to possess the *cbbM* type II isoform of RuBisCo [92]; this sub-type is shared by a only a few microbial species.

Further putative Cbb proteins are encoded by VIMSS581680 and 581688, candidates for CbbR (regulator for the cbb operon) and CbbY (found downstream of RuBisCo in *R. sphaeroides* [93]), respectively.

The presence of the *cbbM* gene suggests the ability to carry out the energetically costly fixation of CO_2, though such functionality has yet to be observed, and CO_2 fixation capability has been found in only a few members of the microbial community.

There is a potential glycolate salvage pathway indicated by the presence of two isoforms of phosphoglycolate phosphatase (*gph*, VIMSS583850 and 581830). In other organisms, phosphoglycolate results from the oxidase activity of RuBisCo in the Calvin cycle, when concentrations of CO_2 are low relative to oxygen. In *Ralstonia (Alcaligenes) eutropha* and *Rhodobacter sphaeroides*, the gph gene (*cbbZ*) is located on an operon along with other Calvin cycle enzymes, including RuBisCo. In *D. aromatica*, the *gph* candidates for this gene (VIMSS583850 and 581830), are removed from the other *cbb* genes on the chromosome in *D. aromatica*; however VIMSS581830 is adjacent to a homolog of Ribulose-phosphate 3-epimerase (VIMSS581829, *rpe*).

The *cco*SNOQP gene cluster codes for a cbb-type cytochrome oxidase that functions as the terminal electron donor to O_2 in the aerobic respiration of *Rhodobacter capsulatus* [94]. These genes are present in a cluster as VIMSS580484-580486 and VIMSS584273-584274; note that these genes are present in a large number of Betaproteobacteria.

Other carbon cycles, such at the reverse TCA cycle and the Wood–Ljungdahl pathways, are missing critical enzymes in this genome, and are not present as such.

Sulfur

Sulfate and thiosulfate transport appear to be encoded in the gene cluster composed of an OmpA type protein (VIMSS581631) followed by orthologs of a sulfate/thiosulfate specific binding protein Sbp (VIMSS581632), a CysU or T sulfate/thiosulfate transport system permease T protein (VIMSS581633), a CysW ABC-type sulfate transport system permease component (VIMSS581634), and a CysA ATP-binding component of sulfate permease (VIMSS581635).

In addition, candidates for the transcriptional regulator of sulfur assimilation from sulfate are present and include: CysB, CysH, and CysI (VIMSS582364, 582360, and 582362, respectively).

A probable sulfur oxidation enzyme cluster is present and contains homologs of *SoxFRCDYZAXB* [95], with a putative SoxCD sulfur dehydrogenase, SoxF sulfide dehydrogenase, and SoxB sulfate thiohydrolase, which is predicted to support thiosulfate oxidation to sulfate (see Figure 5). Functional predictions are taken from Friedrich et al. [95]. A syntenic *sox* gene cluster is also found in *Anaeromyxobacter dehalogens* (although it lacks soxFR) and *Ralstonia eutropha*, but not in *A. aromaticum* EbN1. Thiosulfate oxidation, however, has not been reported under laboratory conditions tested thus far, and experimental support for this physiological capability awaits further investigation.

Sulfur Oxidation

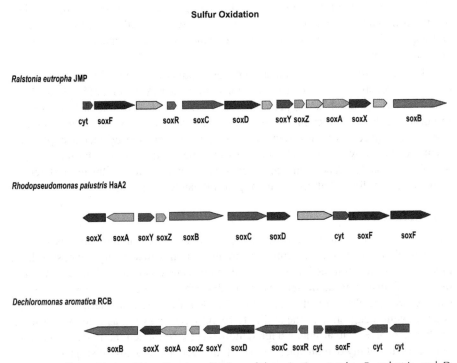

Ralstonia eutropha JMP

cyt soxF soxR soxC soxD soxY soxZ soxA soxX soxB

Rhodopseudomonas palustris HaA2

soxX soxA soxY soxZ soxB soxC soxD cyt soxF soxF

Dechloromonas aromatica RCB

soxB soxX soxA soxZ soxY soxD soxC soxR cyt soxF cyt cyt

Figure 5. Sulfur oxidation (thiosulfate to sulfate) candidates in *R. eutropha, R. palustris,* and *D. aromatica*. Proposed model for this periplasmic complex is as follows: SoxXA, oxidatively links thiosulfate to SoxY; SoxB, potential sulfate thiohydrolase, interacts with SoxYZ (hydrolyzes sulfate from SoxY to regenerate); SoxCD, a sulfur dehydrogenase; oxidizes persulfide on SoxY to cysteine-S-sulfate and potentially yields six electrons per sulfate; SoxC, sulfite oxidase/dehydrogenase with homology to nitrate reductase, induced by thiosulfate; SoxDE, both c-type cytochromes with two heme-type binding sites; and SoxF, a FAD flavoprotein with sulfide dehydrogenase activity. Cyt, cytochrome.

Conversely, the cytoplasmic SorAB complex [96] is not present in *D. aromatica* nor *A. aromaticum* EbN1, although it is found in several other Betaproteobacteria, including *R. metallidurans, R. eutropha, R. solanacearum, C. violaceum,* and *B. japonicum*.

Gene Family Expansion

To determine candidates for recent gene duplication events, extensive phylogenomic profile analyses were conducted for all sets of paralogs in the genome. Flower Power recruitment and clustering against the non-redundant Genbank protein set was done, and the resulting alignments were analyzed using the tree-building SCI-PHY or Belvu based neighbor-joining utilities. The alignment of two or more *D. aromatica* protein sequences in a clade such that they displayed higher % identity to each other than to orthologs present in other species was interpreted as an indication of a probable recent duplication event, either in the *D. aromatica* genome itself or in a progenitor species. Results of this analysis are shown in Table 3.

Potential gene family expansion is indicated in several functional groups, including the following: signaling proteins (including cAMP signaling, histidine kinases, and others), *Mhp*-like aromatic oxidation complexes, nitrogen metabolism proteins, and transport proteins.

Most duplications indicate that a single gene, rather than sets of genes, were replicated. An exception is the Tra/Type IV transport cluster (VIMSS582581-582601 and VIMSS582864-582884) noted previously. In the protein sets for the histidine kinase/response regulator, duplication of histidine kinase appears to occur without duplication of the adjacent response regulator. The paralogs created by recent duplication events are typically found well-removed from one another on the chromosome, although some tandem repeats of single genes were noted. However, the highest percent identity was not found between pairs of genes in tandem repeats.

CONCLUSION

Dechloromonas aromatica strain RCB appears to support a highly complex lifestyle which might involve biofilm formation and interaction with a eukaryotic host. It lacks predicted enzyme families for anaerobic aromatic catabolism, though it supports degradation of several aromatic species in the absence of oxygen. The enzymes responsible for this metabolic function remain to be identified and characterized. It also encodes proteins suggestive of the ability to fix nitrogen and CO_2, as well as thiosulfate oxidation. Converse to aromatic degradation, these enzymatic functionalities have yet to be experimentally demonstrated. In short, this genome was full of surprises.

The utility of TIGRfams and COGs families in these analyses cannot be overstated. New releases of TIGRfams during the course of this analysis provided new insights and identified new functionality (malonate degradation cluster, PEP-Cterm transport and the epsH putative translocon, and urea degradation all were identified in the TIGRfam 7.0 additions). The HMM model building and assessment utilized as the major annotation approach for this study was employed to cover those protein families of interest that are not currently covered by TIGRfams. We utilized K. Sjölander's modeling and analysis tools, which are highly similar to those used to produce TIGRfams models. Overall, the extensive use of HMMs during this analysis allowed high confidence in predicted protein function, as well as certainty that several families of previously characterized anaerobic degradation enzymes for aromatic compounds are not present (e.g., BssABCD and BcrAB).

KEYWORDS

- **Benzylsuccinate synthase**
- **Clusters of orthologs gene**
- ***Dechloromonas aromatic***
- **Open reading frames**
- **Paralogs**
- **Poly-hydroxyalkanoates**

AUTHORS' CONTRIBUTIONS

Alla Lapidus coordinated and oversaw the assembly of the genome. William S. Feil, Helene Feil, and Genevieve Di Bartolo did the initial assembly of the genome. Kennan Kellaris Salinero conducted genome assembly, sequence finishing and gap closure activities, and created the final assembly. Stephan Trong provided internal Joint Genome Institute assembly and analysis tools, and support in their use. Keith Keller was involved in the semi-automated genome annotation, and provided support for the VIMSS dataset and data lists from that set. Kennan Kellaris Salinero conducted all manual annotation work and protein family analyses described here-in. All authors have read and approved the final manuscript.

ACKNOWLEDGMENTS

Kennan Kellaris Salinero sincerely thanks Tanja Woyke for her very helpful suggestions and direction for creation of the tables and figures, Patrick Chain for helpful suggestions on the manuscript, Dan Kirshner for technical help on computational work, Nandini Krishnamurthy for building an internally clustered data-set of *D. aromatica* proteins as well as help with computational tools, Ching Shang for ideas regarding biochemical pathways, Paul Richardson for general support during the finishing phase and Frank W. Larimer for a cogent and extremely helpful critique of the manuscript. The genome finishing portion of this study was performed under the auspices of the US Department of Energy's Office of Science, Biological and Environmental Research Program, and by the University of California, Lawrence Livermore National Laboratory under Contract No. W-7405-Eng-48, Lawrence Berkeley National Laboratory under Contract No. DE-AC02-05CH11231 and Los Alamos National Laboratory under Contract No. W-7405-ENG-36. The majority of the annotation was done as an independent project by Kennan Kellaris Salinero. Considerable intellectual support, computational and data analysis tools were provided by Adam Arkin, Katherine Huang, Morgan Price, Eric Alm, Dan Kirshner, and Kimmen Sjölander—sufficient gratitude cannot be expressed for their generous help.

Chapter 10

Genome-scale Reconstruction of the *Pseudomonas putida* KT2440 Metabolic Network

Jacek Puchałka, Matthew A. Oberhardt, Miguel Godinho,
Agata Bielecka, Daniela Regenhardt, Kenneth N. Timmis,
Jason A. Papin, and Vʜtor A. P. Martins dos Santos

INTRODUCTION

A cornerstone of biotechnology is the use of microorganisms for the efficient production of chemicals and the elimination of harmful waste. *Pseudomonas putida* is an archetype of such microbes due to its metabolic versatility, stress resistance, amenability to genetic modifications, and vast potential for environmental and industrial applications. To address both the elucidation of the metabolic wiring in *P. putida* and its uses in biocatalysis, in particular for the production of non-growth-related biochemicals, we developed and present here a genome-scale constraint-based (CB) model of the metabolism of *P. putida* KT2440. Network reconstruction and flux balance analysis (FBA) enabled definition of the structure of the metabolic network, identification of knowledge gaps, and pin-pointing of essential metabolic functions, facilitating thereby the refinement of gene annotations. The FBA and flux variability analysis (FVA) were used to analyze the properties, potential, and limits of the model. These analyses allowed identification, under various conditions, of key features of metabolism such as growth yield, resource distribution, network robustness, and gene essentiality. The model was validated with data from continuous cell cultures, high-throughput phenotyping data, ^{13}C-measurement of internal flux distributions, and specifically generated knock-out mutants. Auxotrophy was correctly predicted in 75% of the cases. These systematic analyses revealed that the metabolic network structure is the main factor determining the accuracy of predictions, whereas biomass composition has negligible influence. Finally, we drew on the model to devise metabolic engineering strategies to improve production of polyhydroxyalkanoates (PHAs), a class of biotechnologically useful compounds whose synthesis is not coupled to cell survival. The solidly validated model yields valuable insights into genotype–phenotype relationships and provides a sound framework to explore this versatile bacterium and to capitalize on its vast biotechnological potential.

Pseudomonas putida is one of the best studied species of the metabolically versatile and ubiquitous genus of the Pseudomonads [1-3]. As a species, it exhibits a wide biotechnological potential, with numerous strains (some of which solvent-tolerant [4, 5]) able to efficiently produce a range of bulk and fine chemicals. These features, along with their renowned stress resistance, amenability for genetic manipulation and suitability as a host for heterologous expression, make *Pseudomonas putida* particularly attractive for biocatalysis. To date, strains of *P. putida* have been employed to produce phenol, cinnamic acid, cis-cis-muconate, p-hydroxybenzoate, p-cuomarate, and myxochromide [6-12]. Furthermore, enzymes from *P. putida* have been employed in a variety of other biocatalytic processes, including the resolution of D/L-phenylglycinamide into D-phenylglycinamide and L-phenylglycine, production of non-proteinogenic L-amino acids, and biochemical oxidation of methylated heteroaromatic compounds for formation of heteroaromatic monocarboxylic acids [13]. However, most Pseudomonas-based applications are still in infancy largely due to a lack of knowledge of the genotype–phenotype relationships in these bacteria under conditions relevant for industrial and environmental endeavors. In an effort towards the generation of critical knowledge, the genomes of several members of the *Pseudomonads* have been or are currently being sequenced [http://www.genomesonline.org, http://www.pseudomonas.com], and a series of studies are underway to elucidate specific aspects of their genomic programs, physiology and behavior under various stresses (e.g., http://www.psysmo.org, http://www.probactys.org, http://www.kluyvercentre.nl).

The sequencing of *P. putida* strain KT2440, a workhorse of *P. putida* research worldwide and a microorganism Generally Recognized as Safe (GRAS certified) [1, 14], provided means to investigate the metabolic potential of the *P. putida* species, and opened avenues for the development of new biotechnological applications [2, 14-16]. Whole genome analysis revealed, among other features, a wealth of genetic determinants that play a role in biocatalysis, such as those for the hyper-production of polymers (such as PHAs [17, 18]) and industrially relevant enzymes, the production of epoxides, substituted catechols, enantiopure alcohols, and heterocyclic compounds [13, 15]. However, despite the clear breakthrough in our understanding of *P. putida* through this sequencing effort, the relationship between the genotype and the phenotype cannot be predicted simply from cataloguing and assigning gene functions to the genes found in the genome, and considerable work is still needed before the genome can be translated into a fully functioning metabolic model of value for predicting cell phenotypes [2, 14].

The CB modeling is currently the only approach that enables the modeling of an organism's metabolic and transport network at genome-scale [19]. A genome-wide CB model consists of a stoichiometric reconstruction of all reactions known to act in the metabolism of the organism, along with an accompanying set of constraints on the fluxes of each reaction in the system [19, 20]. A major advantage of this approach is that the model does not require knowledge on the kinetics of the reactions. These models define the organism's global metabolic space, network structural properties, and flux distribution potential, and provide a framework with which to navigate through the metabolic wiring of the cell [19-21].

Through various analysis techniques, CB models can help predict cellular phenotypes given particular environmental conditions. The FBA is one such technique, which relies on the optimization for an objective flux while enforcing mass balance in all modeled reactions to achieve a set of fluxes consistent with a maximal output of the objective function. When a biomass sink is chosen as the objective in FBA, the output can be correlated with growth, and the model fluxes become predictive of growth phenotypes [22, 23]. The CB analysis techniques, including FBA, have been instrumental in elucidating metabolic features in a variety of organisms [20, 24, 25] and, in a few cases thus far, they have been used for concrete biotechnology endeavors [26-29].

However, in all previous applications in which a CB approach was used to design the production of a biochemical, the studies addressed only the production of compounds that can be directly coupled to the objective function used in the underlying FBA problem. The major reason for this is that FBA-based methods predict a zero-valued flux for any reaction not directly contributing to the chosen objective. Since the production pathways of most high-added value and bulk compounds operate in parallel to growth-related metabolism, straightforward application of FBA to these biocatalytic processes fails to be a useful predictor of output. Other CB analysis methods, such as extreme pathways and elementary modes analysis, are capable of analyzing non-growth-related pathways in metabolism, but, due to combinatorial explosion inherent to numerical resolution of these methods, they could not be used so far to predict fluxes or phenotypes at genome-scale for guiding biocatalysis efforts [30].

To address both the elucidation of the metabolic wiring in P. putida and the use of P. putida for the production of non-growth-related biochemicals, we developed and present here a genome-scale reconstruction of the metabolic network of Pseudomonas putida KT2440, the subsequent analysis of its network properties through CB modeling and a thorough assessment of the potential and limits of the model. The reconstruction is based on up-to-date genomic, biochemical, and physiological knowledge of the bacterium. The model accounts for the function of 877 reactions that connect 886 metabolites and builds upon a CB modeling framework [19, 20]. Only 6% of the reactions in the network are non-gene-associated. The reconstruction process guided the refinement of the annotation of several genes. The model was validated with continuous culture experiments, substrate utilization assays (BIOLOG) [31], ^{13}C-measurement of internal fluxes [32], and a specifically generated set of mutant strains. We evaluated the influence of biomass composition and maintenance values on the outcome of FBA simulations, and utilized the metabolic reconstruction to predict internal reaction fluxes, to identify different mass-routing possibilities, and to determine necessary gene and reaction sets for growth on minimal medium. Finally, by means of a modified Opt-Knock approach, we utilized the model to generate hypotheses for possible improvements of the production by P. putida of PHAs, a class of compounds whose production consumes resources that would be otherwise used for growth. This reconstruction thus provides a modeling framework for the exploration of the metabolic capabilities of P. putida, which will aid in deciphering the complex genotype–phenotype relationships governing its metabolism and will help to broaden the applicability of P. putida strains for bioremediation and biotechnology.

Highlights of the Model Reconstruction Process

We reconstructed the metabolism of P. putida at the genome-scale through a process summarized in Figure 1. The reconstruction process involved: (1) an initial data collection stage leading to a first pass reconstruction (iJP815^{pre1}); (2) a model building stage in which simulations were performed with iJP815^{pre1} and reactions were added until the model was able to grow in silico on glucose minimal medium (iJP815^{pre2});

and (3) a model completion stage in which BIOLOG substrate utilization data was used to guide model expansion and *in silico* viability on varied substrates. The final reconstruction, named iJP815 following an often used convention [33], consists of 824 intracellular and 62 extracellular metabolites connected by 877 reactions. Eight hundred twenty-one (94%) reactions have at least one assigned gene as delineated in the gene-protein-reaction (GPR) relationships. The GPR relationships are composed of Boolean logic statements that link genes to protein complexes and protein complexes to reactions via combinations of AND and OR operators. An "AND" operator denotes the required presence of two or more genes for a protein to function (as in the case of multi-protein complexes), while an "OR" operator denotes a redundant function that can be catalyzed by any of several genes (as in the case of isozymes). Only 56 reactions, of which nine are non-enzymatic, lack associated genes. The remaining 47 non-gene-associated, enzymatic reactions were added in order to close metabolic network gaps identified during the successive steps of the reconstruction process.

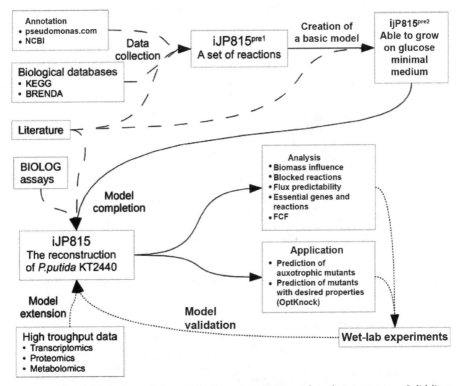

Figure 1. Schematic diagram of the metabolic reconstruction and analysis processes. Solid lines indicate consecutive steps of the reconstruction. Dashed lines represent information transfer. Dotted lines specify planned tasks.

Most network gaps (27) were identified during the second round of the reconstruction and were resolved through detailed literature mining, thereby enabling iJP815 to grow *in silico* on glucose in minimal medium. The remaining gaps identified in the model completion step (Figure 1) were mostly single missing steps in the pathway for which there is experimental evidence of operation (e.g., a compound is consumed but not produced, and no alternative pathways exist). It should be noted that for some gaps, there is more than one combination of reactions with which the gap could be closed [34]. In cases where more than one gap closure method was available, the decision of which to use was made based on similarity queries to related bacteria.

The iJP815 model includes 289 reactions for which non-zero flux values cannot be obtained under any environmental condition while enforcing the pseudo steady-state assumption (PSSA). We term these reactions "unconditionally blocked" meaning that they are unable to function because not all connections could be made with the information available. Three hundred sixty-two metabolites that are only involved in these reactions are classified as "unbalanced metabolites". Another important subset of model reactions is the "weakly annotated" set, which means that all the genes assigned to these 57 reactions are currently annotated as coding for "putative" or "family" proteins. The relationships between all the subsets are shown in Table 1 and Figures 2 and 3.

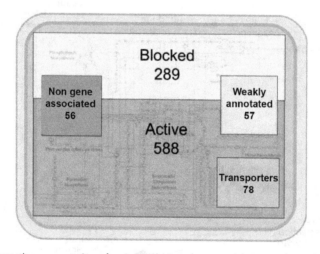

Figure 2. Schematic representation of various reaction classes and their interdependency. The areas of the squares correspond to the sizes of the subsets.

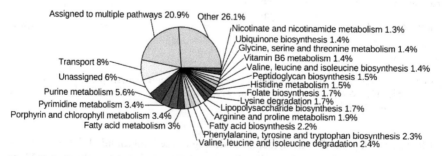

Figure 3. Assignment of the reactions to the particular pathways.

Table 1. Summary of the main characteristics of the iJP815 metabolic model.

System	Parameter	Subset		Size	
P. putida KT2440	Genome size			6.18 Mbp	
	Total ORFs			5446	
iJP815	Reactions	Total		877	
			Potentially active	588 (67.0%)	
			Unconditionally Blocked	289 (33.0%)	
			Well annotated	764 (87.1%)	
			Weakly annotated	57 (6.5%)	
			Non-gene-associated	56 (6.4%)	
			Transport	70 (8.0%)	
	Metabolites	Total		888	
			Internal	824 (92.8%)	
				Balanced	461 (55.9%)
				Unbalanced	363 (44.1%)
			External	64 (7.2%)	
	Genes	Total		815	
			Well annotated	701 (86.0%)	
			Weakly annotated	114 (14.0%)	

The final reconstruction accounts for the function of 815 genes, corresponding to 15% of all genes in the *P. putida* genome and to 65% (1,253) of those currently assigned to the classes "Metabolism" (K01100) and "Membrane Transport" (K01310) in the Kyoto Encyclopedia of Genes and Genomes (KEGG) orthology classification [35]. These figures are consistent with recently published metabolic reconstructions for other prokaryotes.

Model Assessment and Extension Through High-throughput Phenotyping Assays

A high-throughput BIOLOG phenotypic assay was performed on *P. putida* to validate and extend the model. In this assay, *P. putida* was tested for its ability to oxidize 95 carbon substrates in minimal medium. Of these 95 substrates, *P. putida* oxidized 45. We added two other carbon sources to the positive-oxidation group (L-phenylalanine and L-threonine) despite a negative BIOLOG result, since these substrates had been previously shown to be growth substrates [16] and since we confirmed these results experimentally (data not shown), giving altogether 47 compounds utilized *in vivo*. Forty-seven out of the 95 carbon sources tested were accounted for in iJP815^{pre2}, enabling a comparison of these BIOLOG data with FBA simulations of iJP815 grown on *in silico* minimal medium with the respective compound as sole carbon source (see Table 2).

Table 2. Summary of the comparison with the BIOLOG substrate utilization assay.

Compounds tested	95			
Utilized compounds	47			
Reconstruction version	iJP815^{Pre2}		iJP815	
Tested compounds included in 47 51 the model	47		51	
Utilized compounds included in the model	33		37	
Compound supply	Ext	Int	Ext	Int
True positives	14	28	23	33
True negatives	48 (14)	42 (8)	48 (14)	42 (8)
False positives	0	6	0	6
False negatives	33 (19)	20 (6)	24 (14)	14 (4)

Values in brackets indicate onl those compounds that iJP815 accounts for.

The initial working version of the model (iJP815^{pre2}) was able to simulate growth with 14 of the 47 BIOLOG-assayed compounds as sole carbon sources. This version of the reconstruction contained only a few transport reactions, prompting us to identify compounds that could not be utilized *in silico* simply due to the lack of a transporter. This was achieved by allowing the intracellular pool of each compound of interest to be exchanged with environment *in silico*, and by evaluating the production of biomass in each case through FBA simulations. This approach increased the number of utilizable substances to 34 but also produced six false positives (i.e., substances that support *in silico* growth, but which gave a negative phenotype in the BIOLOG assay). These included three metabolites involved in central metabolic pathways (D-glucose 1-phosphate, D-glucose 6-phosphate and glycerol-3-phosphate), an intermediate of the L-histidine metabolism pathway (urocanate), an intermediate of branched amino acids biosynthesis (2-oxobutanoate), and the storage compound glycogen. This analysis suggests that the inability of *P. putida* to utilize these compounds *in vivo* is likely due to the lack of appropriate transport machinery.

The final *P. putida* model (iJP815) grew on 39 of the 51 compounds tested in the BIOLOG assay and that concurrently were accounted for in the model. Of these, 33 were true positives (compounds utilized *in vivo* and allowing for growth *in silico*). The mode of utilization of the remaining 14 *in vivo* oxidized compounds (i.e., false negatives) could not be elucidated. The remaining 42 compounds posed true negatives, eight of which were accounted for in the reconstruction. Ten utilized compounds also lack transport reactions, as nothing is known about their translocation into the cell. Nevertheless, this comparison of *in silico* growth predictions with BIOLOG substrate utilization data indicates that the core metabolism of *P. putida* has been properly reconstructed.

A note of caution when comparing the BIOLOG assays with growth predictions is that this assay evaluates whether an organism is able to oxidize the tested compound and yield energy from it, which is different from growth. However, as *P. putida* is able to grow on minimal medium supplemented with these compounds, we considered the assumption to be justified.

Model-driven Reannotation

The reconstruction process systematizes knowledge about the metabolism of an organism, allowing the identification of errors in, and discrepancies between, various sources of data. A major value of a manual model building effort is the careful revision of the current genome annotation, based on literature evidence encountered during the model building process, BLAST searches, and gap closures. During the reconstruction of the *P. putida* metabolic network, we discovered a number of genes that appear to have been improperly annotated in biological databases (*Pseudomonas* Genome Database, KEGG, NCBI). These mis-annotations arose due to a lack of information at the time of the original annotation or because knowledge that was available in the literature had been overlooked in the original annotation. In a number of other cases, the model building process has also generated new hypotheses for gene functions. For instance, our reconstruction process identified an unlikely gap in the L-lysine degradation pathway of *P. putida*. Extensive literature search and careful reannotation has provided considerable evidence that the genes PP0382 and PP5257, currently annotated as "carbon-hydrogen hydrolase family protein" and "oxidoreductase, FAD binding" respectively, most probably code for a "5-aminopentamidase" and "L-pipecolate oxidase", respectively [36]. Another example is the propanoate degradation pathway: In the iJP815^{pre2} version this pathway was complete except for one enzymatic activity, namely the 2-methylisocitrate dehydratase. Analysis of the enzymes flanking this reaction showed that all of the enzymes are encoded by genes immediately adjacent to the ORF PP2330. Inspection of this region of the genome revealed that PP2336 is annotated as "aconitate hydratase, putative," although the flanking genes are responsible for degradation of propanoate. Analysis of PP2330 via BLAST revealed a homology of more than 99% over the whole length of the protein with the 2-methylisocitrate dehydratase from other bacteria, such as other strains of *P. putida* (GB-1, W619), *Burkholderia prymatum* STM 815, *Burkholderia multivorans* ATCC 17616, *Pseudomonas aeruginosa* PA7, and *Stenotrophomonas maltophilia* R551-3. Consequently the gene was reannotated to code for this function and the gap in propanoate degradation pathway was thus closed by addition of the corresponding GPR. In other cases, discrepancies exist between various databases, as in the case of PP5029, which is annotated in KEGG as "formiminoglutamase" but in NCBI as "N-formylglutamate deformylase". Analysis of network gaps, genomic context, and sequence homology provided a strong indication that "N-formylglutamate deformylase" is the correct annotation. In many other cases the reannotation meant changing the substrate specificity of the enzyme (which corresponds to changing the last part of the EC number). These were mainly identified by BLASTing the protein against protein sequences of other microbes and, whenever available, cross-checking the BLAST results against primary research publications. The full list of reannotations suggested by the reconstruction process is shown in Table 3.

Table 3. List of genes reannotated during the reconstruction process.

Gene	Old Annotation	New Annotation	Reference
PP0213	5uccinate-semialdehyde dehydrogenase; EC:1.2.1 .16	Glutarate-semialdehyde; dehydrogenase EC 1.2.1.20	[36]
PP0214	4-Aminobutyrate aminotransferase; EC:2.6.1.19, EC:2.6.1.22	5-Aminovalerate transaminase; EC 2.6.1.48	[36]
PP0382	Carbon-nitrogen hydrolase family protein	5-Aminopentanamidase; EC 3.5.1.30	[36]
PP0383	Tryptophan 2-monooxygenase, putative	Lysine 2-monooxygenase; EC 1.13.12.2	[36]
PP2336	Aconitate hydratase, putative; EC:4.2.1.3	2-Methylisocitrate dehydratase; EC 4.2.1.99	a
PP2432	Oxygen-insensitive NAD(P)H nitroreductase; EC:1.-.-.	6,7-Dihydropteridine reductase; EC 1.5.1.34	a
PP3591	Malate dehydrogenase, putative; EC:1.1.1.37	Δ^1-Piperideine-2-carboxylate reductase; EC 1.5.1.21	[36]
PP4066	Enoyi-CoA hydratase, putative; EC:4.2.1.17	Methylglutaconyi-CoA hydratase; EC 4.2.1.18	[88]
PP4065	3-Methylcrotonyi-CoA carboxylase, beta subunit, putative EC:6.4.1.3	Methylcrotonoyi-CoA carboxylase; EC 6.4.1.4	[88]
PP4067	AcCoA carboxylase, biotin carboxylase, putative; EC:6.4.1 .3	Methylcrotonoyi-CoA carboxylase; EC 6.4.1.4	[88]
PP4223	Diaminobutyrate-2-oxoglutarate transaminase; EC:2.6.1.76	Putrescine aminotransferase; EC 2.6.1.82	a
PP4481	Acetylornithine aminotransferase; EC:2.6.1.11	Succinylornithine transaminase; EC 2.6.1.81	a
PP5029	Formiminoglutamase; EC:3.5.3.8	N-Formylglutamate deformylase; EC 3.5.1.68	a
PP5036	Atrazine chlorohydrolase	N-Formylglutamate deformylase; EC 3.5.1.68	a
PP5257	Oxidoreductase, FAD-binding	$_\text{L}$-Pipecolate oxidase; EC 1.5.3.7	[36]
PP5258	Aldehyde dehydrogenase family protein; EC:1.2.1.3	$_\text{L}$-Aminoadipate-semialdehyde dehydrogenase; EC 1.2.1.31	[36]

ªAnalysis of the sequence homology and genomic context information.

Comparison of the Predicted and Measured Growth Yields and the Role of Maintenance

After completing the reconstruction, we assessed whether the model was capable of predicting the growth yield of *P. putida*, a basic property of the modeled organism. *In silico* growth yield on succinate was calculated by FBA and compared with *in vivo* growth yield measured in continuous culture [37]. If the *in silico* yield were lower than the experimental, it would indicate that the network may lack important reactions that influence the efficiency of conversion of carbon source into biomass constituents and/ or energy. In fact, the calculated *in silico* yield (0.61 $g_{DW} \cdot g_C^{-1}$) was higher than the experimental yield (0.47 $g_{DW} \cdot g_C^{-1}$), indicating that some of the processes reconstructed in the network might be unrealistically efficient and/or that *P. putida* may be diverting resources into other processes not accounted for in the model. This greater efficiency of the *in silico* model versus *in vivo* growth data is also consistent with recent studies that suggest optimal growth is not necessarily the sole objective (function) of biochemical networks [38, 39].

The *in silico* growth yield is influenced not only by the structure of the metabolic network, but also by other factors including biomass composition and the growth-associated and non-growth-associated energy maintenance factors (GAM and NGAM), the values of which represent energy costs to the cell of "living" and "growing", respectively [22]. Therefore, since both the biomass composition and the GAM/NGAM values were taken from the *E. coli* model [22, 33] due to a lack of organism-specific experimental information, we evaluated the influence of these factors on the predicted growth yield.

First, we analyzed the effects of changes in the ratios of biomass components on the iJP815 growth yield. These analyses indicated that varying any single biomass constituent by 20% up or down has a less than 1% effect on the growth yield of *P. putida*. These results are consistent with results of a previous study on the sensitivity of growth yield to biomass composition [40]. Although it is still possible that some components of *P. putida* biomass are not present in *E. coli* or *vice versa*, we conclude that the use of *E. coli* biomass composition in the *P. putida* model is a justified assumption for the purpose of our application and is probably not a great contributor to the error in our predictions of growth yield.

Subsequently, the effects of changes in the GAM on the *in silico* growth yield were tested. It was found that if GAM was of the same order of magnitude as the value used in the *E. coli* model (13 mmolATPg_{DW}^{-1}), its influence is negligible, as increasing or decreasing it twofold alters the growth yield by merely 5%. A higher GAM value in *P. putida* than in *E. coli* could contribute to the discrepancy between the experimental measurements and *in silico* predictions, but it could not be the only factor unless the *E. coli* and *P. putida* values differ more than twofold, which is unlikely.

Finally, we assessed the effects of changes in the value of NGAM on *in silico* growth yield. The NGAM growth dependency is influenced by the rate of carbon source supply, and thus indirectly by the growth rate. If the carbon intake flux is low (as in the case of the experiments mentioned above, with a dilution rate of 0.05 h^{-1}), the fraction of energy utilized for maintenance purposes is high and therefore so is the influence of the NGAM value on growth yield. Under such low-carbon intake flux conditions, a twofold increase of the NGAM value can decrease the growth yield by about 30%. This indicates that the main cause for the discrepancy between *in vivo* and *in silico* growth yields is that the NGAM value is likely to be higher in *P. putida* than in *E. coli*. Increasing the NGAM value from 7.6 of 12 (mmol$_{ATP}g_{DW}^{-1}$ h^{-1}) would reduce the *in silico* growth yield and lead to a better match with experimental values. Consequently this NGAM value was used in subsequent FBA and FVA [41] simulations.

For a high influx of carbon source the influence of NGAM on the growth yield is low and the influence of the NGAM and GAM values on growth yield are comparable. It should be noted that, while FBA predicts the optimal growth yield, few cellular systems operate at full efficiency. Bacteria tend to "waste" or redirect energy if it is abundant [42], leading to a lower-than-optimal *in vivo* growth yield. It is also worth mentioning that maintenance values may depend on the carbon source used [43] and on environmental conditions [44-46].

Additionally, we computed the growth yields of *P. putida* on sole sources of three other important elements—Nitrogen (N), Phosphorous (P), and Sulfur (S)—and compared these with published experimental data from continuous cultivations [37], as shown in Table 4. Since biomass composition can play a role in the efficiency of *in silico* usage of basic elements, this analysis can aid in assessing how well the biomass equation, which is equivalent to the *E. coli* biomass reaction, reproduces the true biomass composition of *P. putida*. The yield on nitrogen differs only by 10% between *in silico* and *in vivo* experiments, which suggests that the associated metabolic network for nitrogen metabolism is well characterized in the iJP815 reconstruction. The yields on phosphorous and sulfur, however, differ by more than a factor of two between the *in vivo* and *in silico* analyses, suggesting that there may be significant differences between the biomass requirements and the metabolic networks of *P. putida* and *E. coli* for these components. The differences in yields, however, may be also caused by the change of the *in vivo* biomass composition, which decreases the fraction of compounds containing the limited element, when compared to the biomass composition while the bacterium is grown under carbon-limitation. Such changes were observed experimentally in *P. putida* for nitrogen and phosphate limitations [47]. Thus, the biomass composition of *P. putida* needs to be determined precisely in the future. However, for the purpose of this work and since the global effect of the biomass composition on the outcome of the simulations is negligible (as shown above), we considered the use of the original biomass equation to be justified.

Table 4. Comparison of the *in silico* predicted growth yields (in $g_{DW} \cdot g_{Element}^{-1}$) with experimental continuous culture data.

Limiting Element	Yield - Experimental	Yield- Model
C	0.47	0.61
N	5.74	6.67
P	84.95	34.92
S	268.75	130.18

Analysis of Blocked Reactions: The Quest for Completeness

As described above, iJP815 contains 289 unconditionally (i.e., not dependent on external sources) blocked reactions (i.e., reactions unable to function because not all connections are made), corresponding to 33% of the metabolic network. In previously published genome-scale metabolic reconstructions, the fraction of blocked reactions varies between 10 and 70%t [48]. Blocked reactions occur in reconstructions mostly due to knowledge gaps in the metabolic pathways. Accordingly, the blocked-reactions set can be divided into two major groups; (1) reactions with no connection to the set of non-blocked reactions, and (2) reactions that are either directly or indirectly connected to the operating core of the *P. putida* model. The first group of reactions includes members of incomplete pathways that, with increasing knowledge and further model refinement, will gradually become connected to the core. This subset comprises 108 reactions (35% of blocked-reaction set). The second group of reactions comprises also members of incomplete pathways, but many of them belong to pathways that are complete but that lack a transport reaction for the initial or final compound. Examples of pathways lacking a transporter are the degradation of fatty acids and of propanoate.

In addition, there could exist compounds whose production is required only in certain environmental conditions, for example, under solvent stress, and as such are not included in generic biomass equation. Pathways synthesizing compounds that are not included in the biomass equation but that likely are conditionally required include the synthesis of thiamine, various porphyrins and terpenoids. In this case, reactions involved exclusively in the production of such compounds would be blocked if no alternative outlets exist for those pathways. Allowing a non-zero flux through these reactions would require inclusion into biomass of the conditional biomass constituents, which in turn would require having various biomass equations for various conditions. This level of detail, however, is beyond the scope of our initial metabolic reconstruction and investigation.

The high number of blocked reactions in iJP815 clearly indicates that there are still vast knowledge deficits in the model and, thus, in the underlying biochemical and genomic information. Since a genome-scale metabolic model seeks to incorporate all current knowledge of an organism's metabolism, these reactions are integral elements of the metabolic reconstruction and of the modeling scaffold, even if they are not able to directly participate in steady state flux studies. Therefore, the inclusion of these reactions in the model provides a framework to pin-point knowledge gaps, to include novel information as it becomes available and to subsequently study their embedding and function in the metabolic wiring of the cell.

How *P. putida* Allocates its Resources: Evaluating the Prediction of Internal Flux Distributions

The assessment performed as described above by means of high-throughput phenotyping assays, growth experiments, and continuous cultivations, has shown that the model is coherent and that it captures the major metabolic features of *P. putida*. We subsequently used the model to probe the network and to ascertain the distribution of internal fluxes and properties such as network flexibility and redundancy of particular reactions. To this end, we predicted the distribution of reaction fluxes throughout the central pathways of carbon metabolism by FVA, and compared the simulations to internal fluxes computed from experimentally obtained ^{13}C data in *P. putida* [49, 50].

Optimal FVA

Genome-scale metabolic networks are, in general, algebraically underdetermined [41]. As a consequence, the optimal growth rate can often be attained through flux distributions different than the single optimal solution predicted by FBA simulations. Therefore we used FVA to explore the network, as this method provides the intervals inside which the flux can vary without influencing the value of the growth yield (if the flux of the reaction cannot vary then the range is limited to a single value) [41]. The results of the simulations are given in Figure 4. As isotopic (^{13}C) measurements are not able to distinguish which glucose uptake route is being used by *P. putida*, all the fluxes in the ^{13}C experiment and in the FVA simulations were computed assuming that glucose is taken up directly into the cell. For the precise description of the network models used in this comparison (i.e., FBA/FVA vs. ¹³C -Flux analysis).

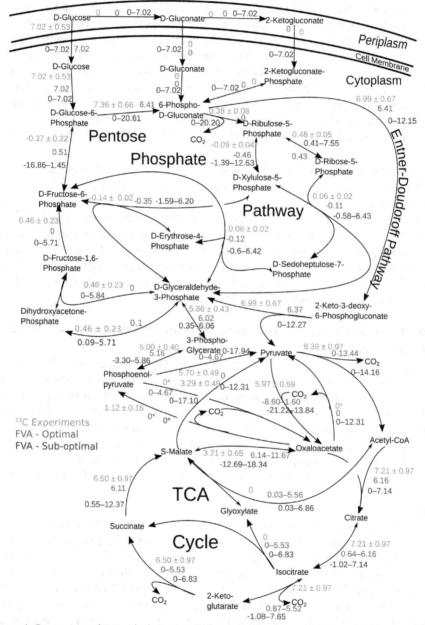

Figure 4. Comparison of FVA calculations with ¹³C experimental flux data. The explanation of color codes is given in the figure. "0*" means that the reaction is not included in the particular metabolic network; double-headed arrows depict reversible reactions, the bigger head shows direction of the positive flux.

Figure 4 shows that the predictions (in red) generally agree well with the measurements (in green) throughout the network, as most of the ^{13}C values fall within the FVA intervals, where intervals were predicted, or both values are close to each other (in absolute values), when a single value was predicted. As *P. putida* lacks phosphofructokinase, glucose can be converted to pyruvate (the entry metabolite of TCA cycle) via the pentose phosphate (PP) or the Entner-Doudoroff (ED) pathways. The ED pathway is energetically more efficient and the ^{13}C measurements indicate that KT2440 uses it preferentially over the PP pathway. Therefore, the FVA yields locally single flux values rather than intervals, which reflects the relative rigidity of this part of the network. In contrast, the energy generating part of the central metabolic network (the TCA cycle and its vicinity) exhibits greater flexibility, as illustrated by the broad flux intervals. First, the conversion of phosphoenylpyruvate into pyruvate can proceed either directly or via oxaloacetate, although the bacterium appears to use the direct route (the [13]C -model assumes, in fact, only the direct route. Second, the conversion of malate to oxaloacetate may also occur directly or via pyruvate. The ^{13}C flux measurements indicate that the bacterium uses the indirect route in addition to the direct one although, according to the FVA, the indirect route is energetically less efficient. Interestingly, our model suggests also that the glyoxylate shunt could be used interchangeably with full TCA-cycle without any penalty on growth yield. However, as the glyoxylate shunt is inactivated in many bacterial species via catabolite repression upon glucose growth [51], it is possible that this alternative is not used in *P. putida*.

Discrepancies Between Model Predictions and Measurements

Despite the general agreement between *in silico* predictions and ^{13}C measurements, there still exist a number of discrepancies. For instance, the [13]C -experiments suggest that the bacterium utilizes the portion of glycolysis between triose-3-phosphate and D-fructose-6-phosphate in the gluconeogenic direction, which is not energetically optimal and as such is not captured in standard FBA (or FVA) simulations. This illustrates one of the possible pitfalls of FBA, which per definition assumes perfect optimality despite the fact that microorganisms might not necessarily allocate their resources towards the optimization function assumed in analysis, and in some cases may not operate optimally at all [52, 53]. Another group of differences concentrates around the pentose phosphate pathway (PPP), although these are relatively minor and are likely due to differences in the quantities of sugar diverted toward biomass in the ^{13}C model versus iJP815. A third group of differences revolves around pyruvate and oxaloacetate, whereby the *in vivo* conversion of malate to oxaloacetate shuttles through a pyruvate intermediate rather than directly converting between the two. The last area where discrepancies exist between *in silico* and ^{13}C data is in the TCA cycle, around which the flux is lower in FVA simulations than in the experiment. This suggests that the *in silico* energetic requirements for growth (maintenance values) are still too low when compared to *in vivo* ones, as the main purpose of the TCA cycle is energy production.

Suboptimal FVA

To investigate further these differences, we carried out a suboptimal FVA (Figure 4, blue values), allowing the production of biomass to range between 90 and 100% of its maximum value. In this suboptimal FVA experiment, the ^{13}C -derived fluxes fall between FVA intervals for every flux value in the ^{13}C network. To filter out artifacts, we re-did all FVA computations using the structure of the network used in the ^{13}C -experiment and found no major differences. We also assessed the influence of the biomass composition on the distribution of internal fluxes and network structure and found that this was negligible on both accounts. The results show that, in principle, the bacterium can use all the alternatives described above and that the penalty on the growth yield is minimal. While this analysis validates the FVA simulation results, the wide breadth of the intervals (i.e., the mean ratio of interval width to mean interval value exceeds three), suggests that the (mathematical) under-determination of central metabolism can be quite high, and indicates that there exist multiple sub-optimal solutions across the network and that is thus difficult to predict exact internal flux and to "pin-point" a particular solution. These results reflect the essence of CB modeling and FBA, which provide only a space of possible flux distributions and not exact values. Therefore, deductions from results of FBA simulations have to be made with great care. This underscores the notion that constraint-based modeling should be seen more as navigation framework to probe and explore networks rather than as an exact predictive tool of cellular metabolism.

Gauging the Robustness of the Network

Essentiality of Genes and Reactions

To assess the robustness of the metabolic network to genetic perturbations (e.g., knock-out mutations), we carried out an *in silico* analysis of the essentiality of single genes and reactions, which enabled us to identify the most fragile nodes of the iJP815 network. Reaction essentiality simulations were performed by systematically removing each reaction from the network and by assessing the ability of the model to produce biomass *in silico* via FBA in minimal medium with a sole carbon source (glucose and acetate). Gene essentiality was assessed by: (i) identifying for each gene the operability of the reaction(s) dependent on this gene, (ii) removing from the network the reactions rendered inoperative by the deletion of that particular gene, and (iii) determining the ability of the model to produce biomass in the same manner as for the reaction essentiality tests. Additionally, we estimated for both carbon sources the smallest possible set of reactions able to sustain *in silico* growth, in order to estimate the number of reactions necessary for biomass synthesis in minimal medium (minimal set). This set encompasses both all reactions that are essential (including those essential regardless of the medium and those "conditionally essential") and the minimal number of non-essential reactions that, together, are able to provide *in silico* growth (see Figure 5). These conditionally essential reactions can be used as a reference for identifying sections of metabolism for which alternative pathways exist. For both glucose and acetate, the minimal sets encompassed approximately 315 reactions. This estimate is consistent with values obtained for other bacteria [54].

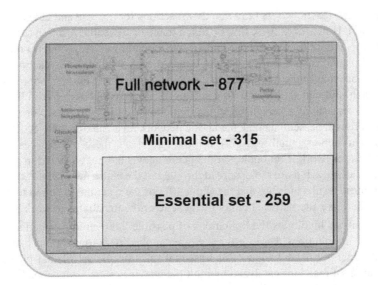

Figure 5. Interdependency between the metabolic network, the minimal set and the set of essential reactions. The set sizes are given for glucose growth conditions.

The sets of essential reactions consists of 259 and 274 reactions for glucose and acetate conditions respectively, constituting 82 and 86% of the minimal set. These numbers indicate that most of the crucial metabolic routes are not duplicated at the level of metabolic network structure. The set of essential reactions under glucose growth is a subset of that under acetate, suggesting that the growth on glucose is more resistant to perturbations (as the smaller number of reactions mean less fragility points in the network). The reactions belonging exclusively to the acetate minimal set are mostly members of glyconeogenic pathway, with ATP synthase, the reactions constituting the glyoxylate shunt, and acetate transport reactions being the exceptions. The inessentiality of ATP synthase under glucose and essentiality of the glyoxylate shunt under acetate conditions are not surprising and similar effects have been reported in *E. coli* [55-57].

The reactions belonging to the non-essential part of the minimal set are mostly members of central metabolic pathways (PPP, TCA cycle, and Pyruvate metabolism), which emphasizes the importance of these pathways for the operation of the metabolism and is in agreement with observations made in other bacteria [58].

Isoenzymes
The metabolic robustness of an organism may also be provided at the genetic level through genes coding for isozymes. Data on gene and reaction essentiality provide insights into this phenomenon. We utilized FBA to generate a list of *in silico* essential gene predictions, including 153 and 159 genes under minimal glucose and acetate growth respectively, in order to determine how gene/pathway redundancy affects network robustness. These values may seem low when compared to the size of the pre-

dicted essential reactions sets (259 and 274 reactions for glucose and acetate growth, respectively). However, it must first be noted that each essential reaction set contains about 25 (26 and 27 for the glucose and acetate essential set, respectively) non-gene-associated reactions, and that elucidating the genes catalyzing these reactions would increase substantially the number of *in silico* essential genes. Further, approximately 20% of each minimal set (78 and 84 genes under glucose and acetate conditions, respectively) consists of essential reactions that can be catalyzed by two or more isozymes and thus are essential at the metabolic network level but not at the genetic level. In contrast to non-essential reactions in the minimal set, these reactions essential at the metabolic network level but not at the genetic level are not clustered in particular metabolic pathways but are rather spread throughout the entire metabolic network. Altogether, these results indicate that for about 40% of the reactions required to produce biomass, there are alternative at either the genetic or the metabolic network level.

This analysis highlights the limitations of possible interventions aimed at reshaping the flux distributions, because these can be applied only to reactions that are not essential (since the inactivation of an essential reaction yields a lethal phenotype). Identification of reactions catalyzed by multiple enzymes shows which reactions may be best avoided when planning mutational strategies as their inactivation may pose additional technical problems, by requiring production of multiple knock-outs.

Flexibility of Flux Distributions
To further investigate these conclusions, we determined the flexibility of fluxes over particular reactions as a measure of metabolic network flexibility during biomass production. We found that the variability of fluxes is similar under either glucose or acetate growth, but that acetate growth instills a slightly higher rigidity to the metabolic network (as observed above). We observed also that the flux of more than a half of the reactions can vary to some degree without influencing biomass output. We next analyzed the pathway-distribution of reactions exhibiting variable flux, and found that biosynthetic pathways are in general more rigid (i.e., the fraction of reactions with flexible flux is relatively lower) than other pathways. This rigidity might reflect the essentiality of these pathways modules for the survival of the cell. A further measure to ascertain network flexibility was the assessment of pairwise couplings between the reactions via flux coupling finder. This analysis indicated that for 90% of the reactions that are unblocked in a given condition, at least one other reaction exists whose flux is proportionally coupled to the flux of the first reaction, and therefore that the great majority of reactions can be inactivated through inactivation of some other reaction. This analysis is helpful in optimizing mutational strategies as it pin-points alternative mutations that exhibit equivalent outcomes.

Prediction of Auxotrophic Mutations and Model Refinement
Assessment of network models through comparison of *in silico* growth-phenotypes with the growth of knock-out strains is a powerful way to validate predictions. This has been done in a number of studies for which knock-out mutant libraries were available [59, 60]. As there is currently no mutant library for *P. putida*, we tested gene knock-out predictions with a set of *P. putida* auxotrophic mutant strains created in our

laboratory that are incapable of growth on minimal medium with acetate as the sole carbon source. First we compared whether the corresponding *in silico* mutants followed the same behavior (lack of growth on minimal medium with acetate, where zero biomass flux during FBA corresponded to a no-growth phenotype). This comparison was performed only for strains whose knocked-out gene is included in iJP815. Thirty-eight out of the 51 strains tested did not grow *in silico*. Of the remaining 13 false positives (i.e., those growing *in silico* but not *in vivo*), four (PP1470, PP1471, PP4679, and PP4680) are mutated in genes considered non-essential *in silico* due to "weakly annotated" gene putatively encoding redundant isozymes. In the case of PP5185 (coding for N-acetylglutamate synthase), its essentiality is removed by PP1346 (coding for bifunctional ornithine acetyltransferase/N-acetylglutamate synthase protein), which is not only an isozyme of PP5185 (the N-acetylglutamate synthase function) but which also catalyses a reaction (ornithine acetyltransferase) that produces N-acetyl-L-glutamate (the product of N-acetylglutamate synthase) and thus renders the activity of PP5185 redundant. It appears either that this is a mis-annotation or that the enzyme is utilized only under different conditions.

In addition, PP0897 (*fumC*) seems to have two paralogues (PP0944, PP1755) coding for isoenzymes of fumarate hydratase, but since the mutant in PP0897 does not grow auxotrophically, they are either non functional or mis-annotated. The enzyme complex that is composed of proteins expressed from the genes knocked-out in the two false positives PP4188 and PP4189 catalyzes the decarboxylation of α-ketoglutarate to succinyl-CoA in the TCA cycle, concurrently producing succinyl-CoA for anabolic purposes. In the model, this functionality is not needed as this part of the TCA cycle can be circumvented by the glyoxylate shunt, whereas succinyl-CoA can be produced by reverse operation of succinate-CoA ligase. Restricting this reaction to be irreversible renders both genes essential. This altogether suggests that either the succinate-CoA ligase is irreversible or the glyoxylate shunt is inactive. The latter solution is, however, impossible, due to the essentiality of the glyoxylate shunt upon growth on acetate.

The false positive PP4782 is involved in thiamine biosynthesis. This cofactor is not included in the biomass, which is why the gene is not *in silico* essential. This suggests thus that the *in-silico P. putida* biomass reaction should be enriched with this cofactor. The remaining false positives (PP1768, PP4909, PP5155) are involved in the serine biosynthesis pathway. We found experimentally that mutants in these genes can grow on acetate if the medium also contains L-serine. These genes can be rendered *in silico* essential by setting glycine hydroxymethyltransferase to operate only unidirectionally from L-serine to glycine. The operation of this enzyme, however, is required for growth of the bacterium on glycine, which is possible; though very slow (results not shown). One of these genes (PP5155) has also a weakly annotated isozyme (PP2335). We found out as well that several of the mutants (PP1612, PP4188-9, PP4191-4) grow *in silico* on glucose, which we confirmed experimentally (results not shown). Altogether, these experimental results assisted us in improving the accuracy of the model.

Albeit limited to a relatively small mutant set, this analysis shows that while CB models are not always able to predict exact flux values, they are very useful in the identification of essential reactions and, through the GPRs, the genes responsible for their catalysis. This enables identification of vulnerable points in the metabolic network.

Model Application—Production of Polyhydroxyalkanoates from Nonalkanoates

To illustrate the utility of a genome-scale model for metabolic engineering, we used iJP815 to predict possible improvements to an industrially relevant process; namely, the production of PHAs from non-alkanoic substrates for biomedical purposes [61-63]. As the production of PHAs uses resources that would be otherwise funneled towards growth, increasing *in silico* PHA production would decrease the growth. Consequently, in classic optimization-based approaches (e.g., FBA), no PHA production would be predicted while optimizing for growth yield. The aim was thus to increase the available pool of the main precursor of PHAs—Acetyl Coenzyme A (AcCoA). This approach was based on the observation that inactivation of isocitrate lyase (ICL) enhances the production of PHAs in *P. putida* due to increased availability of AcCoA that is not consumed by ICL [64]. We therefore searched for other possible intervention points (mutations) in the metabolic network that could lead to the accumulation of AcCoA. This analysis was performed through application of a modified OptKnock approach [28], which allowed for parallel prediction of mutations and carbon source(s) that together provide the highest production of the compound of interest.

Two main methods were employed to model a cellular pooling of AcCoA. The first was the maximization of AcCoA production by pyruvate dehydrogenase (PDH). In the second, an auxiliary reaction was introduced that consumed AcCoA (concurrently producing CoA, to avoid cofactor cycling artifacts) and that would represent the pooling of AcCoA (Figures 6A and B, insets). It is noteworthy that the value of "AcCoA production" predicted by the first method includes AcCoA that is then consumed in other reactions (some of which will lead towards biomass production for instance), whereas the value of "AcCoA pooling" predicted by the second method includes only AcCoA that is taken completely out of the system, and therefore made available for PHA production but unusable for growth or other purposes. Therefore, only with the first method (AcCoA production) can AcCoA fluxes and growth rates be compared directly with the wild type AcCoA flux and growth rate, as the second method (AcCoA pooling) will display lower values for AcCoA fluxes and growth rates but will avoid "double counting" AcCoA flux that is shuttled towards growth, and therefore is not available for PHA production (see plots in Figures 6A and B).

To create the *in silico* mutants, we allowed the OptKnock procedure to block a maximum of two reactions, which corresponds, experimentally, to the creation of a double mutant. To avoid lethal *in silico* strains, the minimal growth yield was limited to a value ranging between 0.83 and 6.67 $g_{DW}mol_C^{-1}$, corresponding to about 5 and 40% of maximum growth yield, respectively.

Six mutational strategies suggested by this approach are presented in Table 5. The first three were generated by the AcCoA production method, and the last three were generated by the AcCoA pooling method. The results provide a range of options for possibly increasing AcCoA production, some of which constrain growth more than others (see Figures 6A and B).

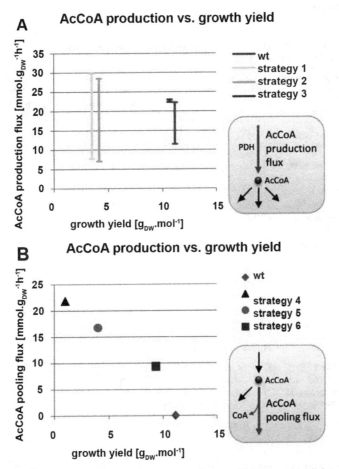

Figure 6. Mutational strategies for increased PHA production. This figure highlights six strategies suggested by the modified Optknock approach for increased production of AcCoA, a precursor for polyhydroxyalkanoates. (A) The AcCoA production ranges versus growth yield of *in silico* strains developed using the "AcCoA production" strategy. (B) The AcCoA pooling versus growth yield of *in silico* strains developed using the "AcCoA pooling" strategy.

One promising hypothesis (strategy 2) generated by the AcCoA production method predicted that a double-mutant devoid of 6-phosphogluconolactonase (pgl/PP1023) and periplasmatic glucose dehydrogenase (gcd/PP1444), would produce 29% more AcCoA than the wild type growing on glucose as a carbon source (Figure 6A). As we are currently still in the process of generating this mutant, we were not yet able to test the prediction. Another promising hypothesis (strategy 1) included knocking-out triose phosphate isomerase (tpiA/PP4715). As the mutant for tpiA was generated in this work, we tested whether it is able to grow on the predicted carbon source (D-fructose), but the observed growth was very weak (only very small colonies grew on agar plates after 3 days). This suggests that growth might be too inhibited by this strategy for it to be of great use.

Table 5. Summary of the characteristics of the *in silico* strains generated in the procedure of optimization of the PHA production.

Strain	Blocked Enzymatic Activity	Loci To Be Blocked	Carbon Source(s)	AcCoA Production [mmol gow⁻¹.h⁻¹]		Growth Yield [gow·mol꜀⁻¹]	
				Min	Max	Limit	Sim
WT	WT	WT	ₗ-Serine	11.47	22.26	0.83	11.16
1	Triose-phosphate isomerase	PP4715	ᴅ-Fructose	7.7	29.74	0.83	3.5
	6-Phosphoglucono lactonase	PP1023					
2	Glucose dehydrogenase (membrane)	PP1444	ᴅ-Glucose	7.05	28.51	0.83	4.17
	6-Phosphoglucono lactonase	PP1023					
3	Isocitrate dehydrogenase	PP4011 or PP4012	ₗ-Serine	22.41	23.01	6.66	10.67
	Formate dehydrogenase	PP0490 or PP0491					
		PP2183 or PP21 84 or PP2185 or PP2186				0.83	1.00
4	Citrate synthase	PP4194	ₗ-V aline	21.85		3.33	4.00
	2-Methylcitrate dehydratase	PP2338					
5	Glycine hydroxymethyl transferase	PP0322	ₗ-Leucine, ₗ-lysine, ₗ-phenylalanine	16.75			
		PP0671					
	Citrate synthase	PP4194					
6	Glycine hydroxymethyl transferase	PP0322	ₗ-Leucine, ₗ-isoleucine	9.35		6.66	9.33
		PP0671					
	Citrate synthase	PP4194					

One strategy suggested by the AcCoA pooling method (strategy 4) called for knocking out 2-methylcitrate dehydratase (prpD/PP2338) and citrate synthase (gltA/PP4194), and supplying *P. putida* with valine. Using this strategy, AcCoA pooling could theoretically reach 21.9 mmolg$_{DW}$⁻¹ h⁻¹, but at a severe expense in bacterial growth (Figure 6B). The other strategies suggested by the AcCoA pooling method highlight a somewhat linear tradeoff between growth and AcCoA pooling, which could be investigated experimentally to determine how much growth disruption is acceptable in a bioengineered production strain of *P. putida* (Figure 6B).

These strategies illustrate the possible approaches to optimizing production of a non-growth-associated compound, and highlight the need for further experimental work to assess the performance of this approach.

DISCUSSION

A primary value of genome-scale metabolic models is their ability to provide a holistic view of metabolism allowing, for instance, for quantitative investigation of dependencies between species existing far apart in the metabolic network [20]. Once experimentally validated, these models can be used to characterize metabolic resource allocation, to generate experimentally testable predictions of cell phenotype, to elucidate metabolic network evolution scenarios, and to design experiments that most effectively reveal genotype–phenotype relationships. Furthermore, owing to their genome-wide scale, these models enable systematic assessment of how perturbations in the metabolic network affect the organism as a whole, such as in determining lethality of mutations or predicting the effects of nutrient limitations. Since these multiple and intertwined relationships are not immediately obvious without genome-scale analysis, they would not be found during investigation of small, isolated circuits or genes as is typical in a traditional reductionist approach [65, 66].

We present here a genome-scale reconstruction and CB model of the *P. putida* strain KT2440, accounting for 815 genes whose products correspond to 877 reactions and connect 886 metabolites. The manually curated reconstruction was based on the most up-to-date annotation of the bacterium, the content of various biological databases, primary research publications and specifically designed functional genomics experiments. New or refined annotations for many genes were suggested during the reconstruction process. The model was validated with a series of experimental sets, including continuous culture data, BIOLOG substrate utilization assays, ^{13}C flux measurements and a set of specifically-generated mutant strains. The FBA and FVA were used to ascertain the distribution of resources in KT2440, to systematically assess gene and reaction essentiality and to gauge the robustness of the metabolic network. Hence, this work represents one of the most thorough sets of analyses thus far performed for an organism by means of CB modeling, providing thereby a solid genome-scale framework for the exploration of the metabolism of this fascinating and versatile bacterium. However, since this modeling endeavor relies upon a number of approximations, the limits, potential and applicability of the analysis must be clearly identified and defined. We address these points below.

Altogether, our results and analyses show that the model accurately captures a substantial fraction of the metabolic functions of *P. putida* KT2440. Therefore, the model was used to generate hypotheses on constraining and redirecting fluxes towards the improvement of production of PHAs, which are precursors for industrially and medically important bioplastics. This is, to our knowledge, the first reported application of CB modeling to direct and improve the yield of a compound of which the production is not directly coupled to the growth of the organism. This opens up novel areas of application for the CB approach. Our approach, based on the OptKnock algorithm, allows for both prediction of mutants with desirable properties and identification of conditions that support the expression of these properties.

Notwithstanding the generally good agreement between experimental results and simulations of our model, several of the discrepancies encountered reflect pitfalls inherent to CB modeling that go beyond the scope of our study:

First, the high number of blocked reactions and the mismatches with the BIOLOG data show that there are still many areas of the metabolism that require thorough exploration. The genes encoding transport-related are particularly relevant, as for most of them, neither the translocated compound nor the mechanism of translocation is known. Furthermore, it should be highlighted that the genome still has 1,635 genes annotated as "hypothetical" or "conserved hypothetical", more than 800 genes annotated as putative, and over 800 for which the functional annotation gives no information beyond the protein family name. It is thus likely that a fraction of the hypothetical and non-specifically annotated genes in the current *P. putida* annotation are responsible for unknown metabolic or transport processes, or that some might code for proteins that add redundancy to known pathways. This observation is common to all genomes sequenced so far and illustrates a major hurdle in the model building process (and hence, its usefulness) that can be overcome only through extensive studies in functional genomics.

Second, although we carefully constrained the *in silico* flux space through FBA and FVA and obtained distribution spaces roughly consistent with those experimentally determined via ^{13}C- flux analysis, these approaches are inherently limited as they assume growth as a sole metabolic objective and ignore any effects not explicitly represented in a CB metabolic model. It has been shown that FBA using objective functions other than growth can improve predictive accuracy under certain conditions [53]. Kinetic limitations also may play a very important role in determining the extent to which a particular reaction or pathway is used. Teusink et al. [52] showed that in the case of *L. plantarum* these factors may lead to false predictions.

Third, the reconstruction includes causal relationships between genes and reactions via GPRs but it lacks explicit information regarding gene regulation. The regulation of gene expression causes that there are many genes in the cell that are expressed only under certain growth conditions. Therefore, the *in silico* flux space is generally larger than the true *in vivo* flux space of the metabolic network. This, in turn, may influence the robustness of the metabolic network and the essentiality of some reactions and genes. The lack of regulatory information and of the genetic interactions involved is likely to be one of the causes for faulty predictions of the viability of mutant strains. Adding this information will be an important step in the further development and improvement of the accuracy of the reconstruction.

Fourth, although our analyses indicated that growth yield is relatively insensitive to changes in biomass composition, these analyses also suggest that factors other than the structure of the metabolic network play an important role in defining the relationship between the growth yield and environmental conditions. The prediction of the exact growth yield requires the precise measurement of maintenance values, which may vary substantially from one condition to the other [44-46]. As the maintenance accounts for 10–30% of the total carbon source provided in unstressed conditions, this may set a limit to the accuracy of the growth yield predictions.

To enhance the usefulness and predictiveness of the model, several avenues could be followed in the future. First, additional constraints can be overlaid on the network to reduce the space of possibilities and increase the accuracy of predictions. In addition to specific knowledge of particular enzymatic or transport processes, such constraints are best based on high-throughput experimental evidence such as transcriptomic and proteomic data, which are instrumental in expanding genotype–phenotype relationships in the context of genome-scale metabolic models [67]. Microarray experiments have guided the discovery of metabolic regulons, and usage of microarray and proteomic data to constrain metabolic models has improved model accuracy for other systems [23]. Second, *P. putida* provides a good opportunity for incorporating kinetic information into a genome-scale model as there are various kinetic models available and under development for small circuits in *P. putida* [68-71]. Incorporating data from these models into the genome-scale reconstruction would provide insights into the relationships of isolated metabolic subsystems within the global metabolism. This synthesis would also improve the flux predictions of the global model, particularly in areas where current FBA-based predictions methods fail due to their inherent limitations.

Experimental validation of a genome-scale model is an iterative process that is performed continuously as a model is refined and improved through novel information and validation rounds. In this work, we have globally validated iJP815 as well as specific parts thereof by using both up-to-date publicly available data and data generated in our lab, but there will be always parts of the model that include blocked reactions and pathways that will require further, specific validation. As more knowledge becomes available from the joint efforts of the large *P. putida* community (e.g., [http://www.psysmo.org]), focus will be put on these low-knowledge areas for future experimental endeavors. We anticipate that this model will be of valuable assistance to those efforts.

The metabolic reconstruction, the subsequent mathematical computation and the experimental validation reported here provide a sound framework to explore the metabolic capabilities of this versatile bacterium, thereby yielding valuable insights into the genotype–phenotype relationships governing its metabolism and contributing to our ability to exploit the biotechnological potential of pseudomonads. By providing the means to examine all aspects of metabolism, an iterative modeling process can generate logical hypotheses and identify conditions (such as regulatory events or conditional expression of cellular functions) that would reconcile disagreements between experimental observations and simulation results. Through a detailed *in silico* analysis of PHA production, we show how central metabolic precursors of a compound of interest not directly coupled to the organism's growth function might be increased via modification of global flux patterns. Furthermore, as the species *Pseudomonas putida* encompasses strains with a wide range of metabolic features and numerous isolates with unique phenotypes, the reconstruction presented provides a basic scaffold upon which future models of other *P. putida* strains can be built with the addition or subtraction of strain-specific metabolic pathways. Due to its applicability across the numerous *P. putida* strains iJP815 provides a sound basis for many future studies towards the elucidation of habitat-specific features, bioremediation applications and metabolic engineering strategies with members of this ubiquitous, metabolically versatile, and fascinating genus.

MATERIALS AND METHODS

Constraint-based Models

The *P. putida* model we present was built using a CB approach. A CB model consists of a genome-wide stoichiometric reconstruction of metabolism and a set of constraints on the fluxes of reactions in the system [19, 20, 24]. The reconstruction represents stoichiometry of the set of all reactions known to act in metabolism of the organism, which can be determined in large part from genomic data since most cellular reactions are catalyzed by enzymes. Thus the model does not require any knowledge regarding the kinetics of the reactions, and the requisite thermodynamic knowledge is limited to the directionality of reactions.

In addition to the reactions, the model includes a set of genes tied via Boolean logic to reactions that their protein products catalyze, which allows for accurate discrimination of the effects of genetic perturbations such as knockouts [33, 72]. These Boolean rules together form the GPRs relationships of the metabolic reconstruction [33].

The second part of the CB model, namely the constraints, constitutes a set of rules that narrow down the interval within which the flux of particular reaction must lie. These constraints rest upon physico-biological knowledge. One of them, the information regarding reaction directionality, has already been mentioned above. Another constraint that is widely applied in biological systems is the PSSA [73], which states that a concentration of a chemical compound stays constant over the simulated time frame. The reactants to which this constraint is applied are usually called internal compounds, and in biological models correspond to the chemical substances located inside the cell or its compartments. Remaining substances, external compounds, correspond to species that can be taken up or secreted and thus exchanged with the environment. Other types of constraints are top and bottom limits that correspond to catalytic capabilities of the enzymes. More detailed description of constraint-based modeling approach can be found in [74].

Analysis Methods

Flux Balance Analysis

The FBA is a primary method for analysis of CB models. Generally, a constraint-based model of metabolism represents an underdetermined system, that is, one in which a range of flux distributions are mathematically possible. The FBA narrows the flux possibilities by determining a point in closed flux space that maximizes a certain linear combination of fluxes. [75]. The FBA poses a linear programming (LP) problem and thus a global maximum always exists, provided that the problem is feasible (i.e., there exists at least one combination of fluxes which fulfills all the constraints). Using the matrix notation the FBA problem can be stated as following:

$$\text{maximize}: c^\tau \bullet v$$

$$\text{subject to}: S_i \bullet v = 0$$

$$v_{min} \leq v \leq v_{max},$$

where S is the stoichiometric matrix containing reaction stoichiometry information, v is a vector of all reaction fluxes in the system, v_{min} and v_{max} represent minimum and maximum constraints on reaction fluxes, respectively, and c^T is a vector containing coefficients for each flux that is to be maximized (for more detail on FBA, refer to [76]).

The FBA optimization yields an optimal value for the objective along with a flux value for every reaction belonging to the metabolic network. Commonly, FBA is used to predict maximal growth or metabolite production yields. Cell growth is simulated by the flux over a special "Biomass" reaction that consumes precursors of cellular components (amino acids, lipids, dNTPs, NTPs, cofactors) and produces a virtual unit of cell biomass. Maximization of this flux is usually set as the FBA objective. This procedure assumes that organisms have been shaped by the evolution towards growth maximization, an assumption that has been validated under a variety of conditions [77].

Flux Variability Analysis

Metabolic networks of living organisms are usually considerably underdetermined [78-80]. The size of the mathematically allowed flux space can vary depending upon the network structure and the constraints. The FVA is a method that allows for rough top estimation of the flux space for a given FBA optimization [41]. The FVA computes for each reaction an interval of values inside of which the flux of the reaction can change without influencing value of the objective function, provided that other fluxes are allowed to vary freely within their constraints.

It is often the case that cells do not operate perfectly optimally when FBA simulations are compared to real data. Therefore, a variant of the FVA approach called suboptimal FVA [41] is sometimes informative, wherein instead of fixing the objective to its optimal value from the initial FBA run (as in standard FVA), the objective value is allowed to vary within a predetermined limit. For every suboptimal FVA presented in this chapter the objective lower limit was chosen at 90% of the initial objective value (assuming that FBA maximized the objective).

OptKnock

The OptKnock is an approach for identification of mutations that selectively increase production of a certain compound of interest, assuming that the mutant would optimize for the same quantity as the wild type (e.g., growth yield) [28]. OptKnock points out reactions (and genes, through GPR logic) that must be blocked in order to maximize a linear combination of target fluxes (outer objective) while simultaneously maximizing for the cell's assumed objective (growth yield; inner objective). OptKnock poses a bi-level optimization approach that is solved via mixed-integer linear programming (MILP).

OptKnock—modification.

In order to enable the choice of the carbon source(s) the original OptKnock procedure was modified as follows:

1. A virtual reaction, with limited flux, was created that sourced the virtual compound "vcarbon"

2. For each carbon source a virtual irreversible reaction that converted the compound "vcarbon" into the respective carbon source was added to the model. The stoichiometry of this virtual reaction corresponded to the number of carbon atoms in the carbon source, for example:

$$6 \text{ vcarbon} \rightarrow \text{D-glucose.}$$

3. For each of those reactions (v_j) a binary variable (z_j) defining its activity was created and following constraint was added to the model: $v_j \leq v_j^{max} \, z_j$, where the v_j^{max} was set to value high enough, so that the whole "vcarbon" could be consumed by each reaction.

This modification allows for the choice of one or more carbon sources that, together with the mutation set identified by OptKnock, provide the highest objective.

Identification of Minimal Growing Reaction Set

The minimal growing set was identified using a MILP approach, by modifying original FBA LP problem. For every non-blocked and non-essential reaction a binary variable was added that reflects the activity of the reaction. When the binary variable takes value of 1 the corresponding reaction is virtually unlimited (or limited by rules of original LP problem). When the variable is set to 0 the corresponding reaction is blocked (non-zero flux is impossible). This was achieved by adding a following set of equations to the original LP problem:

$$-y_i \bullet v_i^{\lim} \leq v_i \leq y_i \bullet v_i^{\lim}$$

for reversible reactions, and

$$v_i \leq y_i \bullet v_i^{\lim}$$

for irreversible reactions. In order to assure that growth was not overly restricted, a minimal flux value was established for the biomass reaction. We set the lower limit on biomass flux to 0.05 when the supply of carbon source was 60 $\text{mmol}_{C} \cdot \text{g}_{DW}^{-1} \text{h}^{-1}$, which corresponds to growth yield of 0.07 $\text{g}_{DW} \cdot \text{g}_{C}^{-1}$, 16 times lower than the wild type. The objective of the problem was set to minimize the sum of all binary variables y_i:

$$\text{minimize} \sum_i y_i.$$

This method searches for a minimal set that is able to sustain growth greater than or equal to to the minimal growth requirement.

Metabolic Network Reconstruction

The main sources of information regarding the composition of the metabolic network of *Pseudomonas putida* KT2440 were various biological databases. Most of the information came from the KEGG [35, 81] and Pseudomonas Genome Database (PGD) [82]. Information regarding *P. putida* contained in these two databases is mainly based on the published genome annotation of the bacterium [14], so there is a large overlap between them. Additionally, substantial information was taken from the BRENDA database, which catalogs reaction and enzyme information [83]. This all was augmented with knowledge coming directly from primary research publications. The reconstruction process was performed in an iterative manner, that is, by adding or removing reactions from the model in between rounds of model testing. First, reaction information for *P. putida* was collected from KEGG and PGD. Reactions supported by sufficient evidence and with specific enough functional annotations were incorporated into the model. For every accepted reaction its reversibility was assessed basing on assignments in KEGG pathways as well as information from BRENDA database. For reactions with inconsistent assignments a decision about reversibility was made basing on analysis of the reaction as well as its reversibility in other organisms. Hereby, a first version of the metabolic model was created (iJP815^{pre1}).

The next step involved assessing whether the reconstructed metabolic network is able to produce energy from glucose. This was achieved by running FBA with ATP production set as the objective function. Subsequently, the ability of the model to grow *in silico* on glucose was tested. Successful *in silico* growth indicates that every chemical compound belonging to the biomass equation can be synthesized from present sources, using the reactions contained in the model. Since the exact cellular composition of *P. putida* is not known, the composition of *E. coli* biomass was used as an approximation. This test was performed by running FBA with production of each biomass constituent set as the objective. If a compound could not be synthesized, the gaps in the pathway leading to it were identified manually and a search was performed for reactions that could fill the gaps. If this approach was unsuccessful, gaps were filled with reactions from the *E. coli* model. This yielded the second version of the reconstruction (iJP815^{pre2}).

The third round of reconstruction consisted of two sub-steps. First, the compounds for which transport proteins exist were identified and appropriate reactions added. Second, the results of BIOLOG carbon-source utilization experiments were compared with *in silico* simulations for growth on those compounds. It was assumed that the ability to grow *in silico* on the particular compound as the sole carbon source approximates the *in vivo* utilization. For those compounds that did not show *in silico* growth, a literature search was performed in order to identify possible pathways of utilization. The results of this search, in the form of reactions and GPRs, were added to the model. The outcome was the final version of the model (iJP815).

Comparison of Growth Yields with the Continuous Culture Experiments

Growth yields on sources of basic elements (C,N,P,S) were compared with experimental values obtained by Duetz et al. [37]. The yields of the model were computed using

FBA, by setting the growth rate to the value of the dilution rate used in experiments and subsequently minimizing for consumption of source of respective element (succinate, ammonia, phosphate, and sulfate).

Computational Methods

The model was created and maintained using ToBiN (Toolbox for Biochemical Networks, http://www.lifewizz.com). The optimizations (FBA, FVA, OptKnock) were computed by free, open source, solvers from the COIN-OR family (COmputational INfrastructure for Operations Research, http://www.coin-or.org) or by the lp_solve version 5.5 (http://lpsolve.sourceforge.net/5.5/) software package. All computations were performed on a Personal Computer with a Intel Core 2 2.40 GHz CPU and 2GB of RAM.

Experimental Methods

Media and Chemicals

Pseudomonas putida KT2440 was grown either on EM-medium (Bacto Trypton – 20 g, Yeast-Extract – 5 g, NaCl – 5 g, Glucose 0.5%, H_2O_{dist} at 1,000 ml; the glucose was as 10% solution autoclaved separately and added in appropriate amount) or SOC-medium (Bacto Trypton – 2%, Yeast-extract – 0.5%, Glucose – 20 nM, NaCl – 10 mM, KCl – 2.5 mM, $MgCl_2$ – 10 mM, $MgSO_4$ – 10 mM, H_2O_{dist} ad 1,000 ml; magnesium salts were autoclaved separately and subsequently merged with the remaining components) or minimal medium (10×; Na_2HPO_4 – 50 g, KH_2PO_4 – 100 g, $MgSO_4 \times 7H_2O$ – 2 g, $(NH_4)_2SO_4$ – 20 g, $CaCl_2$ – 0.01 g, $FeSO_4 \times 7H_2O$ – 0.01 g, H_2O_{dist} ad 1,000 ml; the potassium and sodium salts were dissolved separately and subsequently mixed with other dissolved salts; pH was set to 7.0 by adding 10 mM NaOH) with different compounds as the sole carbon source.

BIOLOG Substrate Utilization Experiments

Pseudomonas putida KT2440 was tested for its ability to utilize various carbon sources using BIOLOG GN2 Microplates [31] (BIOLOG Inc. Hayward, CA, USA). All procedures were performed as indicated by the manufacturer. Bacteria were grown overnight in 28°C on a BIOLOG Universal Growth agar plate. Afterwards they were swabbed from the surface of the plate and suspended in GN inoculating fluid. Each well of the Microplate was inoculated with 150 µl of bacterial suspension and the plate was incubated in 28°C for 24 hr. Subsequently the plate was read by a microplate reader and the read-outs were analyzed with MicroLog3 4.20 software.

Growth Experiments

If not stated differently, cells were grown on agar plates overnight in 30°C.

Transposon Mutagenesis

The mutants of *P. putida* were created using an *in vitro* transposition system (Epicentre Technologies, Madison, Wisconsin, USA) [84]. This system bases on a hyper-reactive Tn5-transposase and Tn5-Transposome that, in the absence of magnesium ions, builds a stable synaptic complex, which can be transmitted into the cell via electroporation.

To render Pseudomonas putida KT2440 electrocompetent, cells were grown in 50 ml of EM-medium to OD600 of 0.6 to 1.0 and subsequently cooled on ice for 15 min. The cells were centrifuged (4,000 g, 4°C) and washed twice with H_2O_{dist}. The cells were washed twice in 0.3 M cold solution of sucrose and resuspended in 0.5–1.0 ml of 0.3 M sucrose solution. The electrocompetent cell were used for transformation by electroporation with Gene Pulser (BioRad, Munich, Germany) using the EZ:TN <Kan-2> Tnp Transposome. The 20–40 µl of cells was mixed with 1–2 µl of DNA in ice-cooled cuvette. The electroporation setting were 25 µF, 200 Ω, and 1.7 or 2.5 for the gap size 0.1 and 0.2 cm, respectively. After 2 hr of incubation in SOC-medium, transformants were selected on EM agar plates with 60 µg/ml of kanamycin. Selection of auxotrophic mutants was performed on minimal medium with acetate as the sole carbon source, by replica-plating P. putida KT2440::Tn5(Kan^r) strains on the minimal and EM media.

Identification of Flanking Sequences

The auxotrophic P. putida KT2440::Tn5(Kan^r) mutants were genotyped by enrichment of either flanking sequences of transposon insertions using PCR [85, 86]. Two rounds of amplifications were performed using primers specific to the ends of transposons and random primers that can anneal to the chromosome. In the first round of amplification the Kan-2 RP1 (5′-GCAATGTAACATCAGAGATTTTGAG-3′) primer complementary to the end of Tn5-element and the arbitrary primer ARB1 (5′-GGCCAC-GCGTCGACTAGTACNNNNNNNNNNNGATAT-3′) were used. A 1 µl of supernatant from a P. putida KT2440 lysate was used as the DNA-template. The PCR-reaction was performed in following mixture (H_2O_{dist} – 28.7 µl, incubation buffer(10×) – 5 µl, dNTPs(5 µM) – 5 µl, primer(10 µM) – 2,5 µl, Taq DNA-polymerase (5U/µl) – 0.2 µl) under following conditions: (i) 5 m at 95°C, (ii) 30 × (30 s at 30°C, 90 s at 72°C), (iii) 30 × (30 s at 95°C, 30 s at 45°C 120 s at 72°C). In the second round of amplification a 5 µl of product of the first PCR-reaction was used as the DNA-template, together with the primers TnINT Rev (5′-GAGACACAATTCATCGATGGTTAGTG-3′) and ARB-2 (5′-GGCCACGCGTCGACTAGTAC-3′). The reaction conditions were following: 30 × (30 s at 95°C, 30 s at 45°C, 120 s at 72°C). The PCR-products were purified with "QIAquick- spin PCR Purification Kit" (Qiagen GmbH, Hilden, Germany) according to manufacturer's instructions. Subsequently, the sequencing procedure was performed. The 200–500 µg of dsDNA in normal sequencing vectors (pBlueskript, pUC18, etc.) with 10 pmol of primer (TnINT Rev) and 6 µl of "Big Dye Terminator v. 2.0 Ready Reaction Mix" were mixed in total volume of 10 µl. The conditions of the reaction were following: 25 × (30 s at 95°C, 30 s at 60°C, 4 m at 60°C]. After the cycle sequencing the remaining dNTP were removed using "Dye Ex Spin Kit" (Qiagen GmbH, Hilden, Germany) according to manufacturer's instructions. To the purified product 50 µl sterile MiliQ-H2O was added and the DNA was precipitated wit 250 µl Ethanol (100% v/v) for 30 min at 16,000 × g in the room temperature. The supernatant was removed and the pellet washed with 250 µl of ethanol (100% v/v), precipitated again by centrifugation (16,000 × g, RT, 10 min) and dried in vacuum-centrifuge. All the DNA-pellets were stored in −20°C in 20 µl Hi-Di Formamide (PE Biosystems) until sequencing. The sequencing was performed with ABI PRISM 377 sequencer [87].

The fluorescence signals were analyzed with ABI PRISM 3100 genetic analyzer and the obtained sequences compared with *P. putida* KT2440 genome sequence.

CONCLUSION

The pseudomonads include a diverse set of bacteria whose metabolic versatility and genetic plasticity have enabled their survival in a broad range of environments. Many members of this family are able to either degrade toxic compounds or to efficiently produce high value compounds and are therefore of interest for both bioremediation and bulk chemical production. To better understand the growth and metabolism of these bacteria, we developed a large-scale mathematical model of the metabolism of *Pseudomonas putida*, a representative of the industrially relevant pseudomonads. The model was initially expanded and validated with substrate utilization data and carbon-tracking data. Next, the model was used to identify key features of metabolism such as growth yield, internal distribution of resources, and network robustness. We then used the model to predict novel strategies for the production of precursors for bioplastics of medical and industrial relevance. Such an integrated computational and experimental approach can be used to study its metabolism and to explore the potential of other industrially and environmentally important microorganisms.

KEYWORDS

- **Constraint-based model**
- **Flux balance analysis**
- **Flux variability analysis**
- **Gene-protein-reaction relationship**
- **Genotype–phenotype relationships**
- ***Pseudomonas putida***

AUTHORS' CONTRIBUTIONS

Conceived and designed the experiments: Jacek Puchałka, Matthew A. Oberhardt, Kenneth N. Timmis, Jason A. Papin, and Vítor A. P. Martins dos Santos. Analyzed the data: Jacek Puchałka and Matthew A. Oberhardt. Wrote the chapter: Jacek Puchałka, Matthew A. Oberhardt, Jason A. Papin, and Vítor A. P. Martins dos Santos. Performed the computational experiments: Jacek Puchałka. Developed the computational platform: Miguel Godinho. Characterized the mutants and carried out wet lab experiments: Agata Bielecka. Produced the mutants: Daniela Regenhardt. Contributed to data interpretation: Kenneth N. Timmis.

ACKNOWLEDGMENTS

We thank Victor de Lorenzo (CSIC, Madrid) and Antoine Danchin (Institute Pasteur, Paris) for their thoughtful comments and valuable contributions to this study. We thank Piotr Bielecki (HZI, Braunschweig) for the help in planning and experimental procedures.

Chapter 11

A New Cold-adapted β-D-galactosidase from the Antarctic *Arthrobacter* Sp. 32c

Piotr Hildebrandt, Marta Wanarska, and Jyzef Kur

INTRODUCTION

The development of a new cold-active β-D-galactosidases and microorganisms that efficiently ferment lactose is of high biotechnological interest, particularly for lactose removal in milk and dairy products at low temperatures and for cheese whey bioremediation processes with simultaneous bio-ethanol production.

In this article, we present a new β-D-galactosidase as a candidate to be applied in the above mentioned biotechnological processes. The gene encoding this β-D-galactosidase has been isolated from the genomic DNA library of Antarctic bacterium *Arthrobacter* sp. 32c, sequenced, cloned, expressed in *Escherichia coli* and *Pichia pastoris*, purified and characterized. 27 mg of β-D-galactosidase was purified from 1 l of culture with the use of an intracellular *E. coli* expression system. The protein was also produced extracellularly by *P. pastoris* in high amounts giving approximately 137 mg and 97 mg of purified enzyme from 1 l of *P. pastoris* culture for the AOX1 and a constitutive system, respectively. The enzyme was purified to electrophoretic homogeneity by using either one step- or a fast two step-procedure including protein precipitation and affinity chromatography. The enzyme was found to be active as a homotrimeric protein consisting of 695 amino acid residues in each monomer. Although, the maximum activity of the enzyme was determined at pH 6.5 and 50°C, 60% of the maximum activity of the enzyme was determined at 25°C and 15% of the maximum activity was detected at 0°C.

The properties of *Arthrobacter* sp. 32cβ-D-galactosidase suggest that this enzyme could be useful for low-cost, industrial conversion of lactose into galactose and glucose in milk products and could be an interesting alternative for the production of ethanol from lactose-based feedstock.

Nowadays low-cost energy bio-industrial processes in biotechnology are highly desired. This has led to increased interest in the production of cold adapted enzymes. One class of such enzymes includes cold-adapted β-D-galactosidases (EC 3.2.1.23) that can find many applications in industrial biotechnology. These enzymes are capable of hydrolyzing 1,4-β-D-galactoside linkages and can sometimes catalyze the synthesis of oligosaccharides. The production of lactose-free milk and synthetic oligosaccharides like lactulose are only examples of this cutting edge enzyme class application.

Currently, commercially available β-galactosidase preparations (e.g., Lactozym-Novo Nordisk, MaxilactDSM Food Specialties) applied for lactose hydrolysis contain

Kluyveromyces lactis β-galactosidase naturally intracellularly biosynthesized by *K. lactis* strains. This enzyme is optimally active at approximately 50°C and displays low activity at 20°C while an ideal enzyme for treating milk should work well at 4–8°C. Besides, the latter enzyme should be optimally active at pH 6.7–6.8 and cannot be inhibited by sodium, calcium, or glucose. Such β-galactosidases are still highly desired. Only several enzymes optimally hydrolyzing lactose at low temperatures have been characterized till now [1-14], however, none of them have been produced on the commercial scale. The β-galactosidases were obtained from different microbial sources, including those from *Arthrobacter* sp. [1, 2, 7, 8, 12], *Arthrobacter psychrolactophilus* [9, 13] *Carnobacterium piscicola* [3], *Planococcus* sp. [4, 14], *Pseudoalteromonas haloplanktis* [5], and *Pseudoalteromonas* sp. [10, 11].

Additionally, in order to make progress in cheaper production of β-D-galactosidases of industrial interest, high efficiency yeast expression systems must be taken into consideration. On the other hand extracellular production must occur to allow easy and fast isolation of target protein. There are several studies in literature related to the extracellular production of the *Aspergillus niger* β-galactosidase by recombinant *Saccharomyces cerevisiae* strains [15-19], although this enzyme is mainly interesting for lactose hydrolysis in acid whey, because of their acidic pH optimum as well as their activity at elevated temperatures. The *S. cerevisiae* expression system was also used for the production of *K. lactis* β-D-galactosidase, the protein of outstanding biotechnological interest in the food industry but in this case the enzyme production was not strictly extracellular. The β-galactosidase was released into the culture medium after osmotic shock of the recombinant *S. cerevisiae* osmotic-remedial thermosensitive-autolytic mutants [20, 21]. To improve the secretion of the *K. lactis* β-D-galactosidase, cytosolic in origin, the hybrid protein from this enzyme and its *A. niger* homologue, that is naturally extracellular, was constructed. The hybrid protein was active and secreted by recombinant *K. lactis* strain, but the amount of extracellular enzyme still remained low [22]. Yeast species especially designated for the production of extracellular proteins are for example *Pichia pastoris* or *Hansenula polymorpha*. There is only one recently published example of an extracellular β-galactosidase production system using *P. pastoris* as a host, however, it concerns thermostable enzyme from *Alicyclobacillus acidocaldarius* [23].

The *S. cerevisiae* is usually the first choice for industrial processes involving alcoholic fermentation but this yeast is unable to metabolize lactose and, therefore, the lactose consuming yeast, *K. fragilis*, has been used in most industrial plants producing ethanol from whey [24]. The engineering of *S. cerevisiae* for lactose utilization has been addressed over the past 20 years by different strategies [25]. However, most recombinant strains obtained displayed no ideal characteristics (such as slow growth, genetic instability, or problems derived from the use of glucose/galactose mixtures) or were ineffective for ethanol production [24, 26, 27]. There is only one published example of efficient ethanol production with a recombinant *S. cerevisiae* strain expressing the *LAC4* (β-galactosidase) and *LAC12* (*lactose permease*) genes of *K. lactis* [28]. Hence, there is still a need for *S. cerevisiae* strains producing new β-galactosidases which may appear to be an interesting alternative for the production of ethanol from lactose-based feedstock.

In this respect, here we report on a new cold-adapted β-D-galactosidase, isolated from psychrothrophic, Antarctic *Arthrobacter* sp. 32c bacterium strain, that possesses low molecular weight of 75.9 kDa of monomer and 195 kDa of native protein. In addition, the presented enzyme is active in the range of temperature 4–8°C that is suitable for milk industry applications and can be produced extracellularly on a large scale using recombinant *P. pastoris* strains cultivated either on methanol or glycerol (a cheap by-product in biodiesel industry).

Characterization of 32c Isolate

Many different colonies were isolated from the Antarctic soil. One isolate, named 32c, that formed yellow colonies was chosen for further study because of its ability to hydrolyze X-Galthe cromogenic analogue of lactose. The cells were gram-negative rods. The optimum growth in LAS medium was observed between 25–27°C. No growth occurred at 37°C. In order to determine the ability of the selected isolate to utilize starch, milk, avicell, or arabinose several plates with different substrates were prepared. It was observed that 32c strain produces enzymes of industrial interest like α-amylase, proteases and has an arabinose utilization pathway. In order to estimate the phylogenetic position of the isolate, we cloned the amplified 16S rRNA gene into pCR-Blunt vector, determined its sequence, and examined its phylogenetic relationships (Figure 1A). The obtained sequence was deposited at GenBank with the accession no. FJ609656. An analysis of the sequence showed that it clustered with other organisms isolated from cold environments, mainly belonging to *Arthrobacter* species. The isolate formed a well-defined cluster with *A. oxidans* (98.59% sequence identity) and *A. polychromogenes* (97.86% sequence identity). Based on 16S rDNA similarity, physiological properties similar to other *Arthrobacter* strains and its presence in the Antarctic soil our isolate was classified as *Arthrobacter* sp. 32c.

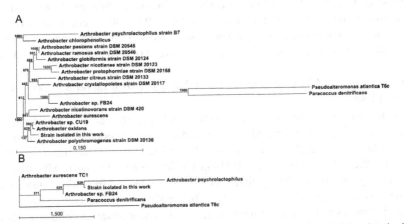

Figure 1. Phylogenetic analysis of the *Arthrobacter* sp. 32c 16S rDNA sequence (A) and *Arthrobacter* sp. 32c β-D-galactosidase gene sequence (B). Sequences were aligned using the sequence analysis softwares: ClustalX 1.5 b and Gene-Doc 2.1.000. Phylogenetic trees were reconstructed with the PHYLIP COMPUTER PROGRAM PACKAGE, using the neighbor-joining method with genetic distances computed by using Kimura's 2-parameter mode. The scale bar indicates a genetic distance. The number shown next to each node indicates the percentage bootstrap value of 100 replicates.

Characterization of the β-D-galactosidase Gene

The psychrotrophic *Arthrobacter* sp. 32c chromosomal library was prepared in *E. coli* TOP10F'. The plasmid pBADmycHisA was used to construct the library, and ampicillin-resistant transformants were selected and screened for the ability to hydrolyze X-Gal. Several transformants out of approximately 5,000 were selected as blue colonies on plates containing X-Gal. Restriction analysis of plasmid inserts from these transformants indicated that they had been derived from the same fragment of chromosomal DNA. Sequence data from the shortest construct, named pBADmycHisALibB32c, contained 5,099 bp insert with an open reading frame (2,085 bp) encoding protein, which shares high homology to a β-D-galactosidase (NCBI Access No. FJ609657). The sequence of *Arthrobacter* sp. 32c β-D-galactosidase was analyzed and found to encode a 694 amino acid protein with a predicted mass of 76.142 kDa and a theoretical pI of 5.59. The analysis of DNA sequence upstream the *Arthrobacter* sp. 32c β-D-galactosidase gene with the promoter prediction tool (BPROM software, http://www.softberry.com webcite) revealed a potential promoter sequence with cttaca and tacaat as -35 and -10 sequences, respectively. A putative ribosomal binding site was apparent eight bases before the initiating methionine codon. The insert fragment and β-D-galactosidase gene had a high G+C content, 67 mol%, and 66 mol%, respectively, which is typical of *Arthrobacter* species.

A comparison of the *Arthrobacter* sp. 32c β-D-galactosidase gene sequence with those from the NCBI database showed that it was most closely related to the *Arthrobacter* sp. *FB24* gene (77.13% sequence identity) and to the *A. aurescens TC1* gene (71.8% sequence identity) (Figure 1B). The deduced amino acid sequence from *Arthrobacter* sp. 32c β-D-galactosidase gene was also used to compare with other amino acid sequences deposited in the NCBI database. The *Arthrobacter* sp. 32c β-D-galactosidase was found to be a member of the glycoside hydrolase family 42 and contained an A4 beta-galactosidase fold. The enzyme shares 84% of identity and 91% of similarity to the sequence of the *Arthrobacter* sp. FB24, 74% identity and 84% similarity to the sequence of the *Arthrobacter aurescens* TC1 and only 51% identity and 65% similarity to the sequence of the *Janibacter* sp. HTCC2649 β-D-galactosidase.

Overexpression and Purification of Recombinant *Arthrobacter* sp. 32c β-D-galactosidase

In order to produce and investigate the biochemical properties of *Arthrobacter* sp. 32c β-D-galactosidase, we constructed bacterial and yeast expression systems. The recombinant arabinose-inducible pBAD-Myc-HisA-β-gal32c plasmid was used for the expression of the *Arthrobacter* sp. 32c β-D-galactosidase gene in *E. coli* LMG194/plysN [29]. The highest enzyme biosynthesis yields were achieved by adding arabinose to the final concentration of 0.02% w/w, at A_{600} 0.5 and by further cultivation for 5 hr. After purification a single protein migrating near 70 kDa was observed following sodium dodecyl sulfate-polyacrylamide gel electrophoresis and staining with Coomassie blue (Figure 2A, lane 3). It was in good agreement with the molecular mass deduced from the nucleotide sequence (75.9 kDa). The applied overexpression system was quite efficient, giving 27 mg (Table 1) of purified β-D-galactosidase from 1 l of induced culture. The relative molecular mass of native enzyme estimated by gel filtration on a column of Superdex 200 HR 10/30, previously calibrated with protein molecular mass standards, was 195,550 Da. Hence, it is assumed that the purified *Arthrobacter* sp. 32c β-D-galactosidase is probably a trimeric protein.

Table 1. Purification of recombinant *Arthrobacter* sp. 32c β-D-galactosidase.

Purification step	Volume (ml)	Protein (mg)	Specific activity (U mg⁻¹)	Total activity (U)	Purification (fold)	Recovery (%)
			$E\ coli$ LMG plysN pBADMyc-HisA-32cβ-gal			
Cell extract	30	580	13.8	8004	1.0	100
Affinity chromatography	3.2	27	155.9	4209	21.0	53
			P. pastoris GS 11S pPICZaA-32cβ-gal			
Broth	1000	3400	28.7	97580	1.0	100
Protein precipitation	54	340	136.1	46274	10.0	47
Affinity chromatography	11	137	154.7	21194	24.8	22
			P. pastoris GS 11S pGAPZaA-32cβ-gal			
Broth	1000	5200	16.2	84240	1.0	100
Protein precipitation	46	450	102.7	46215	11.6	55
Affinity chromatography	10	97	153.1	14851	53.6	18

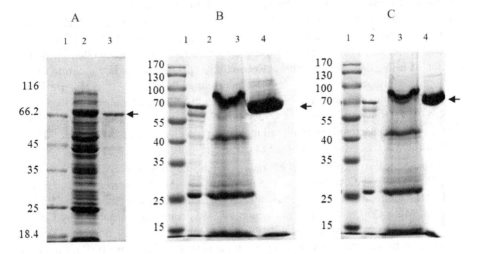

◄   *Arthrobacter* sp. 32c β-**D-galactosidase**

Figure 2. SDS-PAGE analysis of the expression and purification steps of the *Arthrobacter* sp. 32c β-D-galactosidase expressed by *E. coli* host (A), *P. pastoris* GS115 pPICZαA-32cβ-gal methanol induced variant (B) and *P. pastoris* GS115 pGAPZαA-32cβ-gal constitutive variant (C). Lanes 1 protein weight marker. Panel A: lane 2cell extract after expression, lane 3purified β-D-galactosidase after affinity chromatography. Panel B and C: lane 2broth after protein expression, lane 3protein precipitate, lane 4purified β-D-galactosidase after affinity chromatography.

In the *P. pastoris* expression system the methanol induced and constitutive biosynthesis variants for larger scale production of the enzyme were tested. By cloning the gene in the form of translational fusion with the *S. cerevisiae* α-factor leader sequence under the control of either the methanol induced promoter AOX1 or under the constitutive promoter GAP, pPICZαA-32cβ-gal, and pGAPZαA-32cβ-gal recombinant expression plasmids were constructed. *P. pastoris* GS115 strain was transformed with linearized pPICZαA-32cβ-gal or pGAPZαA-32cβ-gal plasmids. The obtained *P. pastoris* GS115 recombinant strains harboring pGAPZαA-32cβ-gal or pPICZαA-32cβ-gal recombinant plasmids were used for extracellular production of the *Arthrobacter* sp. 32c β-D-galactosidase (Figure 2B, lane 2 and Figure 2C, lane 2). The applied overexpression systems were efficient, giving approximately 137 and 97 mg (Table 1) of purified β-D-galactosidase (Figure 2B and C, lanes 4) from 1 l of induced culture for the AOX1 and constitutive system, respectively. Noteworthy is the fact that all attempts in extracellular expression of β-D-galactosidase from *Pseudoalteromonas* sp.22b [10, 11] previously described by us did not succeed (data not shown). The corresponded β-D-galactosidase is a tetramer composed of 115 kDa subunits. All the amount of produced protein with fused secretion signal was accumulated in the cells. We also tried to produce the *Pseudoalteromonas* sp. 22b β-D-galactosidase in the form of fusion protein with other secretion sequences: PHO5 and STA2. All attempts gave negative results. It seems that molecular mass of desired recombinant protein is limited for extracellular production by *P. pastoris* host.

Characterization of *Arthrobacter* sp. 32c β-D-galactosidase

The temperature profiles of the hydrolytic activity of the recombinant *Arthrobacter* sp. 32c β-D-galactosidase showed that the highest specific activity with ONPG was at 50°C (155 U/mg). Lowering or raising temperature from 50°C resulted in the reduction of β-D-galactosidaseactivity. Recombinant β-D-galactosidase exhibited 15% of the maximum activity even at 0°C and approximately 60% at 25°C (Figure 3). In order to determine the optimum pH for recombinant β-D-galactosidase, we measured the enzyme activity at various pH values (pH 4.5–9.5) at 0–70°C, using ONPG as a substrate. β-D-galactosidase exhibited maximum activity in pH 6.5 and over 90% of its maximum activity in the pH range of 6.5–8.5 (Figure 3).

To examine the possible metal ion requirements, the enzyme preparation was treated with EDTA to remove metal ions. No activity was lost during treatment with 100 mM EDTA after 2 h. The activity was not considerably affected by metal ions (5 mM): Na^+, K^+, Mg^{2+}, Co^{2+}, Ca^{2+}. The enzyme activity was completely inhibited by Cu^{2+} or Zn^{2+} (5 mM) and was strongly inhibited by Mn^{2+} (11%), Fe^{2+}(25%), and Ni^{2+} (38%) in comparison to the activity of the enzyme in the absence of cations (100%) (Table 2). The activity of the β-D-galactosidase was not considerably affected by ditiothreitol, β-mercaptoethanol, and L-cysteine, whereas reduced glutathione almost completely inactivated the enzyme (Table 3). The examination of the ethanol influence on the *Arthrobacter* sp. 32c β-D-galactosidaseactivity with ONPG as the substrate shows that addition of ethanol up to 20% still slightly stimulates the enzyme activity (Table 4). The relative enzyme activity was increasing up to 120% in the presence of 8% v/v ethanol at pH 5.5.

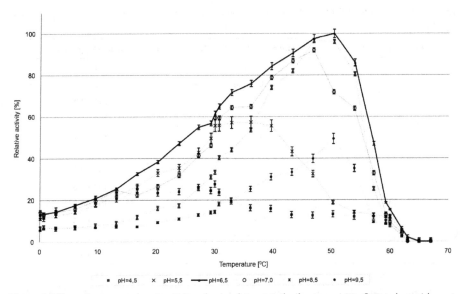

Figure 3. Effect of temperature on activity of recombinant *Arthrobacter* sp. 32c β-D-galactosidase at pH range from 4.5 to 9.5.

Table 2. Effects of metal ions on *Arthrobacter* sp. 32c β-D-galactosidase activity.

Metal ion	Relative activity [%]
None	100
Na^+	97 ± 3
K^+	100 ± 2
Ni^{2+}	38 ± 4
Mg^{2+}	90 ± 2
Fe^{2+}	25 ± 2
Co^{2+}	87 ± 3
Cu^{2+}	0±0
Mn^{2+}	11± 2
Zn^{2+}	0±0
Ca^{2+}	88 ± 2

Table 3. Effects of thiol compounds on recombinant *Arthrobacter* sp. 32c β-D-galactosidase activity.

Compound	Relative activity [%]
None	100
2-mercaptoethanol	92 ± 4
DTT	96 ± 2
Glutathione reduced	6 ± 3
L-cystein	95 ± 2

Table 4. Effect of ethanol concentration on recombinant *Arthrobacter* sp. 32c β-D-galactosidase activity.

Ethanol [% v/v]	Relative activity [%] pH 5.5	Relative activity [%] pH 6.5
0	100	100
1	109 ± 2.0	102 ± 2.4
2	111 ± 2.2	107 ± 3.0
4	114 ± 2.7	109 ± 2.6
6	116 ± 2.5	110 ± 2.4
8	120 ± 2.1	111 ± 2.4
10	119 ± 2.3	109 ± 2.5
12	117± 1.9	107 ± 2.6
14	109 ± 2.2	105 ± 2.4
16	108 ± 2.1	103 ± 2.5
18	105 ± 2.7	102 ± 2.7
20	103 ± 2.9	101 ± 3.1

A study of the substrate specificity of the *Arthrobacter* sp. 32c β-D-galactosidase was performed with the use of various chromogenic nitrophenyl analogues. The recombinant *Arthrobacter* sp. 32c β-D-galactosidase displayed four times higher level of activity with PNPG than with ONPG as substrate. The activities with PNPGlu and ONPGlu were significantly lower with only 1.4% and 0.5% of the activity with ONPG, respectively.

In order to further characterize the biochemical properties of the enzyme the highest specific activity k_{cat}, the K_M values and the catalysis efficiency k_{cat}/K_M in reaction with ONPG and lactose were calculated. The highest observed specific activity with ONPG was 212.4 s^{-1} at 50°C. The half saturation coefficient (K_M) was highest at 10°C (5.75 mM), decreased to 2.62 mM at 50°C and rose again to 5.11 mM at 55°C. The highest catalysis efficiency was achieved at 50°C (81.7 $s^{-1}mM^{-1}$). The same kinetic parameters were also determined with lactose (Table 5). Hereby the half saturation coefficient was significantly higher, the reaction velocity constant was significantly lower and the reaction efficiency was very low. To investigate the reason for such results another test was performed, where glucose was transformed in the reaction mixture by glucose isomerase that converted it to fructose, while galactose remained in the mixture. In this test the reaction efficiency was significantly higher and over 30% from the 5% w/v of lactose was hydrolyzed to glucose and galactose for 12 hr and over 75% of the lactose was found to be hydrolyzed after 72 hr. These results were similar to another test where the recombinant *P. pastoris* strain extracellularly producing *Arthrobacter* sp. 32c β-D-galactosidase (pGAPZαA-32cβ-gal) was cultivated on lactose containing broth. It seems obvious that *Arthrobacter* sp. 32c β-D-galactosidase is inhibited by glucose. Nevertheless this shows that the enzyme might successfully catalyze the conversion of lactose to corresponding monocarbohydrates in a fermentation broth where glucose is consumed by cells of the fermenting strain.

Table 5. Kinetic parameters of *Arthrobacter* sp. 32c β-D-galactosidase.

Substrate	Temperature [°C]	K_m[mM]	k_{cat} [s^{-1}]	k_{cat}/Km [s^{-1}mM^{-1}]
ONPG	10	5.75 ± 0.34	52.4 ± 0.72	9.12 ± 0.71
	20	4.86 ± 0.37	81.0 ± 1.03	16.67 ± 1,60
	30	3.46 ± 0.29	123.9±1.21	35.81 ± 3.66
	40	3.15 ± 0.27	169.9 ± 1.44	53.92 ± 5.56
	50	2.62 ± 0.21	212.4 ± 1.67	81.07 ± 7.76
	55	5.11 ± 0.32	71.2 ± 0.98	13 .93 ± 1.14
lactose	10	77.54 ± 1.77	1.76 ± 0. 11	0.023 ± 0.002
	20	67.82 ± 1.74	2.36 ± 0.14	0.035 ± 0.003
	30	52.67 ± 1.71	4.81 ± 0.22	0.091 ± 0.007
	40	44.31 ± 1.73	5.73 ± 0.21	0.129 ± 0.010
	50	39.73 ± 1.72	6.98 ± 0.23	0.176 ± 0.014

DISCUSSION

The β-D-galactosidase from *Arthrobacter* sp. 32c characterized in this study has interesting industrial properties. It displays optimum activity at pH 6.5 and catalyses the hydrolysis of 1,4-β-D-galactoside linkages at pH 4.5–9.5 with high efficiency. Its optimum activity was observed at about 50°C. Nevertheless it showed over 50% of activity at pH 5.5–7.5 at 30°C and was not considerably inactivated by Ca^{2+} ions what in fact can be of interest in industrial ethanol production from cheese whey by means of brewing *Saccharomyces cerevisiae* strains or by recombinant strains that simultaneously utilize glucose and galactose.

The β-D-galactosidases naturally produced by psychrophilic microorganisms are either intracellular or expressed at low levels. In order to make progress in cheaper production of β-D-galactosidases of industrial interest, we choose highly efficient *P. pastoris* expression systems for consideration to produce enzyme extracellularly. The *P. pastoris* has been successfully used many times in extracellular protein production, however, there are only several examples of cold-adapted proteins and none cold-adapted β-D-galactosidase produced by this host. We have found only one published example of *P. pastoris* extracellular β-D-galactosidase production for a thermostable enzyme from *Alicyclobacillus acidocaldarius* [23].

There are several examples of cold active β-D-galactosidases isolated from *Pseudoalteromonas* strains [5, 10, 11] and *Arthrobacter* strains [7-9, 12, 13] with molecular mass above 110 kDa of monomer and forming an active enzyme of over 300 kDa. Most of them belong to the family 42 β-D-galactosidases. However, the β-D-galactosidase belonging to family two obtained from the Antarctic *Arthrobacter* isolate appears to be one of the most cold-active enzymes characterized to date [8]. All of the known cold-adapted β-D-galactosidases, except two of them isolated from *Planococcus* sp. strains [4, 14] and from *Arthrobacter* sp. 32c (this study), form very large oligomers and therefore are of minor interest in industrial application probably because of many problems in effective overexpression. The β-D-galactosidases isolated from psychrophilic *Planococcus* sp. strains have low molecular weight of about 75 kDa of monomer and about 155 kDa of native protein. The β-D-galactosidase isolated from *Planococcus* sp. L4 is particularly thermolabile, loosing its activity within only 10 min at 45°C [14] and therefore larger scale production of this enzyme by recombinant yeast strains cultivated at 30°C might be economically not feasible. Only the β-D-galactosidase from *Planococcus* sp. isolate SOS orange [4] displays interesting activity and might be considered in biotechnological production on a larger scale.

In comparison with known β-D-galactosidases, the *Arthrobacter* sp. 32c β-D-galactosidase is a protein with a relatively low molecular weight. Molecular sieving revealed that the active enzyme is a trimmer with a molecular weight of approximately 195 ± 5 kDa. Relatively low molecular weight of the protein did not interfere with extracellular production of the protein by *P. pastoris*. Therefore the constructed recombinant strains of *P. pastoris* may serve to produce the protein extracellularly with high efficiency and in a cheap way. The calculated production cost of 1 mg of purified β-D-galactosidase was estimated at 0.03 €.

The same *Pichia pastoris* expression systems had been unsuccessfully used for extracellular expression of previously reported β-D-galactosidase from *Pseudoalteromonas* sp. 22b [10, 11]. This enzyme is much bigger than *Arthrobacter* sp. 32c β-D-galactosidase and forms a tetramer of approximately 490 kDa. It is worth noting that we have tried to secrete this enzyme with three different secretion signals (α-factor from *Saccharomyces cerevisiae*, glucoamylase STA2 from *Saccharomyces diastaticus* or phosphatase PHO$_5$ from *S. cerevisiae*) with no success. It seems that the molecular mass of the desired recombinant protein is limited to extracellular production by *P. pastoris* host, whereas the used secretion signal is without any influence. Based on our experience with *Pichia pastoris* expression systems we assert that the larger protein the lower expression yield can be achieved.

In comparison with the known β-D-galactosidase from *Planococcus* sp. isolate SOS orange [10], β-D-galactosidase from *Arthrobacter* sp. 32c is more thermostable and it has a similar activity profile. Moreover, as shown in this study, it can be produced extracellularly in high amounts by yeast strain. The displayed activity profile of the *Arthrobacter* β-D-galactosidase, especially the activity at pH range from 5.5 to 7.5, over 50% of relative activity at 30°C and enhancement of the activity by the presence of ethanol suggest that this enzyme is compatible with the industrial process conditions for ethanol production by yeast. The construction of corresponding *S. cerevisiae* recombinant strains and fermentation tests for the production of ethanol from cheese whey by the application of this β-D-galactosidase are pending.

The *Arthrobacter* β-D-galactosidase was strongly inhibited by glucose and therefore the catalysis efficiency was very low. Removal of this product resulted in 75% hydrolysis of a solution containing 5% of lactose after 72 hr in a combined enzyme assay. These results clearly indicate that the enzyme can be used for the production of sweet lactose free milk where hydrolysis of lactose to glucose and galactose is performed by simultaneous isomerisation of glucose to fructose by glucose isomerase.

MATERIALS AND METHODS

Isolation, Characterization, and Identification of the 32c Isolate

A 5 g of Antarctic soil was dissolved in 45 ml of water containing 1% of sea salt (Sigma-Aldrich). After decantation 100 μl of the supernatant was spread out on LAS agar plates that contained 1% lactose, 0.1% pepton K, 0.1% yeast extract, 1% of marine salt, 1.5% agar and 20 μg/ml of X-gal. Pure cultures of microorganisms were isolated. One of them was found to be a producer of β-D-galactosidase and also exhibited amylolytic and proteolytic activities. This strain was primarily classified as 32c isolate and used for further analyzes. The bacterium 32c was cultured in the liquid LAS medium containing 1% lactose, 1% pepton K, 0.5% yeast extract, and 1% artificial sea salt at 15°C for 2 days at 150 rpm in air shaker. The temperature profile of growth was determined in the range 0–37°C, by means of stationary cultures in the LAS medium.

16S rDNA Gene Amplification

Genomic DNA from isolate 32c was used as a template to amplify *16S rDNA* gene using primers: 16S For 5' AGAGTTTGATCCTGGCTCAG 3' and 16S Rev 5' ACG-GCTACCTTGTTACGACTT 3'. Reaction was performed in mixture containing: 0.2 μM of each primer, 0.2 μg of chromosomal DNA, 250 μM of each dNTP, 1 U of DNA polymerase (Hypernova, DNA-Gdańsk, Poland) in 1 × PCR buffer (20 mM Tris-HCl pH 8.8, 10 mM KCl, 3.4 mM $MgCl_2$, 0.15% Triton X-100). The reaction mixture was incubated for 3 min at 95°C, followed by 30 cycles at 95°C for 1 min, 55°C for 1 min, 72°C for 1.5 min, and a final incubation for 5 min at 72°C using a Mastercycler Gradient (Eppendorf, Germany). PCR product was purified from an agarose gel band using DNA Gel-Out kit (A&A Biotechnology, Poland), and cloned directionally into pCR-Blunt vector (Invitrogen). The *16S rDNA* insert was sequenced using ABI 3730 xl/ABI 3700 sequencing technology (Agowa DE, Germany).

Genomic DNA Library Construction

The chromosomal DNA from 32c strain cells was isolated using a genomic DNA Prep Kit (A&A Biotechnology, Poland) according to protocol for gram-negative bacteria. The DNA was digested using the 20 U of SalI and 20 U of BglII endonucleases (Fermentas, Lithuania) for 2 hr at 37°C in 1× buffer O+ (Fermentas), and 2- to 8-kb fragments were purified from a 0.8% agarose gel using the DNA Gel Out kit (A&A Biotechnology, Poland). Then DNA fragments were ligated with T4 DNA ligase (Epicentre, USA) for 1 hr at 16°C into pBAD/Myc/HisA vector (Invitrogen) pre-cutted with the same restriction enzymes. The *E. coli* TOP10F' cells were transformed to give the genomic library by incubation at 37°C on LA agar (10 g pepton K, 5 g yeast extract, 10 g NaCl, and 15 g agar) containing 100 μg/ml ampicillin, 1 mM IPTG, and 20 μg/ml X-gal. After 12 hr incubation, plates were transferred to 20°C and incubated further for 16 hr. Blue colonies were taken for analysis. These *E. coli* TOP10F' cells were transformed with plasmid containing the *Arthrobacter* sp. 32c β-galactosidase gene. Plasmid DNA was extracted from these recombinant strains. The insert of the smallest recombinant plasmid (pBADmycHisALibB32c) was sequenced using ABI 3730 xl/ABI 3700 sequencing technology (Agowa DE, Germany).

β-D-Galactosidase Gene Amplification and Cloning to Bacterial Expression System

Based on the known β-D-galactosidase gene sequence of *Arthrobacter* sp. 32c (GenBank Accession No. FJ609657), the specific primers for PCR amplification were designed and synthesized. The gene was amplified using two separate reactions. The first DNA fragment was amplified using the forward primer: F1Nc-β-gal CATGGGCAAGCGTTTTCCAAG, and reverse primer: R32c-β-gal CCCCGTC-GACTTTTCTAGATCAGTCCTCCGCGATCAC (containing *Sal*I and *Xba*I recognition sites, underlined). The second DNA fragment was amplified using the forward primer: F2Nc-β-gal GGCAAGCGTTTTCCAAGCGG, and reverse primer: R32c-β-gal CCCCGTCGACTTTTCTAGATCAGTCCTCCGCGATCAC (containing *Sal*I recognition site, underlined). The start and stop codons are given in bold. For the NcoI sticky end generation the second forward F2Nc-β-gal primer contains only one nucleotide of the start codon. Each PCR reaction mixture contained: 0.2 μM of each primer, 0.2 μg of pBADmycHisALibB32c DNA, 250 μM of each dNTP, 1 U of DNA polymerase (*Hypernova*, DNA-Gdańsk, Poland) in 1 × PCR buffer (20 mM Tris-HCl pH 8.8, 10 mM KCl, 3.4 mM MgCl2, 0.15% Triton X-100). The reaction mixtures were incubated for 3 min at 95°C, followed by 5 cycles at 95°C for 1 min, 50°C for 1 min, 72°C for 2 min and 25 cycles at 95°C for 1 min, 60°C for 1 min, 72°C for 2 min, and a final incubation for 5 min at 72°C using a Mastercycler Gradient (Eppendorf, Germany). Both amplification reaction products were purified and mixed together at ratio 1:1. This mixture was denatured at 95°C for 3 min and cooled down to room temperature at 0.2°C/s. Afterwards DNA were purified by ethanol precipitation, digested with *Sal*I endonuclease and cloned into pBAD/Myc/HisA (Invitrogen) vector pre-cutted with NcoI and *Sal*I endonucleases. The resulting recombinant plasmid pBAD/Myc/HisA-β-gal32c containing the *Arthrobacter* sp. 32c β-D-galactosidase gene under control of the pBAD promoter was used to transform chemically competent *E. coli* LMG194 plysN cells [29].

Expression of the Recombinant β-D-galactosidase Gene in *E. coli*

The recombinant plasmid pBAD/Myc/HisA-32cβ-gal was used for the expression of the putative β-D-galactosidase gene in *E. coli* LMG 194 plysN under the control of pBAD promoter. The cells were grown overnight at 37°C in LB medium containing chloramphenicol (34 μg/ml) and ampicillin (100 μg/ml) in air shaker at 220 rpm. The preculture was inoculated (1%) into fresh 1 l of LB medium containing the same antibiotics and cultivation was continued at 37°C to OD600 of 0.5. The culture was then supplemented with 0.02% (w/w) arabinose (final concentrations) and grown for 4 hr at 37°C to achieve the overexpression of β-D-galactosidase gene.

Pichia Pastoris Expression Plasmids Construction

The primers used for amplification of the *Arthrobacter* sp. 32c β-D-galactosidase gene were: F32c-β-gal ATGGGCAAGCGTTTTCCAAGCGGC and R32c-β-gal CCCC-GTCGAC TTTTCTAGATCAGTCCTCCGCGATCAC (containing *Sal*I and *Xba*I recognition sites, underlined) (reaction A). The start and stop codons are given in bold. The second PCR reaction was performed to obtain a linear form of DNA vectors using primers: Phos-alfa-factor phos-TCTTTTCTCGAGAGATACCCCTTCTTCTTTAG-CAGCAATGC and AOX1-res-insert-ATTTGAATTCTCTAGACTTAAGCTTGTTT-GTAGCCTTAGACATGACTGTT CCTCAGTTCAAGTTG and pPICZαA (reaction B) or pGAPZαB (reaction C) plasmid DNA as DNA template. Each PCR reaction mixture contained: 0.2 μM of each primer, 0.2 μg of recombinant plasmid, 250 μM of each dNTP, 1 U of DNA polymerase (Hypernova, DNA-Gdańsk, Poland) in 1 × PCR buffer (20 mM Tris-HCl pH 8.8, 10 mM KCl, 3.4 mM MgCl2, 0.15% Triton X-100). Reaction A was performed using following conditions: 95°C–3 min, (95°C–1 min, 53°C–1 min, 72°C–2 min; 5 cycles), (95°C–1 min, 65°C–1 min, 72°C–2 min; 25 cycles), 72°C–5 min. Reaction B and C were performed at conditions: 95°C–3 min, (95°C–1.5 min, 66°C–1 min, 72°C–4 min; 5 cycles), (95°C–1.5 min, 68°C–1 min, 72°C–4 min; 25 cycles), 72°C–10 min. The PCR products were purified from an agarose gel bands using DNA Gel-Out kit (A&A Biotechnology, Poland), digested with XbaI endonuclease and ethanol precipitated. The DNA fragments from reaction A and B and from reaction A and C were ligated with each other and chemically competent E. coli TOP10F' (Invitrogen) cells were transformed with those ligation mixtures, spread out on LA plates containing 12.5 μg/ml zeocine (Invitrogen) and incubated at 37°C for 16 hr. Afterwards recombinant plasmids were isolated, linearized by SacI or XmaJI endonuclease and used to transform P. pastoris GS115 competent cells using Pichia EasyComp™ Transformation Kit (Invitrogen). The obtained P. pastoris GS115 recombinant strains harboring pGAPZαA-32cβ-gal or pPICZαA-32cβ-gal recombinant plasmids were used to extracellular production of the Arhrobacter sp. 32c β-D-galactosidase.

Expression of the β-D-galactosidase Gene in *Pichia pastoris*

The P. pastoris GS115 recombinant strains harboring pGAPZαA-32cβ-gal or pPICZαA-32cβ-gal plasmid were used to extracellular expression of the Arhrobacter sp. 32c β-D-galactosidase either constitutively or after methanol induction, respectively. For both expression systems 900 ml of YPG medium (Yeast extract 1%, Pepton K 2%, 2% glycerol) was inoculated with 100 ml of YPG medium cells cultures of the P. pastoris pGAPZαA-32cβ-gal or P. pastoris pPICZαA-32cβ-gal. In case of the constitutive β-D-galactosidase expression the inoculated culture was grown with agitation at 30°C for 4 days. After 2 days additional carbon source in form of glycerol was added to final concentration of 3% v/v to the broth. In case of the methanol induced variant, 100 ml overnight culture of the P. pastoris pPICZαA-32cβ-gal was centrifugated at 1,500 × g for 10 min. The supernatant was discarded, cells were dissolved in 100 ml of BMMY medium (1% yeast extract, 2% peptone, 0.004% L-histidine, 100 mM potassium phosphate, pH 6.0, 1.34% YNB, 4×10^{-5}% biotin, 0.5% methanol) and added to 900 ml of the same medium. The cultivation was performed for 4 days, where methanol was added to final concentration of 0.65%, 0.8%, and 1% after first, second and third day, respectively.

β-D-galactosidase Purification

After protein expression in E. coli host, the cells were disrupted according to protocol described earlier with some modifications [29]. Cells were harvested by centrifugation at 5,000 × g for 20 min and the cell pellet was resuspended in 30 ml of buffer A (20 mM K_2HPO_4-KH_2PO_4, pH 7.5) and frozen at 20°C for 15 min. After thawing at room temperature, the samples were centrifuged at 10,000 × g. The supernatant containing the desired protein was applied onto affnity matrix of agarose coupled with *p*-aminobenzyl-1-thio-β-D-galactopyranoside (PABTG-agarose, Sigma) (10 ml column) equilibrated with four volumes of buffer A. The column was washed with 300 ml of the buffer A, and the recombinant β-D-galactosidase was eluted three times with 10 ml of 0.05 M sodium borate (pH 10.0) buffer at a flow rate of 0.5 ml/min. Active fractions containing the β-D-galactosidase were collected and dialyzed three times against 3 l of buffer D (100 mM NH_4HCO_3).

In case of the purification of the extracellular produced β-D-galactosidase in *P. pastoris* cultures, the yeast cells were separated from the post-culture medium through centrifugation. Next, the ammonium sulfate was added to the post-culture medium to 60% w/w, at 4°C. The precipitated proteins were centrifugated at 20,000 × g, dissolved in buffer A and dialyzed overnight against the same buffer. For β-D-galactosidase purification the dissolved sample was applied further directly onto affnity matrix of agarose coupled with *p*-aminobenzyl-1-thio-β-D-galactopyranoside and purified as described above for bacterial system. The concentration of purified protein was determined by the Bradford method using bovine serum albumin (BSA) as a standard.

β-D-galactosidase Activity Assays

The activity of purified *Arthrobacter* sp. 32c β-D-galactosidase was determined by the use of chromogenic substrates as described elsewhere [4, 14]. The o-nitrophenol released from 10 mM of o-nitrophenyl-β-*D*-galactopyranoside (ONPG) by β-D-galactosidase at 0–70°C and pH range 4.5–9.5 (0.02 M citrate buffer for pH 4.5 and 5.5; 0.02 M K_2HPO_4-KH_2PO_4 for pH 6.5 and 7.0 and 0.02 M Tris-HCl for pH 8.5 and 9.5) was measured at 405 nm. The reaction was stopped after 10 min with 1 M Na_2CO_3. One unit is defined as one micromolar of o-nitrophenol released per minute.

Substrate specificity was estimated using 1 mM solution of chromogenic substrates: The ONPG, *p*-nitrophenyl-β-D-galactopyranoside (PNPG), o-nitrophenyl-β-D-glucopyranoside (ONPGlu), and *p*-nitrophenyl-β-D-glucopyranoside (PNPGlu). Activity determination was carried out under standard conditions in 0.02 M K_2HPO_4-KH_2PO_4 (pH 6.5) buffer at 10, 20, 30, 40, or 50°C. The activity of the β-D-galactosidase towards lactose was monitored by HPLC analysis (column Bio-rad, Aminex HPX-87H) where 1% solutions of lactose, glucose, fructose, and galactose were used as standards.

In the combined enzyme assay glucose isomerase from *Streptomyces murinus* (Sigma G4166) was used in the amount of 0.01 g/ml of 5% w/v solution of lactose (0.02 M K_2HPO_4-KH_2PO_4, pH 6.5). The *Arthrobacter* sp. 32c β-D-galactosidase was used at concentration of 200 U/ml of the mixture. The reaction mixture was set at 37°C for 72 hr and products were analyzed by HPLC every 12 hr.

Effects of 5 mM dithiothreitol, 5 mM of 2-mercaptoethanol, 5 mM of L-cysteine, 5 mM of reduced glutathione, and metal ions (Na^+, K^+, Mn^{2+}, Mg^{2+}, Ca^{2+}, Fe^{2+}, Zn^{2+}, Cu^{2+}, Co^{2+} and Ni^{2+}; each at concentration of 5 mM) on *Arthrobacter* sp. 32c β-D-galactosidase activity were determined under standard conditions.

All measurements and/or experiments were conducted five times. Results are presented as mean SD. Relative activities were estimated in above experiments by comparison to highest activity (100%).

CONCLUSION

In this study we present the purification and characterization of a new β-D-galactosidase from *Arthrobacter* sp. 32c. From the sequence analyses it is obvious that the protein is a member of the family 42 β-D-galactosidases. The protein weight deduced from the 695 amino acid sequence was 75.9 kDa. Molecular sieving revealed that the active enzyme has a molecular weight of approximately 195 ± 5 kDa and therefore it is probably a trimmer. The new characterized β-D-galactosidase is of industrial interest and can be produced extracellularly in its economically feasible variant by the constructed *P. pastoris* strain.

The constructed *P. pastoris* strain may be used in co-fermentation of lactose from cheese whey by a consortium of microorganisms with industrial strains of brewing yeast *S. cerevisiae*, where the *P. pastoris* produces β-D-galactosidase in the oxygen phase and accelerates the shift between the oxidative and reductive conditions.

KEYWORDS

- *β-D-galactosidase*
- *Arthrobacter* sp.
- *Escherichia coli*
- **Half saturation coefficient**
- *Pichia pastoris*
- *Saccharomyces cerevisiae* strains

AUTHORS' CONTRIBUTIONS

Piotr Hildebrandt carried out the molecular genetic studies, participated in the design of the study and drafted the manuscript. Marta carried out the molecular genetic studies, participated in drafted the manuscript. Józef Kur conceived of the study, and participated in its design and coordination. All authors read and approved the final manuscript.

ACKNOWLEDGMENTS

This work was supported by the Polish State Committee for Scientific Research Grant 2 P04B 002 29 to J.K.

This research work was supported by the European Social Fund, the State Budget and the Pomeranian Voivodeship Budget in the framework of the Human Capital Operational Programme, priority VIII, action 8.2, under-action 8.2.2 Regional Innovative Strategies, the system project of the Pomorskie Voivodeship "Innodoktorant Scholarships for PhD students, I edition."

Chapter 12

Global Transcriptional Response to Natural Infection by Pseudorabies Virus

J. F. Yuan, S. J. Zhang, O. Jafer, R. A. Furlong, O. E. Chausiaux, C. A. Sargent, G. H. Zhang, and N. A. Affara

INTRODUCTION

Pseudorabies virus (PRV) is an alphaherpesviruses whose native host is pig. The PRV infection mainly causes signs of central nervous system (CNS) disorder in young pigs, and respiratory system diseases in the adult.

In this chapter, we have analyzed native host (piglets) gene expression changes in response to acute PRV infection of the brain and lung using a printed human *oligonucleotide* gene set from Illumina. A total of 210 and 1,130 out of 23,000 transcript probes displayed differential expression respectively in the brain and lung in piglets after PRV infection (p-value < 0.01), with most genes displaying up-regulation. Biological process and pathways analysis showed that most of the up-regulated genes are involved in cell differentiation, neurodegenerative disorders, the nervous system and immune responses in the infected brain whereas apoptosis, cell cycle control, and the *mTOR* signaling pathway genes were prevalent in the infected lung. Additionally, a number of differentially expressed genes were found to map in or close to quantitative trait loci for resistance/susceptibility to PRV in piglets.

This is the first comprehensive analysis of the global transcriptional response of the native host to acute alphaherpesvirus infection. The differentially regulated genes reported here are likely to be of interest for the further study and understanding of host viral gene interactions.

The PRV, is a member of the alphaherpesvirus subfamily and has multiple closely related family members, such as the herpes simplex virus1 (HSV-1), varicellovirus (VZV), avian herpes viruses, bovine herpesviruses (BHV-1), equine herpesviruses (EHV-1 and EHV-4), feline herpesvirus type 1, and canine herpesvirus type [1, 2]. Thus, PRV has served as a useful model organism for the study of herpesvirus biology [1]. Owing to its remarkable propensity to infect synaptically connected neurons, PRV is also studied as a "live" tracer of neuronal pathways [1]. Finally, while vaccination strategies to eradicate PRV in the US and Europe have shown great progress, they fail to eradicate completely viral infection from a population. Thus outbreaks in swine populations result in substantial economic losses. These include restrictions on animal movement and trade for affected countries, with disease and infection control measures increasing production costs owing to antibody testing, vaccination programs, and extra labor.

Although PRV has been widely studied (especially its agricultural impact, its viral pathogenesis, its molecular biology, its use as a neuronal tracer, and in DNA vaccine exploration [1]) how the native host responds globally after infection with wild type PRV is still poorly understood. Clinically, infection in older pigs ranges from asymptomatic to severe respiratory disease but with limited mortality. Young piglets exhibit more serious clinical signs and often succumb to fatal encephalitis preceded by typical behaviors consistent with infection of the CNS. In recent years, microarray technology has proven useful to assess the cellular transcriptional responses to herpesvirus infections in human and mouse cell lines [3-5]. It has been used to study host gene expression after PRV infection of rat embryo fibroblasts [5], and the CNS in rodent brain at various times post infection *in vivo* [6]. However few porcine genome-wide expression studies have been published. Most experiments have used "in-house" cDNA arrays to study transcriptional events in pig tissues, such as the stress-genes related to early weaning of piglets [7]. The down side of these cDNA-based clone libraries is that the genes represented on the array are often very focused on a given biological system or process and lack a whole genome overview.

In this study, piglet samples were hybridized onto an Illumina Human Refset Chip (Illumina Inc. San Diego), corresponding to 23,000 transcript probes. This cross-species comparison potentially allows the study of the whole transcriptome. There are now porcine arrays available from commercial suppliers (e.g., Affymetrix and Qiagen), but these are not all representative of the entire pig genome and were not widely available at the time of this study. In the absence of a comprehensive species-specific array deeper interrogation of the pig gene complement was afforded by the use of the better annotated human geneset. Although the use of this approach can only be partially informative when there are no confirmed pig orthologs in the public databases, we have identified host cellular genes whose mRNA levels change during natural PRV infection of piglet brain and lung. The resulting data define key pathways of *host*-gene expression that characterize the host response to an acute CNS and respiratory infection.

MATERIALS AND METHODS

Experimental Pigs and Housing

The experimental animals were sourced from an outbreak of PRV that occurred in the farrowing house of a local commercial farmer due to a reduced level of protection via maternal antibody. Clinical signs were described as follows: suckling piglets were listless, febrile, and uninterested in nursing. Within 24 hr of exhibiting these clinical signs, some piglets progressively developed indications of CNS infection including trembling, excessive salivation, lack of coordination, ataxia, and seizures. Infected piglets sat on their haunches in a "dog-like" position, lay recumbent and paddled, or walked in circles. The appearance of the dissected organs in selected piglets was typical of PRV infection: bleeding in meninges, oedema in the brain, bleeding spots in the lung, and on the adenoids [1, 8].

Three strict criteria were imposed for the selection of piglets included in this study: (1) piglets exhibited the typical clinical signs described above; (2) piglets exhibited the expected pathology, especially in brain, and lung; (3) virus isolation, antibody identification, or detection of viral antigen-positive tissues were used to confirm the organic infection by PRV, and diseases including swine fever (SF), Porcine Reproductive, and Respiratory Syndrome Virus (PRRSV) and other potential bacterial infections which could be clinically and pathologically confused with PRV infection were excluded by viral antigen, antibody identification, and polymerase chain reaction (PCR) detection.

Six piglets aged from 2 to 4 days (commercial breed Landrace X Yorkshire) which were infected by PRV but not by the other tested diseases (see above) and three healthy piglets (not infected, and negative for all tests under the strict criteria used above), matched for age and breed from the same farm were used in this experiment. All experiments were carried out in strict accordance with accepted HuaZhong Agricultural University, China, and governmental policies.

Microarray Experimental Design

Total mRNA samples from the brains and lungs of the three normal piglets were pooled for the reference mRNA. Ten independent RNA samples (six biological replicates for brain and four biological replicates of lung) from the six infected piglets were paired with the reference sample for hybridization on two-color microarrays. Using a dye-swap configuration, comparing each sample provides technical replicates to adjust for dye bias [9]. A total of 20 slides were used in this study.

RNA Purification

Total mRNA was prepared using Qiazol reagent (Qiagen, Crawley, West Sussex, UK) following the manufacturer's instructions. A second purification step was performed immediately post extraction on the isolated total mRNA using the RNeasy Midi kit (Qiagen Inc., Valencia, CA) and each sample was treated with DNase (20 U of grade I DNase; Roche, Lewes, UK) to remove any genomic contamination following the manufacturer's instructions. With a cut-off of 150 bp, 5S rRNA, and tRNAs were removed from the samples by the columns, limiting interference in downstream experiments. The RNA concentration and integrity were assessed on the Nanodrop ND-1000 spectrophotometer (Nanodrop, USA) and on the Agilent 2,100 bioanalyzer system (Agilent Technologies, Palo Alto, CA), using an RNA 6000 Nano LabChip kit.

SMART Amplification and Labeling of the Samples

The extracted RNA was amplified using the SMART amplification protocol (BD Smart TM Amplification Kit, UK) and labeled with Cy5 or Cy3 using Klenow enzyme as described by Petalidis et al. 2003 [10] with two modifications; (a) a constant number of 14 cycles was used, and (b) for the labeling step, 1 μl of Cy3 or Cy5-dCTP was used with 22 μl (250 ng) of second strand cDNA. The labeled products were purified using G50 columns, according to manufacturer's instructions (Amersham Biosciences, UK). Labeled samples were combined and precipitated for at least 2 hr at 20°C with 2 μl of human Cot-1 DNA, 1 μl PolyA (8 μg/μl), 1 μl yeast tRNA (4 μg/μl), 10 μl Na acetate (3 M, pH 5.2), and 250 μl 100% ethanol.

Microarray Hybridization and Scanning

The labeled product was re-suspended in 40 μl hybridization buffer (40% deionised formamide, 5 × SSC, 5 × Denhart's, 1 mM Na Pyrophosphate, 50 mM Tris pH 7.4 and

0.1%SDS) and hybridized onto a microarray slide containing 23,000 human oligo-nucleotides (Illumina Inc. San Diego), printed in-house on to Codelink slides using a BioRobotics Microgrid II arrayer. After over-night hybridization of the slides at 48°C in a water bath, they were washed in $2 \times$ SSC, $0.1 \times$ SSC, 0.05% Tween 20, and $0.1 \times$ SSC sequentially for 5 min each and scanned using an Axon 40001A scanner. Signal quantification was performed using Bluefuse software (2.0) (BlueGnome, Cambridge, UK).

Analysis of the Data

Data exported from Bluefuse was analyzed using the R package [http://www.r-project.org/library] FSPMA [11], which is based on the mixed model ANOVA library YASMA [12]. Expression values in both channels were converted to log ratios and normalized by subtracting a M/A (i.e., log ratio/log amplitude) loess fit and adjusting the within-slide scale of the data. The ANOVA model used a nested design with spot-replication (1) as the innermost effect, nested inside biological replication (six for brains; four for lungs), with dye-swap (2) as the outermost effect. Spot-replication was considered to be a random effect and biological replication and dye-swap fixed effects. Genes were considered to be up- or down-regulated, if the average channel log ratios relative to the control were found to be highly significantly different from zero, using a p-value threshold of 0.05. The p-values were calculated within the ANOVA model, using FSPMA's VARIETY option and a correction for multiple comparisons by false discovery rate. This analysis takes into consideration the variance across samples and excludes those genes with a high level of variance. We can, therefore, be confident that the smaller fold changes observed are real.

The 70-mer human oligonucleotide sequences from differentially expressed probe sets with a p-value < 0.01 were used to BLAST search pig sequences in the public databases [http://www.ncbi.nlm.nih.gov/BLAST/] including Unigene and ESTs [13]. For matches to Unigene clusters, Homologene was used to indicate orthology to the human probe sets. With novel ESTs, pig data were matched against the human genomic and transcript database to confirm that the best matches were to orthologs sequences. Hits were considered to be reliable if there was a putatively orthologs match of 6070 bp, and oligonucleotides with fewer matches, in the range of 5059 bp, were also selected if p-values were significant in this study. Probe sets that could not be verified by BLAST as described above are not reported in this chapter. Analysis of the signal intensity distribution of the cross-species hybridizations for both the lung and brain experiments showed a normal distribution similar to that obtained when homologous human RNA is hybridized to the chip. The proportion of the approximately 23K probes showing a signal greater than 100 signal value (i.e., above background) in the cross-hybridization is 22,300 from the 22,800 probes on the chip (~97%). The microarray data (accession number E-MEXP-2376) is available through ArrayExpress.

Functional Annotation of Gene Expression Data

In order to understand the biological phenomena studied here and reduce the interpretive challenge that is posed by a long list of differentially expressed genes. Onto-Express was used to classify our lists of differentially regulated genes into functional profiles characterizing the impact of the infection on the two different tissues [http://

vortex.cs.wayne.edu/ontoexpress/] [14]. Initial analysis used the non-filtered dataset, that is, all differentially regulated probe sets against the full human oligonucleotide geneset. We then looked at differentially expressed probes (p-value < 0.01) identified from our microarray analysis, and statistical significance values were calculated for each category using the binomial test available in Onto-Express [15]. This makes no assumptions about those probesets with good matches to known pig sequences. However, only those probesets for which we could confidently assume orthology are reported in the tables in this chapter. Here we present categories of gene ontology based on a maximum pairwise p-value of 0.05 for the "biological processes." To gain a better understanding of the gene interactions (pathways) involved in the disease, Pathway-Express was also applied to our data. In order to quantify the over/under representation of each category, the library composition has been taken into account in the presentation of the results.

Quantitative RNA Analyses Using Real-time PCR Methodology (qRT-PCR)

Quantitative real-time reverse transcriptase polymerase chain reaction (qRT-PCR) analysis using SYBR green and selected primers was carried out following the manufacturer's protocol (QIAGEN, QuantiTect SYBR Green RT-PCR) to confirm the microarray results. All probes and primers were designed using Express Primer 3 software developed by the Whitehead Institute for Biomedical Research. The nucleotide sequences of selected genes were obtained from GenBank, and the primer information is shown in Table 1. The *PSMD2* (primers kindly provided by Ms. Gina Oliver and Dr. Claire Quilter) was selected for use as the reference gene because it was previously shown to be a good control for pig brain (personal communication from Ms. Gina Oliver and Dr. Claire Quilter) and was also shown to be one of the most constant housekeeping genes in a human tissue study. The qRT-PCR was performed on 300 ng RNA equivalents in 25 μl/reaction/well on an Icycler (Bio-Rad Laboratories Ltd, USA) (50°C for 60 min; 95°C for 15 min; 40 cycles of 95°C for 15 sec, 58°C for 30 sec, and 72°C for 30 sec). For each gene reactions were performed in triplicate to allow statistical evaluation of the data. The average Ct (threshold cycle) was used for the analysis. Relative expression levels were calculated by using the $2^{-(\Delta\Delta Ct)}$ method as previously described [16].

Table 1. Validation of array data by real-time PCR.

Gene name	Pig homologene	Primer sequences (5'-3')	Microarray data		qRT-PCR data	
			Brain (n-fold change)	Lung (n-fold change)	Brain (n-fold change)	Lung (n-fold change)
PSMD2	Ssc.1642	F: tggggagaataagcgttttg R: tattcatgaccccatgatgc	Ref	Ref	Ref	Ref
AKTI	Ssc.29760	F: tgggcgacttcatccttg R: tggaagtggcagtgagca	ND[a]	1.68	ND	2.19
CDC42	Ssc.6687	F: aaagtgggtgcctgagata R: ctccacatacttgacagcc	-[b]	2.03	-	7.38
LY96	Ssc.25550	F:cattgcacgaagagacataca R: tgtattcacagtctctcccttc	1.37	3.32	6.91	9.23

Table 1. (Continued)

Gene name	Pig homologene	Primer sequences (5'-3')	Microarray data		qRT-PCR data	
			Brain (n-fold change)	Lung (n-fold change)	Brain (n-fold change)	Lung (n-fold change)
PIK3RI	Ssc.49949	F: cccaggaaatccaaatga R: ggtcctcctccaaccttc	-	-	0.61	0.45
SERPI-NEI	Ssc.9781	F: ccagcagcagatccaaga R: cggaacagcctgaagaagt	-1.66	2.36	-0.64	4.28

[a]ND, not done;
[b]-, not changed or absent.

Microarray Analysis of Gene Expression Profiles in Brain and Lung

Six brain samples and four lung samples were used for microarray hybridization and qRT-PCR, and two of the lung samples were excluded as they were found to be degraded. Table 2 shows the number of differentially expressed human probe sets initially identified in brain and lung tissues (p-value < 0.01 and p-value < 0.05). Based on BLAST analysis, those probes with putative pig gene homologues have been considered for further analysis and numbers are shown in Table 2. This avoids making assumptions about other probes that detect expression changes but have weaker matches to pig ESTs. Most probes with porcine homologues remained unchanged, and few showed a reduction in transcription level by microarray analysis. For example, expression of only four (6070 bp human match category) and one (5059 bp human match category) were decreased in infected lung tissue (p-value < 0.01). In contrast, a large number of host transcripts were induced in response to wild type PRV infection (Table 2). Here we identified 120 and 866 up-regulated transcripts in brain and lung (p-value < 0.01) with pig: human matches ≥ 60 bp, and 42, and 259 genes with matches of 5059 bp for further gene ontology and pathway classification (Table 2).

Table 2. Number of probe sets and pig gene homologues in brain and lung tissues affected by wild type PRV infection.

p-value	Up/down regulated	Brain				Lung			
		A	B	C	D	A	B	C	D
p-value < 0.01	down	253	35(34)	14(14)	17	195	4(4)	1 (1)	11
	up	528	132(120)	44(42)	115	2283	888(866)	261(259)	424
p-value < 0.05	down	588	77(76)	26(26)	43	1657	25(24)	4(4)	51
	up	879	209(196)	69(67)	173	3284	1122(1075)	357(355)	545

A= Total number of differentially expressed human probes.
B =Total number of pig Unigene matches of 60-70 basepairs (subset of verified gene or thologues).
C =Total number of pig Unigene matches of 50-59 basepairs (subset of verified gene or thologues).
D =Total number of EST matches >50 basepairs with no assigned Unigene ID.

Of the transcripts with matches ≥ 60 bp, 76, corresponding to 74 unique pig gene homologues, are up-regulated in common between the two tissues. Forty-four probe sets corresponding to 41 unique pig gene homologues with matches of 5059 bp also displayed increased expression in both tissues after infection by wild type PRV.

Gene Ontology and Bioinformatics Analysis

To characterize the sets of functionally related genes that are differentially expressed between the infected and uninfected group, we used the Onto-Express tool to classify up-regulated genes in each tissue according to their biological process. Table 3 summarizes the largest classes identified on the basis of biological process. Twelve defined biological processes with matches $\geq$ 60 bp, and 10 with matches of 5059 bp, are observed in brain at least two fold more often than expected. In comparison, nine processes with matches $\geq$ 60 bp, and only four with matches of 5059 bp are over-represented in lung, although the total number of up-regulated genes in lung is more than that in brain tissue (Table 2).

Table 3. Classes of biological processes involving up-regulated pig gene homologues (*p*-value < 0.01) in brain and lung tissues infected with wild type PRV.

Biological Process	Library	Brain Pig Unigene Matches over 60 base-pairs (gene homologues)	Brain Pig Ungene Matches between 50-59b base-pairs (gene homologues)	Lung Pig Unigene Matches over 60 base-pairs (gene homologues)	Lung Pig Unigene Matches between 50 59b base-pairs (gene homologues)
Apoptosis	230	3*	0	30*	4
Biological function unknown	472	4	0	31	10
Cation transport	139	2*	3*	0	2
Cell adhesion	429	8*	5*	11	4
Cell cycle	303	3	0	31*	3
Cell differentiation	230	4*	1*	7	2
Immune response	255	0	0	7	3
Intracellular protein transport	135	5*	0	16*	5*
Intracellular signaling cascade	285	4*	0	14	3
Ion transport	304	5*	1	6	4
Metabolism	280	3*	3	15*	9*
Nervous system development	239	9*	3*	9	3
Protein amino acid dephosphorylation	108	1	2*	14*	1
Protein amino acid phosphorylation	412	1	2*	29	7
Protein folding	165	4*	0	27*	5*
Protein transport	217	2	2*	33*	4
Proton transport	47	1*	2*	3	7*
Regulation of progression through cell cycle	215	2	2*	20*	4

Table 3. *(Continued)*

Biological Process	Library	Brain Pig Unigene Matches over 60 base-pairs (gene homologues)	Brain Pig Ungene Matches between 50-59b base-pairs (gene homologues)	Lung Pig Unigene Matches over 60 base-pairs (gene homologues)	Lung Pig Unigene Matches between 50 59b base-pairs (gene homologues)
Regulation of transcription, DNA dependent	1285	9	2	73	7
Signal transduction	1110	6	1	40	6
Synaptic transmission	164	4*	4*	5	1
Transcription	945	7	1	63	7
Ubiquitin cycle	217	1	0	30*	4

* Biological processes with at least two times the expected number of genes (calculated from the library composition).

Pathways Affected by Wild-type PRV Infection in Brain and Lung

One indication that the observed transcript differences (p-value < 0.01) may have biological relevance is that sets of genes in known pathways show coordinated regulation. Accordingly, the functionally classified genes were mapped to known cellular pathways. Fifteen pathways with at least five times the expected number of genes (matches ≥ 60 bp) have been highlighted with pathway-express in the infected brain. Interestingly, most of them belong to neurodegenerative disorders, nervous system, and immune system pathways. Twelve pathways (including the calcium signaling pathway, the phosphatidylinositol signaling system, and the TGFβ signaling pathway) with at least five times the expected number of genes (matches of 5059 bp) were also highlighted in the infected brain. However only four pathways (ubiquitin mediated proteolysis and prion disease, matches ≥ 60 bp; ALS, and mTOR signaling pathway, matches of 5059 bp) showed at least five times the expected number of genes in the infected lung (Table 4). Interestingly, ubiquitination of PRV glycoproteins for vaccination has been shown to be related to decreased cellular immune responses following wild type infection.

Table 4. Cellular pathways involving up-regulated (p-value < 0.01) pig gene homologues in brain and lung tissues infected with wild type PRV.

Pathway Name	Library	Brain Pig Unigene Matches over 60 base-pairs (gene homologues)	Brain Pig Unigene Matches between 50-59b basepairs (gene homologues)	Lung Pig Unigene Matches over 60 base-pairs (gene homologues)	Lung Pig Unigene Matches between 50-59b base-pairs (gene homologues)
Behavior					
Circadian rhythm	17	0	0	1	0
Cancers					
Colorectal cancer	73	2	0	9	0
Cell Communication					
Adherens junction	72	2	0	9	2

Table 4. (Continued)

Pathway Name	Library	Brain Pig Unigene Matches over 60 base-pairs (gene homologues)	Brain Pig Unigene Matches between 50-59b basepairs (gene homologues)	Lung Pig Unigene Matches over 60 base-pairs (gene homologues)	Lung Pig Unigene Matches between 50-59b base-pairs (gene homologues)
Focal adhesion	187	4	1	11	4
Gap junction	91	3	0	5	0
Tight junction	106	2	0	13	3
Cell Growth and Death					
Apoptosis	81	0	0	3	1
Cell cycle	105	1	1	0	5
Cell Motility					
Regulation of actin cytoskeleton	195	6	0	12	2
Development					
Axon guidance	119	2	1	7	3
Endocrine System					
Adipocytokine signaling pathway	68	1	0	2	2
GnRH signaling pathway	94	3	2	6	1
Insulin signaling pathway	125	2	1	8	1
Folding, Sorting and Degradation					
Regulation of autophagy	24	0	0	2	0
SNARE interactions in vesicular transport	28	0	1	3	1
Ubiquitin mediated proteolysis	41	0	1	11	1
Immune System					
Antigen processing and presentation	80	0	0	3	0
B cell receptor signaling pathway	61	2	0	6	1
Complement and coagulation cascades	60	0	0	1	0
Fe epsilon Rl signaling pathway	73	2	0	4	1
Leukocyte transendothelial migration	111	1	0	6	3
Natural killer cell mediated cytotoxicity	119	2	0	4	2

Table 4. *(Continued)*

Pathway Name	Library	Brain Pig Unigene Matches over 60 base-pairs (gene homologues)	Brain Pig Unigene Matches between 50-59b basepairs (gene homologues)	Lung Pig Unigene Matches over 60 base-pairs (gene homologues)	Lung Pig Unigene Matches between 50-59b base-pairs (gene homologues)
T cell receptor signaling pathway	87	3	0	5	3
Toll-like receptor signaling pathway	87	1	0	6	0
Infectious Diseases					
Epithelial cell signaling in Helicobacter pylori infection	45	1	0	5	1s
Metabolic Disorders					
Type I diabetes mellitus	42	1	1	2	0
Nervous System					
Long-term depression	72	2	0	5	0
Long-term potentiation	65	**2**	**2**	**5**	**3**
Neurodegenerative disorders					
Neurodegenerative disorders	33	23	1	0	1
Alzheimer's disease	18	0	0	2	0
Amyotrophic lateral sclerosis (ALS)	17	3	0	0	2
Dentatorubropallidoluysian atrophy (DRPLA)	12	0	0	1	0
Huntington's disease	26	2	1	4	0
Parkinson's disease	15	1	0	0	0
Prien disease	10	1	0	3	0
Sensory System					
Olfactory transduction	30	0	2	2	0
Taste transduction	51	1	0	1	0
Signal Transduction					
Calcium signaling pathway	173	0	4	3	3
Hedgehog signaling pathway	54	0	0	3	0
Jak-STAT signaling pathway	147	0	0	6	2
MAPK signaling pathway	267	5	2	18	7
mTOR signaling pathway	44	0	0	4	3

Table 4. *(Continued)*

Notch signaling pathway	39	0	0	1	0
Phosphatidylinositol signaling system	77	0	1	0	0
TGF-beta signaling pathway	70	1	1	11	0
VEGF signaling pathway	68	3	0	5	2
Wnt signaling pathway	138	0	1	12	2
Signaling Molecules and Interaction					
Cell adhesion molecules (CAMs)	123	2		3	2
Cytokine-cytokine receptor interaction	242	0		3	2
ECM-receptor interaction	85	1		3	2
Neuroactive ligand-receptor interaction	275	1		1	0

* Cellular pathways with at least five times the expected number of genes (calculated from the library composition).

Ten genes up-regulated in both tissues by wild-type PRV infection segregated into known pathways. Most of them are involved in multiple pathways, such as *SPP1* in the immune response pathway, the ECM-receptor interaction and focal adhesion pathway, and *FOS* and *CDC42* in the T cell receptor signaling pathway and MAPK signaling pathway. Moreover, it is also interesting to note that a few genes such as *SERPINE1* and *LCP2* respond differently in the two tissues studied, and while some of the pathways responding to the infection are ubiquitous, others appear to be tissue specific.

qRT-PCR Analysis for Validation of Microarray Results

In order to verify the data obtained in the microarray experiment, we confirmed the expression profile of five selected genes with different patterns of expression: *LY96* is differentially expressed in the same direction in both tissues; *SERPINE1* is down-regulated in the brain but up-regulated in the lungs after infection; *CDC42* and *AKT1* are significantly up-regulated in lung tissue only, and *PIK3R1* is not significantly differentially expressed. Results from qRT-PCR confirmed the direction of expression (up- or down-regulated) obtained by microarray analysis in the five genes tested (Table 1). The magnitude of the fold change is not the same. This is most probably due to the fact that the array analysis is based on a cross-species hybridization whereas the RT-PCR has been performed using species homologous primers. It is likely that the RT-PCR analysis reflects more accurately the fold change in expression.

DISCUSSION

The virus replication cycle involves a series of host-virus interactive processes causing changes in expression of cellular genes, and an infected host activates both innate and adaptive immune responses to eliminate the invading virus [17]. The pig is an ideal animal model for studying human diseases, so the identification of pig model biomarkers for viral diseases is an important step towards identification of human counterparts. The identification of biomarkers has already been proposed as a way to create new diagnostic tools for specific microbial infection [18, 19].

Previous studies have shown the value of using cross-species hybridization [20]. Here, using the Illumina human oligonucleotide Refset in a cross-species study we identified hundreds of probes with expression levels that were altered in brain and lung following wild type PRV infection of young piglets, which typically have more severe clinical manifestations than the adult. In adult pigs one observes mainly, or exclusively, the respiratory symptoms, whereas in piglets and rodent hosts there is invariably invasion of the CNS [21, 22]: piglets exhibit signs in the form of tremor, trembling, and incoordination. Thus piglets permit the potential identification of a wider spectrum of genes involved in the disease processes in different tissues.

Classification of the genes that are differentially expressed in piglet brain into functional groups revealed that several genes are also implicated in human neurodegenerative disorders. These include genes in the pathways for amyotrophic lateral sclerosis (*NEF3, NEFL, NEFH*), Huntington's disease (*CALM3, CLTC, CLTB*), neurodegenerative disorders (*APLP1, NEFH, FBXW7*), Parkinson's disease (*GPR37*) and prion disease (*APLP1, NFE2L2*). It is not known if these transcriptional changes are primary or secondary effects of the PRV infection.

Several members of the immune response pathways (e.g., the B cell receptor signaling pathway, the Fc epsilon RI signaling pathway, natural killer cell mediated cytotoxicity, and the T cell receptor signaling pathway) were also transcriptionally regulated by PRV infection in brain. This is in agreement with the results from PRV or HSV-1 infection in primary cultures of rat embryonic fibroblasts [5]. In addition, similar changes to immune response pathway (e.g., antigen processing and presentation, complement, and coagulation cascades), cell differentiation and metabolism pathway genes have been described in the host following PRV infection in rat CNS [6]. Our experiment not only identified pathways, but also several genes in common with these previous studies: *FOS* and *LCP2*, both involved in T cell receptor signaling pathways; the *TGFβ* signal transduction pathway components *ID4* and *THBS4*, highlighted in the study of PRV infection of primary cultures of rat embryonic fibroblasts [5, 6]; and *SERPINE-1*, identified in both earlier rat studies. These genes may be potential diagnostic and therapeutic targets for viral encephalitis and other neurodegenerative or neuroinflammatory diseases.

Several genes of the TGFβ pathway were also identified here in the infected lung tissue (e.g., *PPP2CA, PPP2CB, ID2, ID3*, and *ID4*). After PRV infection, most older swine exhibit signs of respiratory disease, and the study of the lung is therefore important for understanding what genes may be involved in the disease process. We identified 1,130 differentially expressed probes as a result of wild-type PRV infection; this is five times higher than in the brain. The lung may be more transcriptionally active, or have a more pronounced immune response that might involve more immune cell types than the brain. In addition, we have identified five possible viral receptors, normally necessary for the spread of virus between cells, up-regulated in the infected lung: *HveC* (*PVRL1*), *PVRL3*, *HveD* (*PVR*, *CD155*), *HS3ST4*, and *HS3ST5* [23, 24]. Finally, a number of members of the TNF receptor family, usually involved in apoptosis, were identified (*TNFRSF10, 21, 25, 9, 17, 8, 1α*). This apoptotic pathway was also described in the study of HSV infection of glial cell types [25]. However, the result is interesting as the family member *TNFRSF14* has been shown to be involved in some cases of viral entry, but we do not know whether these other family members are involved in viral entry and cell fusion, or only have a downstream role.

Numerous other genes involved in cellular proliferation (*YWHAB, BUB1, PCNA, GADD45, MCM7, CDK4, CDK7*) and apoptosis (*PRKACA, PDCD8, AKT1, PPP3CA*), were identified. These pathways were previously described following PRV and HSV infection in several models [5, 25] and might reflect the proliferation of immune cells. A number of other genes differentially expressed in the lung, such as *HSPD1, HSPB2, SERPINE-1*, are in common with human and mouse models infected by HSV-1 [5, 26].

Recently, Flori et al. [27] have published a time course transcription profiling study (based on the *Qiagen 8,541* gene porcine oligonucleotide array and a 1,789 porcine and PRV cDNA array) investigating both the PRV transcriptome and the host transcriptome responses of PK15 (porcine kidney) cells in culture. This study reports the early down-regulation of many cellular genes in contrast to the data in this chapter. This difference most probably arises from the artificial cell culture study where there is a homogeneous cell population, whereas our present study is an *in vivo* investigation of complex tissues. It is entirely possible that minor tissue cell types exhibit down-regulation of many of the same genes, however, their contribution to the overall signal renders these changes undetectable. This may also explain the differences in gene expression changes for shared genes between lung and brain. In general, fold changes are lower in brain which probably reflects the complexity of cell types in the tissue, not all of which may respond equally to infection. Nevertheless, it is clear that the Flori et al. study has also observed changes in gene expression in the main categories of cellular functions described in this chapter; most notably genes involved in immune responses and cell proliferation and apoptosis.

Genetic differences have been reported in the susceptibility to PRV between European Large White and Chinese Meishan pigs, with differences in cell-mediated and humoral immunity, as well as the outward clinical signs in young pigs [28]. In this study we identified several differentially expressed genes located at or close to the QTL regions previously reported. Two genes (*CD36* and *NPL*) up-regulated in the infected brain and lung are located near the SW749 marker, which is associated with changes in body temperature and neurological signs. The ETA1 (alias *SPP1*), which is involved in the recruitment of T-lymphocytes [29, 30], was up-regulated in both tissues after natural PRV infection, and is linked to the QTL region of chromosome 8. One of the PRV receptors, PVRL3, which is differentially expressed in infected lung, is linked to a QTL on chromosome 13. The CLDN7, which is involved with cell communication, was down-regulated in the infected brain and is linked to a QTL on chromosome 13 associated with neurological signs.

CONCLUSION

By combining the array data presented here with the information from the previous QTL study, it may be possible to identify the best candidates for the clinical features and increased resistance to PRV infection. In addition, further studies and functional analysis of these candidates will broaden the scientific understanding of PRV infection, provide biomarkers to use as diagnostic tools, and may also lead to the development of novel antiviral treatments and/or the application of marker assisted selection for disease resistance.

KEYWORDS

- **Central nervous system**
- **Oligonucleotide**
- **Pseudorabies virus**
- **Varicellovirus**

AUTHORS' CONTRIBUTIONS

J. F. Yuan, S. J. Zhang, and O. Jafer performed the microarray experiments. R. A. Furlong and O. E. Chausiaux contributed towards the data analysis. G. H. Zhang carried out animal experiments and sample collection. C. A. Sargent and N. A. Affara contributed intellectually to the study, and to manuscript preparation. All authors have read and approved the final manuscript.

ACKNOWLEDGMENTS

We thank Anthony Brown, Peter Ellis, Gina Oliver, Claire Quilter, Junlong Zhao, and Rui Zhou for their skilled technical assistance. Financial assistance from the 863 High Technology and Development Project of China (2006AA10Z195, 2007AA10Z152), Chinese projects (2006BAD14B08-02, 2006BAD04A02-11), Hubei project (2006CA023), Wuhan project (20067003111-06), and National Project of China (04EFN214200206) is greatly appreciated.

Chapter 13

Proteomics of *Porphyromonas gingivalis*

Masae Kuboniwa, Erik L. Hendrickson, Qiangwei Xia, Tiansong Wang, Hua Xie, Murray Hackett, and Richard J. Lamont

---

INTRODUCTION

Porphyromonas gingivalis is a periodontal pathogen that resides in a complex multi-species microbial biofilm community known as dental plaque. Confocal laser scanning microscopy (CLSM) showed that *P. gingivalis* can assemble into communities *in vitro* with *Streptococcus gordonii* and *Fusobacterium nucleatum*, common constituents of dental plaque. Whole cell quantitative proteomics, along with mutant construction and analysis, were conducted to investigate how *P. gingivalis* adapts to this three species community.

The 1,156 *P. gingivalis* proteins were detected qualitatively during comparison of the three species model community with *P. gingivalis* incubated alone under the same conditions. Integration of spectral counting and summed signal intensity analyses of the dataset showed that 403 proteins were down-regulated and 89 proteins up-regulated. The proteomics results were inspected manually and an ontology analysis conducted using Database for Annotation, Visualization and Integrated Discovery (DAVID). Significant decreases were seen in proteins involved in cell shape and the formation of the cell envelope, as well as thiamine, cobalamin, and pyrimidine synthesis and DNA repair. An overall increase was seen in proteins involved in protein synthesis. The hmuR, a TonB dependent outer membrane receptor, was up-regulated in the community and a hmuR deficient mutant was deficient in three species community formation, but was unimpaired in its ability to form mono- or dual-species biofilms.

Collectively, these results indicate that *P. gingivalis* can assemble into a heterotypic community with *F. nucleatum* and *S. gordonii*, and that a community lifestyle provides physiologic support for *P. gingivalis*. Proteins such as hmuR, that are up-regulated, can be necessary for community structure.

The microbial communities that exist on oral surfaces are complex and dynamic biofilms that develop through temporally distinct patterns of microbial colonization [1, 2]. For example, initial colonizers of the salivary pellicle on the coronal tooth surface are principally commensal oral streptococci such as *S. gordonii* and related species. Establishment of these organisms facilitates the subsequent colonization of additional gram-positives along with gram-negatives such as *Fusobacterium nucleatum*. As the biofilm extends below the gum line and becomes subgingival plaque, further maturation is characterized by the colonization of more pathogenic gram-negative anaerobes including *Porphyromonas gingivalis* [2-4]. While organisms such as *P. gingivalis* are considered responsible for destruction of periodontal tissues, pathogenicity is only expressed in the context of mixed microbial communities. Periodontal diseases, therefore, are essentially microbial community diseases, and the interactions among the constituents of these communities and between the communities and host cells and tissues, are of fundamental importance for determining the health or disease status of the periodontium.

Oral biofilm developmental pathways are driven by coadhesive, signaling, and metabolic interactions among the participating organisms. Pioneer bacteria provide a substratum and appropriate metabolic support for succeeding organisms. Complex consortia then accumulate through recognition and communication systems. These interbacterial signaling processes can be based on cell–cell contact, short range soluble mediators, AI-2, or nutritional stimuli [2, 5-8]. In general, bacterial adaptation to the community lifestyle is accompanied by distinct patterns of gene and protein expression [9, 10]. In *S. gordonii* for example, arginine biosynthesis genes are regulated in communities with *Actinomyces naeslundii* which enables aerobic growth when exogenous arginine is limited [11]. Over 30 genes are differentially regulated in *P. gingivalis* following community formation with *S. gordonii* but not with *S. mutans* [12], whereas in mono-species *P. gingivalis* biofilm communities there are changes in abundance of over 80 envelope proteins [13].

While over 700 species or phylotypes of bacteria can be recovered from the oral cavity, in any one individual there are closer to 200 species [14] and the diversity of bacteria assembled in dense consortia will be further limited by nutritional and other compatibility constraints. *Porphyromonas gingivalis* can accumulate into single species biofilms and mixed species consortia with *S. gordonii* and related oral streptococci [15-17]. Moreover, introduction of *P. gingivalis* into the mouths of human volunteers results in almost exclusive localization in areas of streptococcal-rich plaque [18]. Development of more complex multi-species communities in aerated environments such as supragingival tooth surfaces may require oxygen scavenging by *F. nucleatum* [19]. *Fusobacterium nucleatum* is also able to coaggregate with *P. gingivalis* and with oral streptococci [19-21]. Hence communities of *S. gordonii*, *F. nucleatum*, and *P. gingivalis* are likely to be favored *in vivo*; however, community formation by these three organisms has not been investigated. The aim of this study was to examine the ability of *S. gordonii*, *F. nucleatum,* and *P. gingivalis* to form multi-species communities *in vitro*, and to utilize a global proteomic approach to investigate differential protein expression in *P. gingivalis* in response to presence of these organisms.

RESULTS AND DISCUSSION

Assembly of *P. gingivalis–F. nucleatum–S. gordonii* Communities *In Vitro*

The CLSM was used to investigate the ability of *P. gingivalis* to assemble into communities with *S. gordonii* and *F. nucleatum*. In order to mimic the temporal progression of events *in vivo*, *S. gordonii* cells were first cultured on a glass surface and this streptococcal substratum was then reacted in succession with *F. nucleatum* and *P. gingivalis*. The *F. nucleatum* and *P. gingivalis* cells were maintained in the absence of growth media in order to be able to detect any metabolic support being provided by the other organisms in the community. A 3D reconstruction of the heterotypic community is shown in Figure 1. Both *P. gingivalis* and *F. nucleatum* formed discrete accumulations and could be either separate from each other or interdigitated, consistent with the concept that the later gram-negative colonizers such as *P. gingivalis* and *F. nucleatum* initially establish themselves on the streptococcal rich supragingival plaque [4, 18]. The results demonstrate the mutual compatibility of these three organisms for heterotypic community development, an early step in the overall process of plaque biofilm accumulation. Participation in multi-species communities may provide a basis for synergistic interactions in virulence. For example, mixed infections of *P. gingivalis* and *F. nucleatum* are more pathogenic in animal models than either species alone [22], and *F. nucleatum* can enhance the ability of *P. gingivalis* to invade host cells [23].

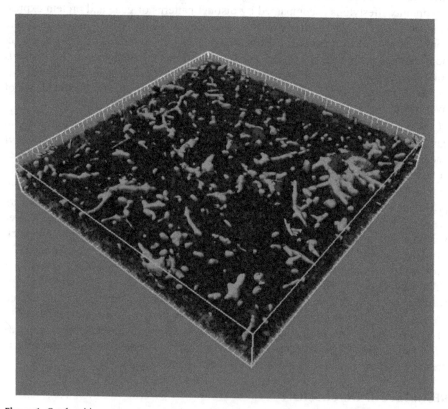

Figure 1. Confocal laser scanning microscopy of *P. gingivalis–F. nucleatum–S. gordonii* community. The *S. gordonii* cells (red, stained with hexidium iodide) were cultured on a glass plate. The FITC-labeled *F. nucleatum* cells (green), followed by DAPI labeled *P. gingivalis* cells (blue), were reacted sequentially with the *S. gordonii* substratum. Bacterial accumulations were examined on a Bio-Rad Radiance 2,100 confocal laser scanning microscope. A series of fluorescent optical x-y sections in the z-plane to the maximum vertical extent of the accumulation were collected with Laser Sharp software. Images were digitally reconstructed with Imaris software. Image is representative of three independent experiments.

Proteome of *P. gingivalis* in a Three Species Community

To begin to investigate the mechanisms of adaptation of *P. gingivalis* to a community environment, the proteome of non-growing *P. gingivalis* cells incorporated into a community with *F. nucleatum* and *S. gordonii* was compared to the proteome of non-growing *P. gingivalis* cells alone. The expressed proteome of *P. gingivalis* in a community consisted of 1,156 annotated gene products detected qualitatively. Based on spectral counting, 271 gene products showed evidence of relative abundance change at a *q*-value of 0.01: 109 proteins at higher relative abundance and 162 at lower relative abundance, using *P. gingivalis* alone as a reference state. Spectral counting is a conservative measure of protein abundance change that tends to generate low FDRs [24-26] but that often suffers from high false negative rates (FNRs) in studies of the kind described here [27]. Less conservative calculations based on intensity measurements [27] found 458 gene products with evidence of relative abundance change at a *q*-value of 0.01: 72 proteins at higher relative abundance, and 386 proteins at lower relative abundance. Spectral counting and protein intensity measurements were examined for common trends. Trends tended to be consistent across both biological replicates, but the magnitudes of the abundance ratios showed significant scatter, similar to most published expression data at either the mRNA or protein level [27]. In most cases the abundance ratio trends were the same, using both quantitation methods, although not necessarily significantly so. In only eight cases were the spectral counting trend and summed intensity trend significantly in opposite directions for the same protein (PGN 0329, 0501, 1094, 1341, 1637, 1733, 2065). The integrated relative abundance trends found 403 gene products with evidence of lower relative abundance change and 89 at higher relative abundance. For purposes of examining the totals for combined trends, if an abundance change was called as significant in one measurement, it was considered significant for the above combined totals only if the ratio of the other measurement showed the same direction of abundance change, with a $\log_2$ ratio of $\pm$ 0.1 or greater regardless of the *q*-value in the second measurement. The experimental data for differential protein abundance are shown in Figure 2 as a pseudo M/A plot [28, 29] with a locally weighted scatterplot smoothing (LOWESS) curve fit [30]. The same data are plotted in Figure 3 as open reading frames according to PGN numbers from the ATCC 33277 genome annotation [31].

To assess global sampling depth, average spectral counts were calculated by summing all spectral count numbers for all *P. gingivalis* proteins in the FileMaker script output described under Materials and Methods and dividing by the total number of *P. gingivalis* proteins in that file. The average redundant spectral count number for peptides unique to a given ORF for *P. gingivalis* alone was 80, for *P. gingivalis* in the community it was 64. The lower number of counts observed for *P. gingivalis* proteins in the community is consistent with the added sampling demands placed on the analytical system by sequence overlaps in the proteomes of all three microbes and thus the smaller number of unique proteolytic fragments predicted.

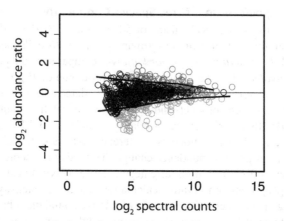

Figure 2. Pseudo M versus A plot [28, 29] of the average protein abundance ratios over all replicates for the *P. gingivalis–F. nucleatum–S. gordonii/P. gingivalis* comparison versus total abundance as estimated by spectral counting. Color codes: red, *P. gingivalis* protein is over-expressed in the *P. gingivalis–F. nucleatum–S. gordonii* community relative to *P. gingivalis* alone; green, *P. gingivalis* protein is under-expressed in the community relative to *P. gingivalis* alone; black, no significant abundance change. Solid black lines represent a LOWESS curve fit [30] to the biological replicates of *P. gingivalis* alone, and represent the upper and lower boundaries of the experimentally observed error regions or null distributions associated with the relative abundance ratio calculations. Proteins coded as either red or green were determined to be significantly changed at the *q*-value [24] cut-off value of 0.01. Thus, the G-test predictions [56] were in good agreement with the curve fitting procedure. Details regarding hypothesis testing procedures can be found in Materials and Methods and in the explanatory notes to the data tables

Figure 3. Genomic representation of the *P. gingivalis* proteome, showing changes in relative abundance for the *P. gingivalis–F. nucleatum–S. gordonii/P. gingivalis* comparison by spectral counting. Each dot represents a PGN ORF number in the order followed by the ATCC 33,277 strain annotation. Color codes: red, over-expression in the *P. gingivalis–F. nucleatum–S. gordonii* community relative to *P. gingivalis* alone; green, under-expression in the community relative to *P. gingivalis* alone; yellow, protein was detected qualitatively, but did not change in abundance; gray, proteins that were qualitative non-detects; gaps indicate ORFs that were not common to both the ATCC 33277 and W83 annotations according to a master cross-reference compiled by LANL (G. Xie, personal communication).

Proteins and Functions Differentially Regulated by *P. gingivalis* in a Community

Cell Envelope and Cell Structure

In bacterial communities significant surface-surface contact occurs both within and among accumulations of the constituent species, as was also observed in the *P. gingivalis–F. nucleatum–S. gordonii* consortia. Regulation of outer membrane constituents of *P. gingivalis* would thus be predicted in the context of a community and this was borne out by the proteomic results. Overall, 84 proteins annotated as involved in the cell envelope were detected, and 40 of these showed reduced abundance in the three species community, indicating an extensive change to the cell envelope. Only four proteins showed increased abundance, two OmpH proteins (PGN0300, PGN0301) and two lipoproteins (PGN1037, PGN1998). The MreB (PGN0234), a bacterial actin homologue that plays a role in determining cell shape, showed almost a 2-fold decrease in community derived *P. gingivalis*. Expression of MreB has been found to decrease under stress or during stationary phase in *Vibrio paraheamolyticus* [35]. However, stress-related proteins were generally reduced in *P. gingivalis* cells in the community (see below) so stress is an unlikely explanation for the change in MreB. Rather, the decrease in MreB abundance may be due to the *P. gingivalis* cells entering a state resembling stationary phase or responding in a previously unseen way to the formation of the three species community.

Protein Synthesis

Extensive changes were observed in ribosomal proteins and in translation elongation and initiation proteins. While overall more proteins showed reduced abundance in the three species community, the changes to the translational machinery were almost exclusively increases in abundance. Of 49 ribosomal proteins detected, 27 showed increased abundance, while only one showed decreased abundance. Of nine translation elongation and initiation proteins detected, none showed significant abundance decreases but five showed increased abundance (EfG (PGN1870), putative EfG (PGN1014), EfTs (PGN1587), EfTu (PGN1578), and If2 (PGN0255)). This represents not only a substantial portion of the translational machinery but also a large portion, 36%, of the proteins showing increased abundance. It is well known that ribosomal content is generally proportional to growth rate [36]; however, given that the cells were not in culture medium during the assay, rapid growth is an unlikely explanation for these results. The increased ribosomal content presumably indicates increased translation, consistent with the community providing physiologic support to *P. gingivalis* and allowing higher levels of protein synthesis.

Vitamin Synthesis

Pathways for synthesizing several vitamins showed reduced protein abundance in the three species community. Most of the proteins involved in thiamine diphosphate (vitamin B1) biosynthesis were down-regulated (Figure 4). Thiamine is a cofactor for the 2-oxoglutarate dehydrogenase complex that converts 2-oxoglutarate to succinyl-CoA and for the transketolase reactions of the anaerobic pentose phosphate pathway [37]. However, transketolase (PGN1689, Tkt) showed no abundance change while of the three components of the 2-oxoglutarate dehydrogenase complex (PGN1755, KorB) only the beta subunit showed an abundance increase.

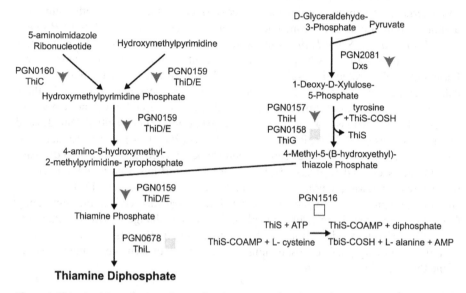

Figure 4. Thiamine biosynthetic pathway, showing protein abundance changes for the *P. gingivalis–F. nucleatum–S. gordonii/P. gingivalis* comparison. Proteins catalyzing each step in the pathway are shown by their *P. gingivalis ATCC 33277* gene designation (PGN number) and protein name, where applicable. Green downward arrows indicate decreased abundance in the three species community. Yellow squares indicate no statistically significant abundance change. Empty squares indicate that the protein was not detected in the proteomic analysis. Thiamine diphosphate is shown in bold.

Only incomplete pathways have been identified for many of the other vitamin biosynthesis activities in *P. gingivalis*. However, cobalamin (vitamin B12) synthesis [38] can be predicted to be decreased in the community, with five (PGN0010, CobC; PGN0316, CbiG; PGN0317, CobL; PGN0318, CobH/CbiG; PGN0735, CobU) of the seven identified proteins having statistically significant reductions. Less complete population of pathways was observed for pyridoxal phosphate (vitamin B6) and biotin synthesis. Only two of the four detected proteins for vitamin B6 synthesis showed reduced abundance (PGN1359, PdxB and PGN2055, PdxA). For biotin synthesis, three of the six detected proteins showed reduced abundance (PGN0133, BioA; PGN1721, BioF; PGN1997, BioD). None of the vitamin/cofactor synthesis pathways showed any indication of increased protein levels in the three species community.

The decrease in several vitamin/cofactor pathways could be due to a decreased utilization of those cofactors. However, in the case of thiamine, the proteins that utilize this cofactor showed no decrease, and a possible increase in abundance, implying that demand for vitamin B1 was unchanged. A more likely explanation for the reduced cofactor pathways is therefore nutrient transfer. Either one or both of the other organisms in the three species community could be providing *P. gingivalis* with cofactors, allowing reduced cofactor synthesis without reducing expression of the cofactor dependent pathways. Nutritional cross-feeding among members of oral biofilms is well established [5], and indeed *P. gingivalis* has been found to utilize succinate produced by *T. denticola* [39].

Nucleotide Synthesis

Pyrimidine biosynthesis appeared to be reduced in the three species community (Figure 5) as many of the proteins leading to the production of finished pyrimidine nucleotides have decreased abundance. However, the proteins responsible for incorporating finished ribonucleotides into RNA show unchanged or increased abundance. As with vitamin biosynthesis this may be the result of nutrient transfer from the other organisms in the community. *Porphyromonas gingivalis* can acquire nucleosides and nucleobases and it has even been suggested that they may represent an important nutrient source for *P. gingivalis* [40]. Consistent with uptake of nucleosides and their precursors, uracil permease (PGN1223) shows increased expression in the three species community.

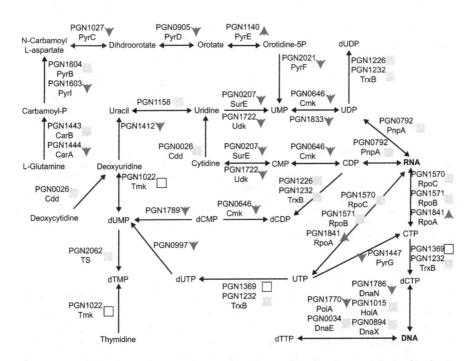

Figure 5. Pyrimidine biosynthetic pathway, showing protein abundance changes for the *P. gingivalis–F. nucleatum–S. gordonii/P. gingivalis* comparison. The protein names follow the same conventions as in Figure 4. Green downward arrows indicate decreased abundance in the three species community. Red upward arrows indicate increased abundance. Yellow squares indicate no statistically significant abundance change. Empty squares indicate that the protein was not detected in the proteomic analysis. The RNA and DNA are shown in bold.

Purine biosynthesis does not appear to be significantly effected in the three species community (Figure 6). A few proteins showed reduced abundance, but the central biosynthesis pathway was primarily unchanged.

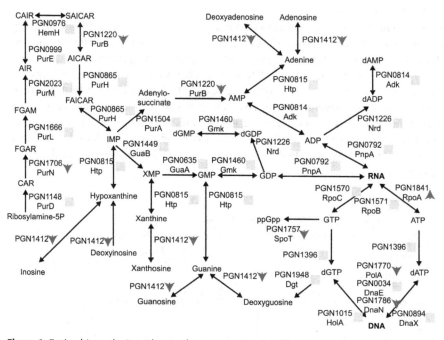

Figure 6. Purine biosynthetic pathway, showing protein abundance changes for the *P. gingivalis–F. nucleatum–S. gordonii/P. gingivalis* comparison. The protein names and arrows/squares follow the same conventions as in Figure 5. The RNA and DNA are shown in bold. GAR: 5-Phosphoribosyl glycinamide; FGAM: 5-phosphoribosyl-N-formylglycineamidine; FGAR: 1-(5′-Phosphoribosyl)-N-formylglycinamide; AICAR: 5′-phosphoribosyl-4-(N-succinocarboxamide)-5-aminoimidazole; AIR: 1-(5′-Phophoribosyl)-5-aminoimidazole; CAIR: 5′P-Ribosyl-4-carboxy-5-aminoimidazole; SAICAR: 5′P-Ribosyl-4-(N-succinocarboximide)-5-aminoimidazole; FAICAR: 1 (5′-Phosphoribosyl)-5-formamido-4-imidazole carboxamide.

Stress Proteins

The ability of the community to provide physiologic support to constituent species might result in *P. gingivalis* experiencing lower levels of environmental stress than occurs in monoculture. Consistent with this concept, community derived *P. gingivalis* showed a significant reduction in abundance of DNA repair proteins (PGN0333, RadA; PGN0342, Ung; PGN0367, Xth; PGN1168, MutS; PGN1316, UvrA; PGN1388, LigA; PGN1567, RecF; PGN1585, UvrB; PGN1712, Nth; PGN1714, Mfd; PGN1771, Pol1). The DNA repair genes are generally induced in the presence of damaged DNA [41], and lower abundance of DNA repair proteins is consistent with the monoculture experiencing more DNA damage than *P. gingivalis* in the three species community where the presence of the partner organisms provides protection against DNA damage.

Only two stress proteins showed increased abundance, and then only 30% increases, the molecular chaperone DnaK (PGN1208) and a PhoH family protein possibly involved in oxidation protection (PGN0090).

Role of the Differentially Regulated *P. gingivalis* Protein HmuR

To begin to test the functional relevance of proteins identified as differentially regulated in the three species community, we undertook a mutational analysis. For this purpose it was important to target a protein that directly effectuates a biological function and lacks homologs in the genome. The hmuR, a major hemin uptake protein, and potential adhesin [42], was selected. As shown in Figure 7A, while wild type *P. gingivalis* cells are abundant within a *S. gordonii–F. nucleatum–P. gingivalis* community, *P. gingivalis* cells lacking hmuR are deficient in community formation. Biovolume analysis showed a 70% reduction in community formation by the hmuR mutant (Figure 7C). Furthermore, this effect was specific for the three species community as a decrease in accumulation by the hmuR deficient mutant was not observed in monospecies biofilms, or in two species communities of *P. gingivalis* with either *S. gordonii* or *F. nucleatum* (Figure 7B, D–G). Hence loss of hmuR, that is up-regulated by *P. gingivalis* when the organism is associated with *S. gordonii* and *F. nucleatum*, results in a phenotype that is restricted to three species community formation. *Porphyromonas gingivalis* cells were first cultured in hemin excess, under which conditions the hmu operon is expressed at a basal level [42]. As the three species model system involves metabolically quiescent *P. gingivalis* cells in buffer, it is unlikely that the role of hmuR is related to its hemin uptake capacity. However, TonB dependent receptors can exhibit functions distinct from transport across the outer membrane. For example, in *E. coli* the TonB dependent catecholate siderophore receptor Iha confers an adhesin function and contributes to colonization and virulence in the mouse urinary tract [43]. Hence, hmuR may have a cohesive function in community formation by *P. gingivalis* although further studies are necessary to resolve this issue.

MATERIALS AND METHODS

Bacteria and Culture Conditions

Fusobacterium nucleatum subsp. nucleatum American type culture collection (ATCC) 25586 and *Porphyromonas gingivalis* ATCC 33277 were grown anaerobically (85% N_2, 10% H_2, 5% CO_2) at 37°C in trypticase soy broth supplemented with 1 mg/ml yeast extract, 1 µg/ml menadione, and 5 µg/ml hemin (TSB). *Streptococcus gordonii* DL1 was grown anaerobically at 37°C in Todd-Hewitt broth (THB).

CHEMICALS

The HPLC grade acetonitrile was from Burdick & Jackson (Muskegon, MI, USA); high purity acetic acid (99.99%), and ammonium acetate (99.99%), from Aldrich (Milwaukee, WI, USA). High purity water was generated with a NANOpure UV system (Barnstead, Dubuque, IA, USA).

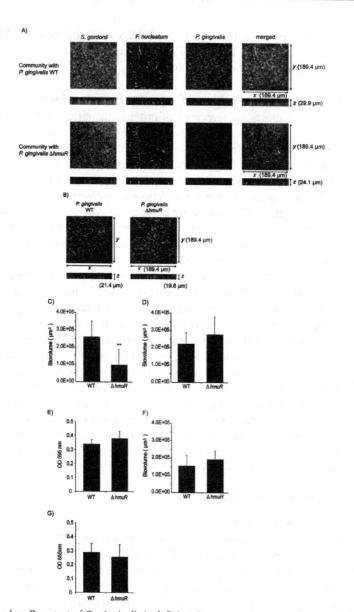

Figure 7. The hmuR mutant of *P. gingivalis* is deficient in community accumulation. (A) Confocal microscopy showing x-y and x-z projections of communities of *S. gordonii* (red), *F. nucleatum* (green) and *P. gingivalis* (blue) wild type (WT) or ΔhmuR mutant strains. Representative image from three independent experiments. (B) Confocal microscopy showing x-y and x-z projections of single species *P. gingivalis* WT or ΔhmuR mutant accumulations. Representative image from three independent experiments. (C) Biovolume analysis of *P. gingivalis* WT or ΔhmuR mutant accumulation in the *P. gingivalis–F. nucleatum–S. gordonii* communities shown in A. (D) Biovolume analysis of *P. gingivalis* WT or ΔhmuR single species accumulations shown in B. (E) Biomass of *P. gingivalis* WT or ΔhmuR single species accumulations measured by crystal violet staining and release. (F) Biovolume analysis of *P. gingivalis* WT or ΔhmuR accumulation in two species *P. gingivalis–S. gordonii* communities. (G) Biomass of *P. gingivalis* WT or ΔhmuR two species accumulation with *F. nucleatum* measured with *P. gingivalis* antibodies. ** denotes $p < 0.01$ (n = 3) compared to WT.

Proteomics of Model Bacterial Communities

High density bacterial communities were generated by the method of Merritt et al. [44]. Bacteria were cultured to mid-log phase, harvested by centrifugation and resuspended in pre-reduced PBS (rPBS). 1×10^9 cells of *P. gingivalis* were mixed with an equal number of *S. gordonii* and *F. nucleatum* as a combination of the three species. *Porphyromonas gingivalis* cells alone were also used as a control. Two independent biological replicates from separate experiments comprised of at least two technical replicates were analyzed. Bacteria were centrifuged at 3,000 g for 5 min, and pellets were held in 1 ml rPBS in an anaerobic chamber at 37°C for 18 hr. The bacterial cells remain viable under these conditions, as determined by both colony counts and live/dead fluorescent staining. Supernatant and bacterial cells were separated and processed separately. Bacterial cells were lyzed with ice cold sterile distilled water and proteins were digested with trypsin as previously described for *P. gingivalis* [33], then fractionated on a 2.0 mm × 150 mm YMC polymer C18 column. There were five pre-fractions collected for each cellular sample, with a final volume of 50 μl for each fraction. The 2D capillary High Performance Liquid Chromatography/ Mass Spectrometry/Mass Spectrometry (HPLC/MS/MS) analyses [32, 45, 46] were conducted using an in-house fabricated semi-automated system, consisting of a Thermo LTQ mass spectrometer (Thermo Fisher Corp. San Jose, CA, USA), a Magic 2002 HPLC (Michrom BioResouces, Inc., Auburn, CA, USA), a Pump 11 Plus syringe pump (Harvard Apparatus, Inc., Holliston, MA, USA), an Alcott 718 autosampler (Alcott Chromatography, Inc., Norcross, GA, USA) and a micro-electrospray interface built in-house. About 2 μl of sample solution was loaded into a 75 μm i.d. × 360 μm o.d. capillary column packed with 11 cm of AQUA C18 (5 μm, Phenomenex, Torrance, CA, USA) and 4 cm of polysulfoethyl aspartamide SCX (strong cation exchange) resin (PSEA, 5 μm, Michrom BioResouces, Inc.). The peptides were eluted with a seven step salt gradient (0, 10, 25, 50, 100, 250, and 500 mM ammonium acetate) followed by an acetonitrile gradient elution (Solvent A: 99.5% water, 0.5% acetic acid. Solvent B: 99.5% acetonitrile, 0.5% acetic acid), 5% B hold 13 min, 5–16% B in 1 min, hold 6 min, 16–45% B in 45 min, 40–80% B in 1 min, hold 9 min, 80–5% B in 5 min, then hold 10 min. For the secreted proteins in the supernatant no pre-fractionation or SCX was performed, and 4 μl of digested sample was loaded into a 75 μm i.d. × 360 μm o.d. column packed with 11 cm AQUA C18 for a single dimension of capillary HPLC/ tandem MS analysis. After 20 min of flushing with 5% acetonitrile, peptides were eluted by an acetonitrile gradient (5–12% B in 1 min, hold 9 min, 12–40% B in 50 min, 40–80% B in 1 min, hold 10 min, 80–5% B in 5 min, hold 14 min). The MS^1 (First stage of tandem mass spectrometry) scan range for all samples was 400–2,000 m/z. Each MS^1 scan was followed by 10 MS^2 (Second stage of tandem mass spectrometry) scans in a data dependent manner for the 10 most intense ions in the MS^1 scan. Default parameters under Xcalibur 1.4 data acquisition software (Thermo Fisher) were used, with the exception of an isolation width of 3.0 m/z units and a normalized collision energy of 40%.

Data Processing and Protein Identification

Raw data were searched by SEQUEST [34] against a FASTA protein ORF database consisting of the Ver. 3.1 curation of *P. gingivalis* W83 (2006, TIGR-CMR [47]), *S.*

gordonii Challis NCTC7868 (2007, TIGR-CMR [48], *F. nucleatum* ATCC 25586 (2002, TIGR-CMR [49]), bovine (2005, UC Santa Cruz), nrdb human subset (NCBI, as provided with Thermo Bioworks ver. 3.3) and the MGC (Mammalian Gene collection, 2004 curation, NIH-NCI [50]) concatenated with the reversed sequences. After data processing, the genome sequence for strain 33277 became available [31] and the data were subsequently cross-referenced to PGN numbers from the 33,277 specific FASTA database provided by Los Alamos National Laboratory (LANL) (personal communication with G. Xie). Although Naito et al. [31] reported extensive genome re-arrangements between W83 and ATCC 33277, the actual protein amino acid sequences are sufficiently similar across the proteome that the use of a database based on W83 was not expected to greatly impact the analysis. Our proteomic methods are not sensitive to genome re-arrangements, only to changes in amino acid sequence for a given protein. The reversed sequences were used for purposes of calculating a peptide level qualitative False discovery rate (FDR) using the published method [51, 52]. The SEQUEST peptide level search results were filtered and grouped by protein using DTASelect [53], then input into a FileMaker script developed in-house [32, 33] for further processing. The DTASelect Version 1.9 filter parameters were: peptides were fully tryptic; ΔCn/Xcorr values for different peptide charge states were 0.08/1.9 for + 1, 0.08/2.0 for + 2, and 0.08/3.3 for +3; all spectra detected for each sequence were retained (t = 0). Only peptides that were unique to a given ORF were used in the calculations, ignoring tryptic fragments that were common to more than one ORF or more than one organism, or both. In practice this had the consequence of reducing our sampling depth from what we have achieved with single organism studies [27, 32, 33], because the gene sequence overlap among the three organisms is significant. A bioinformatic analysis (data not shown) of inferred protein sequence overlaps between *P. gingivalis* and *S. gordonii,* or *F. nucleatum* suggested the reduction in the number of predicted tryptic fragments unique to *P. gingivalis* would not be sufficient to impact the analysis of more than a small number of proteins. The qualitative peptide level FDR was controlled to approximately 5% for all conditions by selecting a minimum non-redundant spectral count cut-off number appropriate to the complexity of each condition, *P. gingivalis* alone or the *P. gingivalis–F. nucleatum–S. gordonii* community.

Protein Abundance Ratio Calculations

Protein relative abundances were estimated on the basis of summed intensity or spectral count values [27, 32, 33] for proteins meeting the requirements for qualitative identification described above. Summed intensity refers to the summation of all processed parent ion (peptide) intensity measurements (MS^1) for which a confirming CID spectrum (MS^2) was acquired according to the DTASelect filter files. For spectral counts, the redundant numbers of peptides uniquely associated with each ORF were taken from the DTAselect filter table (t = 0). Spectral counting is a frequency measurement that has been demonstrated in the literature to correlate with protein abundance [54]. These two ways of estimating protein relative abundance, that avoids the need for stable isotope labeling, have been discussed in a recent review [27] with specific reference to microbial systems. To calculate protein abundance ratios, a normalization scheme was applied such that the total spectral counts or total intensities for all *P. gingivalis* proteins in each condition were set equal for each comparison. This normalization also had the effect of zero centering the $\log_2$ transformed relative abundance ratios, see Figure 2. The normalized data for each abundance ratio comparison was tested for significance using either a global G-test or a global paired t-test for each condition, the details of which have been published for this type of proteomics data in which all biological replicates are compared against each other [55, 56], and are also described in the explanatory notes. Both of these testing procedures weigh deviation from the null hypothesis of zero abundance change and random scatter in the data to derive a probability or *p*-value that the observed change is a random event, that is the null hypothesis of no abundance change is true. Each hypothesis test generated a *p*-value that in turn was used to generate a *q*-value as described [24, 32], using the R package QVALUE [26]. The *q*-value in this context is a measure of quantitative FDR [25] that contains a correction for multiple hypothesis testing.

Ontology Analysis

An overall list of detected proteins as well as lists of proteins that showed increased or decreased levels in the three species community were prepared using *Entrez* gene identifiers. Ontology analyses were then conducted using the DAVID [57] functional annotation clustering feature with the default databases. Both increased and decreased protein level lists were analyzed using the overall list of detected proteins as the background. Potentially interesting clusters identified by DAVID were then examined manually.

Construction of *P. gingivalis HmuR* Mutant

A mutation in the *hmuR* gene was generated using ligation-independent cloning of PCR mediated mutagenesis (LIC-PCR) [58]. A 2.1-kb ermF-ermAM cassette was introduced into the *hmuR* gene by three steps of PCR to yield a hmuR-erm-hmuR DNA fragment as described previously [59]. The fragment was then introduced into *P. gingivalis* 33,277 by electroporation. The hmuR deficient mutant (ΔhmuR) was generated via a double crossover event that replaces hmuR with the hmuR-erm-hmuR DNA fragment in the 33,277 chromosome. The mutants were selected on TSB plates containing erythromycin (5 μg/ml), and the mutation was confirmed by PCR analysis. Growth rates of mutant and parent strains were equivalent.

Quantitative Community Development Assays

(i) Crystal violet assay. Homotypic community formation by *P. gingivalis* was quantified by a microtiter plate assay [60], as adapted for *P. gingivalis* [61]. Parental and mutant strains in early log phase (2×10^8 cells) were incubated at 37°C anaerobically for 24 hr. Wells were washed, stained with 1% crystal violet and destained with 95% ethanol. Absorbance at 595 nm was determined in a Benchmark microplate reader. (ii) ELISA. *F. nucleatum* was incubated at 37°C anaerobically for 36 hr in microtiter plate wells. After washing, parental and mutant *P. gingivalis* strains (2×10^6 cells) were incubated with the fusobacterial biofilm at 37°C anaerobically for 24 hr. *Porphyromonas gingivalis* accumulation was detected with antibodies to whole cells (1:10,000) followed by peroxidase-conjugated secondary antibody (1:3,000), each for 1 hr at 37°C. Antigen-antibody binding was determined by a colorimetric reaction using the 3,3',5,5'-tetramethylbenzidine (TMB) liquid substrate, and absorbance at 655 nm. *Porphyromonas gingivalis* antibody binding to the fusobacterial biofilm alone was subtracted as background. (iii) Confocal microscopy assay. (A) Single species. *Porphyromonas gingivalis* was stained with 4',6-diamidino-2-phenylindole (50 µg ml⁻¹) and 2 × 10⁶ cells in rPBS incubated anaerobically at 37°C for 16 hr with rocking in individual chambers of the CultureWell coverglass system (Grace Bio Labs). Chambers were washed three times in rPBS. (B) Dual-species. Heterotypic *P. gingivalisS. gordonii* communities were generated as described previously [15]. *Streptococcus gordonii* cells were labeled with hexidium iodide (15 µg ml⁻¹), then cultured anaerobically at 37°C for 16 hr with rocking in CultureWell chambers. *Porphyromonas gingivalis* was stained with 5-(and-6)-carboxyfluorescein, succinimidyl ester (10 µg ml⁻¹), and 2 × 10⁶ cells in rPBS were reacted with the surface attached *S. gordonii* for 24 hr anaerobically at 37°C with rocking. (C) Three species. Surface attached hexidium iodide-stained *S. gordonii* were generated as above. *Fluorescein* stained *F. nucleatum* (2 × 10⁶ cells in rPBS) reacted with *S. gordonii* for 24 hr anaerobically at 37°C with rocking. The coverglass was then washed with rPBS to remove non-attached bacteria. *Porphyromonas gingivalis* was stained with 4',6-diamidino-2-phenylindole (50 µg ml⁻¹) and 2 × 10⁶ cells in rPBS were added and further incubated for 24 hr anaerobically at 37°C with rocking. Communities were observed on a Bio-Rad Radiance 2,100 confocal laser scanning microscope (Blue Diode/Ar/HeNe) system with a Nicon ECLIPSE TE300 inverted light microscope and 40 × objective using reflected laser light of combined 405, 488, and 543 nm wavelengths where appropriate. A series of fluorescent optical x-y sections were collected to create digitally reconstructed images (z-projection of x-y sections) of the communities with Image J V1.34s (National Institutes of Health) or Laser Sharp software (Bio-Rad). Z stacks of the x-y sections of CLSM were converted to composite images with "Iso Surface" functions of the "Surpass" option on Imaris 5.0.1 (Bitplane AG; Zurich, Switzerland) software. Iso Surface images of *P. gingivalis* were created at threshold of 20 and smoothed with Gaussian Filter function at 0.5 width, and *P. gingivalis* biovolume was calculated.

Biofilm assays were repeated independently three times with each strain in triplicate. Crystal violet results were compared by t-tests. Biovolume calculations were compared with a t-test using the SPSS statistics software.

CONCLUSION

Complex multi-species biofilms such as pathogenic dental plaque accumulate through a series of developmental steps involving attachment, recruitment, maturation, and detachment. Choreographed patterns of gene and protein expression characterize each of these steps. In this study we developed a model of the early stages of plaque development whereby three compatible species accreted into simple communities. *Porphyromonas gingivalis* increased in biomass due to attachment and recruitment, and this allowed us to catalog differential protein expression in *P. gingivalis* consequent to contact dependent interbacterial signaling and communication through short range soluble mediators. The proteomic analysis indicated that around 40% of *P. gingivalis* proteins exhibit changes in abundance in a community with *F. nucleatum* and *S. gordonii*, implying extensive interactions among the organisms. The proteomic results were consistent with the formation of a favorable environment in a *P. gingivalis–F. nucleatum–S. gordonii* community, wherein *P. gingivalis* showed evidence of increased protein synthesis and decreased stress. Moreover, nutrient transfer may occur among the constituents of the community. As evidenced by hmuR, these proteins may have a functional role in the development of multi-species communities and ultimately shape the pathogenic potential of plaque.

KEYWORDS

- **Bacterial adaptation**
- *Fusobacterium nucleatum*
- *Porphyromonas gingivalis*
- **Protein relative abundances**
- *Streptococcus gordonii*

AUTHORS' CONTRIBUTIONS

Masae Kuboniwa carried out the community construction and analysis by confocal microscopy; Erik L. Hendrickson did the pathway analysis; Qiangwei Xia and Tiansong Wang performed the protein biochemistry, separations, and mass spectrometry; Hua Xie constructed the hmuR mutant; Murray Hackett and Richard J. Lamont conceived the experiments.Murray Hackett, Erik L. Hendrickson, Masae Kuboniwa, and Richard J. Lamont wrote the manuscript. Masae Kuboniwa and Erik L. Hendrickson contributed equally.

ACKNOWLEDGMENTS

This work was supported by NIDCR research grants DE14372, DE12505, and DE11111, and by a Grant-in-Aid for Scientific Research (C)(20592453) from the Ministry of Education, Culture, Sports, Science, and Technology of Japan. We thank the Institute for Systems Biology and Nittin Baliga for the use of Gaggle and assistance with the pathway analysis. We thank Fred Taub for the FileMaker database and assistance with the figures. We thank LANL (Los Alamos National Laboratory) and Gary Xie in particular for bioinformatics support.

Permissions

Chapter 1: Microfabricated Microbial Fuel Cell Arrawas originally published as "Microfabricated Microbial Fuel Cell Arrays Reveal Electrochemically Active Microbes" in *PLoS ONE 8:10, 2009*. Reprinted with permission under the Creative Commons Attribution License or equivalent.

Chapter 2: Quorum Sensing Drives the Evolution of Cooperation in Bacteria was originally published as "Microbial Communication, Cooperation and Cheating: Quorum Sensing Drives the Evolution of Cooperation in Bacteria" in *PLoS ONE 8:17, 2009*. Reprinted with permission under the Creative Commons Attribution License or equivalent.

Chapter 3: Resveratrol as an Antiviral Against Polyomavirus was originally published as "Resveratrol exhibits a strong cytotoxic activity in cultured cells and has an antiviral action against polyomavirus: potential clinical use" in *BioMed Central 7:1, 2009*. Reprinted with permission under the Creative Commons Attribution License or equivalent.

Chapter 4: Novel Anti-viral Peptide Against Avian Influenza Virus H9N2was originally published as "Identification and characterization of a novel anti-viral peptide against avian influenza virus H9N2" in *BioMed Central 6:5, 2009*. Reprinted with permission under the Creative Commons Attribution License or equivalent.

Chapter 5: Prion Disease Pathogenesis was originally published as "Prion Protein Modulates Cellular Iron Uptake: A Novel Function with Implications for Prion Disease Pathogenesis" in *PLoS ONE 2:12, 2009*. Reprinted with permission under the Creative Commons Attribution License or equivalent.

Chapter 6: Activity and Interactions of Liposomal Antibiotics was originally published as "Activity and interactions of liposomal antibiotics in presence of polyanions and sputum of patients with cystic fibrosis" in *PLoS ONE 5:28, 2009*. Reprinted with permission under the Creative Commons Attribution License or equivalent.

Chapter 7: Discovery of Novel Inhibitors of *Streptococcus pneumonia* was originally published as "Discovery of novel inhibitors of *Streptococcus pneumoniae* based on the virtual screening with the homology-modeled structure of histidine kinase (VicK)" in *BioMed Central 6:22, 2009*. Reprinted with permission under the Creative Commons Attribution License or equivalent.

Chapter 8: Archaeosomes Made of *Halorubrum tebenquichense* Total Polar Lipids was originally published as "Archaeosomes made of *Halorubrum tebenquichense* total polar lipids: a new source of adjuvancy" in *BioMed Central 8:13, 2009*. Reprinted with permission under the Creative Commons Attribution License or equivalent.

Chapter 9: Soil Microbe *Dechloromonas aromatica* Str. RCB Metabolic Analysis was originally published as "Metabolic analysis of the soil microbe *Dechloromonas aromatica* str. RCB: indications of a surprisingly complex life-style and cryptic anaerobic pathways for aromatic degradation" in *BioMed Central 8:3, 2009*. Reprinted with permission under the Creative Commons Attribution License or equivalent.

Chapter 10: Genome-scale Reconstruction of the *Pseudomonas putida* KT2440 Metabolic Network was originally published as "Genome-Scale Reconstruction and Analysis of the Pseudomonas putida KT2440 Metabolic Network Facilitates Applications in Biotechnology" in *Puchałka J, Oberhardt MA, Godinho M, Bielecka A, Regenhardt D, et al. (2008) Genome-Scale Reconstruction and Analysis of the Pseudomonas putida KT2440 Metabolic Network Facilitates Applications in Biotechnology. PLoS Comput Biol 4(10): e1000210. Doi: 10.1371/journal.pcbi.10002*. Reprinted with permission under the Creative Commons Attribution License or equivalent.

Chapter 11: A New Cold-adapted B-D-galactosidase from the Antarctic *Arthrobacter* Sp. 32c was originally published as "A new cold-adapted β-D-galactosidase from the Antarctic *Arthrobacter* sp. 32c – gene cloning, overexpression, purification and properties" in *BioMed Central 7:27, 2009*. Reprinted with permission under the Creative Commons Attribution License or equivalent.

Chapter 12: Global Transcriptional Response to Natural Infection by Pseudorabies Virus was originally published as "Global transcriptional response of pig brain and lung to natural infection by Pseudorabies virus" in *BioMed Central 12:1, 2009*. Reprinted with permission under the Creative Commons Attribution License or equivalent.

Chapter 13: Proteomics of *Porphyromonas gingivalis* was originally published as "Proteomics of *Porphyromonas gingivalis* within a model oral microbial community" in *BioMed Central 5:19, 2009*. Reprinted with permission under the Creative Commons Attribution License or equivalent.

References

1

1. Lovley, D. R. (2008). The microbe electric: Conversion of organic matter to electricity. *Curr. Opin. Biotechnol.* **19**, 564–571.

2. Debabov, V. G. (2008). Electricity from microorganisms. *Microbiol.* **77**, 123–131.

3. Lovley, D. R. (2006). Bug juice: Harvesting electricity with microorganisms. *Nat. Rev. Microbiol.* **4**, 497–508.

4. Logan, B. E., Hamelers, B., Rozendal, R., Schrorder, U., Keller, J., et al. (2006). Microbial fuel cells: Methodology and technology. *Environ. Sci. Technol.* **40**, 5181–5192.

5. Rabaey, K. and Verstraete, W. (2005). Microbial fuel cells: novel biotechnology for energy generation. *Trends. Biotechnol.* **23**, 291–298.

6. Logan, B. E. and Regan, J. M. (2006). Microbial fuel cells–challenges and applications. *Environ. Sci. Technol.* **40**, 5172–5180.

7. Rabaey, K., Boon, N., Hofte, M., and Verstraete, W. (2005). Microbial phenazine production enhances electron transfer in biofuel cells. *Environ. Sci. Technol.* **39**, 3401–3408.

8. Marsili, E., Baron, D. B., Shikhare, I. D., Coursolle, D., Gralnick, J. A., et al. (2008). *Shewanella* Secretes flavins that mediate extracellular electron transfer. *Proc. Nat. Acad. Sci. USA* **105**, 3968–3973.

9. Gorby, Y. A., Yanina, S., McLean, J. S., Rosso, K. M., Moyles, D., et al. (2006). Electrically conductive bacterial nanowires produced by *Shewanella oneidensis* strain MR-1 and other microorganisms. *Proc. Nat. Acad. Sci. USA* **103**, 11358–11363.

10. El-Naggar, M. Y., Gorby, Y. A., Xia, W., and Nealson, K. H. (2008). The Molecular Density of States in Bacterial Nanowires. *Biophys. J.* **95**(1), L10–12.

11. Reguera, G., Nevin, K. P., Nicoll, J. S., Covalla, S. F., Woodard, T. L., et al. (2006). Biofilm and nanowire production leads to increased current in *Geobacter sulfurreducens* fuel cells. *Appl. Environ. Microbiol.* **72**, 7345–7348.

12. Du, Z. W, Li, H. R., and Gu, T. Y. (2007). A state of the art review on microbial fuel cells: A promising technology for wastewater treatment and bioenergy. *Biotechnol. Adv.* **25**(5), 464–482.

13. Rabaey, K. and Keller, J. (2008). Microbial fuel cell cathodes: From bottleneck to prime opportunity? *Water Sci. Technol.* **57**, 655–659.

14. Shantaram, A., Beyenal, H., Raajan, R., Veluchamy, A., and Lewandowski, Z. (2005). Wireless sensors powered by microbial fuel cells. *Environ. Sci. Technol.* **39**, 5037–5042.

15. Tender, L. M., Gray, S. A., Groveman, E., Lowy, D. A., Kauffman, P., et al. (2008). The first demonstration of a microbial fuel cell as a viable power supply: Powering a meteorological buoy. *J. Power Sources* **179**, 571–575.

16. Wilkinson, S. (2000). "Gastrobots"Benefits and challenges of microbial fuel cells in food powered robot applications. *Autonomous Robots* **9**, 99–111.

17. LaVan, D. A., McGuire, T., and Langer, R (2003). Small-scale systems for *in vivo* drug delivery. *Nat. Biotech.* 21, 1184–1191.

18. Zuo, Y., Maness, P. C., and Logan, B. E. (2006). Electricity production from steam-exploded corn stover biomass. *Energy Fuels* **20**, 1716–1721.

19. Clauwaert, P., Aelterman, P., Pham, T. H., De Schamphelaire, L., Carballa, M., et al. (2008). Minimizing losses in bio-electrochemical systems: The road to applications. *Appl. Microbiol. Biotechnol.* **79**, 901–913.

20. Zuo, Y., Xing, D. F., Regan, J. M., and Logan, B. E. (2008). Isolation of the exoelectrogenic bacterium *Ochrobactrum anthropi* YZ-1 by using a U-tube microbial fuel cell. *Appl. Environ. Microbiol.* **74**, 3130–3137.

21. Xing, D., Zuo, Y., Cheng, S., Regan, J. M., and Logan, B. E. (2008). Electricity generation by *Rhodopseudomonas palustris* DX-1. *Environ. Sci. Technol.* **42**, 4146–4151.

22. Izallalen, M., Mahadevan, R. Burgard, A., Postier, B., Didonato, R., et al. (2008). *Geo-*

bacter sulfurreducens strain engineered for increased rates of respiration. *Metab. Eng.* **10**, 267–275.

23. Tang, Y. J. J., Laidlaw, D., Gani, K., and Keasling, J. D. (2006). Evaluation of the effects of various culture conditions on Cr(VI) reduction by *Shewanella oneidensis* MR-1 in a novel high-throughput mini-bioreactor. *Biotechnol. Bioeng.* **95**, 176–184.

24. Jadhav, G. S. and Ghangrekar, M. M. (2009). Performance of microbial fuel cell subjected to variation in pH, temperature, external load and substrate concentration. *Bioresour. Technol.* **100**, 717–723.

25. Logan, B. E. (2009). Exoelectrogenic bacteria that power microbial fuel cells. *Nat. Rev. Microbiol.* **7**(5), 375381.

26. Logan, B. E. and Regan, J. M. (2006). Electricity-producing bacterial communities in microbial fuel cells. *Trends Microbiol.* **14**, 512–518.

27. Rabaey, K., Boon, N., Siciliano, S. D., Verhaege, M., and Verstraete, W. (2004). Biofuel cells select for microbial consortia that self-mediate electron transfer. *Appl. Environ. Microbiol.* **70**, 5373–5382.

28. Hoheisel, J. D. (2006). Microarray technology: Beyond transcript profiling and genotype analysis. *Nat. Rev. Genet.* **7**, 200–210.

29. Siu, C. P. B. and Chiao, M. (2008). A Microfabricated PDMS Microbial Fuel Cell. *J. Microelectromech. Syst.* **17**, 1329–1341.

30. Biffinger, J., Ribbens, M., Ringeisen, B., Pietron, J., Finkel, S., et al. (2009). Characterization of Electrochemically Active Bacteria Utilizing a High-Throughput Voltage-Based Screening Assay. *Biotechnol. Bioeng.* **102**, 436–444.

31. Weibel, D. B., DiLuzio, W. R., and Whitesides, G. M. (2007). Microfabrication meets microbiology. *Nat. Rev. Microbiol.* **5**, 209–218.

32. Xia, Y. N. and Whitesides, G. M. (1998). Soft lithography. *Annul. Rev. Mater. Scie.* **28**, 153–184.

33. Lowy, D. A., Tender, L. M., Zeikus, J. G., Park, D. H., and Lovley, D. R. (2006). Harvesting energy from the marine sediment-water interface II Kinetic activity of anode materials. *Biosens. Bioelectron.* **21**, 2058–2063.

34. Park, D. H. and Zeikus, J. G. (2002). Impact of electrode composition on electricity generation in a single-compartment fuel cell using *Shewanella putrefaciens*. *Appl. Microbiol. Biotechnol.* **59**, 58–61.

35. Richter, H., McCarthy, K., Nevin, K. P., Johnson, J. P., Rotello, V. M., et al. (2008). Electricity generation by *Geobacter sulfurreducens* attached to gold electrodes. *Langmuir* **24**, 4376–4379.

36. Ringeisen, B. R., Henderson, E., Wu, P. K., Pietron, J., Ray, R., et al. (2006). High Power Density from a Miniature Microbial Fuel Cell Using *Shewanella oneidensis* DSP10. *Environ. Sci. Technol.* **40**, 2629–2634.

37. Fredrickson, J. K., Romine, M. F., Beliaev, A. S., Auchtung, J. M., Driscoll, M. E., et al. (2008). Towards environmental systems biology of *Shewanella*. *Nat. Rev. Microbiol.* **6**, 592–603.

38. Crittenden, S. R., Sund, C. J., and Sumner, J. J. (2006). Mediating electron transfer from bacteria to a gold electrode via a self-assembled monolayer. *Langmuir* **22**, 9473–9476.

39. Pearce, C. I., Christie, R., Boothman, C., von Canstein, H., Guthrie, J. T., et al. (2006). Reactive azo dye reduction by *Shewanella* strain j18 143. *Biotechnol. Bioeng.* **95**, 692–703.

40. Pham, T. H., Boon, N., De Maeyer, K., Hofte, M, Rabaey, K., et al. (2008). Use of *Pseudomonas* species producing phenazine-based metabolites in the anodes of microbial fuel cells to improve electricity generation. *Appl. Microbiol. Biotechnol.* **80**, 985–993.

41. Pham, T. H., Boon, N., Aelterman, P., Clauwaert, P., De Schamphelaire, L., et al. (2008). Metabolites produced by *Pseudomonas* sp enable a Gram-positive bacterium to achieve extracellular electron transfer. *Appl. Microbiol. Biotechnol.* **77**, 1119–1129.

42. Bretschger, O., Obraztsova, A., Sturm, C. A., Chang, I. S., Gorby, Y. A., et al. (2007). Current production and metal oxide reduction by *Shewanella oneidensis* MR-1 wild type and mutants. *Appl. Environ. Microbiol.* **73**, 7003–7012.

43. Ishii, S., Watanabe, K., Yabuki, S., Logan, B. E., and Sekiguchi, Y. (2008). Compari-

son of Electrode Reduction Activities of *Geobacter sulfurreducens* and an Enriched Consortium in an Air-Cathode Microbial Fuel Cell. *Appl. Environ. Microbiol.* **74**, 7348–7355.

2

1. Hardin, G. (1968). The tragedy of the commons. *Science* **162**, 1243–1244.

2. Hamilton, W. D. (1964). The genetical evolution of social behaviour, I & II. *J. Theor. Biol.* **7**, 1–52.

3. West, S. A., Griffin, A. S., and Gardner, A. (2007). Evolutionary explanations for cooperation. Curr. Biol. **17**, R661–72.

4. Fletcher, J. A. and Doebeli, M. (2008). A simple and general explanation for the evolution of altruism. Proceedings.

5. Queller, D. C. (1992). Quantitative genetics, inclusive fitness, and group selection. *American Naturalist* **139**, 540–558.

6. Frank, S. A. (1998). *Foundations of social evolution..* Princeton University Press, Princeton, NJ.

7. Crespi, B. J. (2001). The evolution of social behavior in microorganisms. *Trends ecol. evol. (Personal edition)* **16**, 178–183.

8. Velicer, G. J. (2003). Social strife in the microbial world. *Trends Microbiol.* **11**, 330–7.

9. Parsek, M. R. and Greenberg, E. P. (2005). Sociomicrobiology: The connections between quorum sensing and biofilms. *Trends Microbiol.* **13**, 27–33.

10. West, S. A., Griffin, A. S., Gardner, A., and Diggle, S. P. (2006). Social evolution theory for microorganisms. *Nat. rev.* **4**, 597–607.

11. Shapiro, J. A. (1998). Thinking about bacterial populations as multicellular organisms. *Annu. Rev. Microbiol.* **52**, 81–104.

12. Henke, J. M. and Bassler, B. L. (2004). Bacterial social engagements. *Trends cell bio.* **14**, 648–56.

13. Schaber, J. A., Carty, N. L., McDonald, N. A., Graham, E. D., Cheluvappa, R., et al. (2004). Analysis of quorum sensing-deficient clinical isolates of *Pseudomonas aeruginosa. J. med. microbio.* **53**, 841–53.

14. Smith, E. E., Buckley, D. G., Wu, Z., Saenphimmachak, C., Hoffman, L. R., et al. (2006). Genetic adaptation by *Pseudomonas aeruginosa* to the airways of cystic fibrosis patients. *Proceedings of the National Academy of Sciences of the United States of America* **103**, 8487–92.

15. Brookfield, J. F.Y. (1998). Quorum sensing and group selection. *Evolution* **52**, 1263–1269.

16. Brown, S. P. and Johnstone, R. A. (2001). Cooperation in the dark: Signalling and collective action in quorum-sensing bacteria. *Proceedings* **268**, 961–5.

17. Czaran, T. and Hoekstra, R. F. (2007). A spatial model of the evolution of quorum sensing regulating bacteriocin production. *Behav. Ecol.* **18**, 866–873.

18. Diggle, S. P., Griffin, A. S., Campbell, G. S., and West, S. A. (2007). Cooperation and conflict in quorum-sensing bacterial populations. *Nature* **450**, 411–4.

19. Toffoli, T. and Margolus, N. (1987). *Cellular Automata Machines: A New Environment For Modeling.* MIT Press, Cambridge, MA.

20. Maynard, S. J. (1982). *Evolution and the Theory of Games.* Cambridge University Press, Cambridge, UK.

21. Czaran, T. L., Hoekstra, R. F., and Pagie, L. (2002). Chemical warfare between microbes promotes biodiversity. *Proceedings of the National Academy of Sciences of the United States of America* **99**, 786–90.

22. Keller, L. and Surette, M. G. (2006). Communication in bacteria: An ecological and evolutionary perspective. *Nat. rev.* **4**, 249–58.

23. Williams, P., Winzer, K., Chan, W. C., and Camara, M. (2007). *Look who's talking: communication and quorum sensing in the bacterial world.* Philosophical transactions of the Royal Society of London **362**, 1119–34.

24. West, S. A. and Buckling, A. (2003). Cooperation, virulence and siderophore production in bacterial parasites. *Proceedings* **270**, 37–44.

25. Griffin, A. S., West, S. A., and Buckling, A. (2004). Cooperation and competition in pathogenic bacteria. *Nature* **430**, 1024–7.

26. Buckling, A., Harrison, F., Vos, M., Brockhurst, M. A., Gardner, A., et al. (2007).

Siderophore-mediated cooperation and virulence in *Pseudomonas aeruginosa*. FEMS *microbiol. ecol.* **62**, 135–41.

27. Denervaud, V., TuQuoc, P., Blanc, D., Favre-Bonte, S., Krishnapillai, V., et al. (2004). Characterization of cell-to-cell signaling-deficient *Pseudomonas aeruginosa* strains colonizing intubated patients. *J. clin. microbiol.* **42**, 554–62.

28. Lee, B., Haagensen, J. A., Ciofu, O., Andersen, J. B., Hoiby, N., and Molin, S. (2005). Heterogeneity of biofilms formed by non-mucoid *Pseudomonas aeruginosa* isolates from patients with cystic fibrosis. *J. clin. microbiol.* **43**, 5247–55.

29. Harrison, F., Browning, L. E., Vos, M., and Buckling, A. (2006). Cooperation and virulence in acute *Pseudomonas aeruginosa* infections. *BMC biology* 4, 21.

30. Sandoz, K. M., Mitzimberg, S. M., and Schuster, M. (2007). Social cheating in *Pseudomonas aeruginosa* quorum sensing. *Proceedings of the National Academy of Sciences of the United States of America* **104**, 15876–15881.

31. Rumbaugh, K. P., Diggle, S. P., Watters, C. M., Ross-Gillespie, A., and Griffin, A. S., et al. (2009). Quorum Sensing and the Social Evolution of Bacterial Virulence. *Curr. Biol.* **19**, 341–345.

32. Slater, H., Alvarez-Morales, A., Barber, C. E., Daniels, M. J., and Dow, J. M. (2000). A two-component system involving an HD-GYP domain protein links cell-cell signaling to pathogenicity gene expression in *Xanthononas campestris. Mol. Microbiol.* **38**, 986–1003.

33. Chatterjee, S., Wistrom, C., Lindow, S. E. (2008). A cell-cell signaling sensor is required for virulence and insect transmission of *Xylella fastidiosa. Proceedings of the National Academy of Sciences of the United States of America* **105**(7), 2670–2675.

3

1. Tooze, J., (Ed.) (1982). *Molecular biology of tumor viruses: DNA Tumor Viruses. second edition.* Cold Spring Harbor Laboratory Press, New York, USA.

2. Howley, P. M. and Livingston, D. M. (2009). Small DNA tumor viruses: Large contributors to biomedical sciences. *Virology* **384**, 2569.

3. Yaniv, M. (2009). Small DNA tumour viruses and their contributions to our understanding of transcription control. *Virology* **384**, 369374.

4. Moens, U. and Johannessen, M. (2008). Human polyomaviruses and cancer: Expanding repertoire. *J. Dtsch. Dermatol. Ges.* **6**, 704708.

5. Jiang, M., Abend, J. R., Johnson ,S. F., and Imperiale, M. J. (2009). The role of polyomaviruses in human disease. *Virology* **384**, 26673.

6. zur Hausen, H. (2008). Novel human polyomavirusesRe-emergence of a well known virus family as possible human carcinogens. *Int. J. Cancer* **123**, 247250.

7. Khalili, K., Sariyer, I. K., and Safak, M. (2008). Small tumor antigen of polyomaviruses: Role in viral life cycle and cell transformation. *J. Cell Physiol.* **215**, 309319.

8. Iacoangeli, A., Melucci-Vigo, G., and Risuleo, G. (2000). Mechanism of the inhibition of murine polyomavirus DNA replication induced by the ionophore monensin. *Biochimie* **82**, 3539.

9. Campanella, L., Delfini, M., Ercole, P., Iacoangeli, A., and Risuleo, G. (2002). Molecular characterization and action of usnic acid: A drug that inhibits proliferation of mouse polyomavirus *in vitro* and its main target is RNA transcription. *Biochimie* **84**, 329334.

10. Pastore, D., Iacoangeli, A., Galati, G., Izzo, L., Fiori, E., Caputo, M., Castelli, M., and Risuleo, G. (2004). Variations of telomerase activity in cultured mouse fibroblasts upon proliferation of polyomavirus. *Anticancer Res.* **24**, 791794.

11. Pillich, R. T., Scarsella, G., Galati, G., Izzo, L., Iacoangeli, A., Castelli, M., and Risuleo, G. (2005). The diimide drug PIPER has a cytotoxic dose-dependent effect *in vitro* and inhibits telomere elongation in HELA cells. *Anticancer Res* **25**, 33413346.

12. Pillich, R. T., Scarsella, G., and Risuleo, G. (2005). Reduction of apoptosis through the mitochondrial pathway by the administra-

tion of acetyl-L-carnitine to mouse fibroblasts in culture. *Exp. Cell Res.* **306**, 18.

13. Di Ilio, V., Pasquariello, N., Esch, S. A., Cristofaro, M., Scarsella, G., and Risuleo, G. (2006). Cytotoxic and antiproliferative effects induced by a non terpenoid polar extract of A. indica seeds on 3T6 murine fibroblasts in culture. *Molec. Cell Biochem.* **287**, 6977.

14. Piccioni, F., Borioni, A., Delfini, M., Del Giudice, M. R., Mustazza, C., Rodomonte, A., and Risuleo, G. (2007). Metabolic Alterations in Cultured Mouse Fibroblasts Induced by an Inhibitor of the Tyrosine Kinase Receptor FGFR-1. *Analyt.l Biochem.* **367**, 111121.

15. Calandrella, N., Risuleo, G., Scarsella, G., Mustazza, C., Castelli, M., Galati, F., Giuliani, A., and Galati, G. (2007). Reduction of cell Proliferation induced by PD166866: An Inhibitor of the basic fibroblast growth factor. *J. Exp. Clin. Cancer Res.* **26**, 405409.

16. Schmutterer, H. (2002). *The neem tree and other meliaceous plants*. The neem Foundation, Mumbai, India.

17. Brahmachari, G. (2004). Neem-an omnipotent plant: A retrospection. *Chembiochem.* **5**, 40821.

18. Ricci, F., Berardi, V., and Risuleo, G. (2008). Differential cytotoxicity of MEX: A component of Neem oil whose action is exerted at the cell membrane level. *Molecules* **14**, 122132.

19. Bonincontro, A., Di Ilio, V., Pedata, O., and Risuleo, G. (2007). Dielectric properties of the plasma membrane of cultured murine fibroblasts treated with a nonterpenoid extract of *Azadirachta indica* seeds. *J. Membr. Biol.* **215**, 7579.

20. Parida, M. M., Upadhyay, C., Pandya, G., and Jana, A. M. (2002). Inhibitory potential of neem (*Azadirachta indica* Juss) leaves on dengue virus type-2 replication. *J. Ethnopharmacol.* **79**, 273278.

21. López-Vélez, M., Martínez-Martínez, F., and Del Valle-Ribes, C. (2003). The study of phenolic compounds as natural antioxidants in wine. *Crit. Rev. Food Sci. Nutr.* **43**, 233244.

22. Palamara, A. T., Nencioni, L., Aquilano, K., De Chiara, G., Hernandez, L., Cozzolino, F., Ciriolo, M. R., and Garaci, E. (2005). Inhibition of influenza A virus replication by resveratrol. *J. Infect. Dis.* **191**, 17191729.

23. Docherty, J. J., Sweet, T. J., Bailey, E., Faith, S. A., and Booth, T. (2006). Resveratrol inhibition of varicella-zoster virus replication in vitro. *Antiviral. Res.* **72**, 171177.

24. Docherty, J. J., Fu, M. M., Hah, J. M., Sweet, T. J., Faith, S. A., and Booth, T. (2005). Effect of resveratrol on herpes simplex virus vaginal infection in the mouse. *Antiviral. Res.* **67**, 15562.

25. Faith, S. A., Sweet, T. J., Bailey, E., Booth, T., and Docherty, J. J. (2006). Resveratrol suppresses nuclear factor-kappa B in herpes simplex virus infected cells. *Antiviral. Res.* **72**, 242251.

26. Hirt, B. (1969). Replicating molecules of polyoma virus DNA. *J. Mol. Biol.* **40**, 141144.

27. Mosmann, T. (1983). Rapid colorimetric assay for cellular grow and survival: application to proliferation and cytotoxixity assay. *J. Immunol. Methods.* **65**, 5563.

28. Delmas, D., Lançon, A., Colin, D., Jannin, B., and Latruffe, N. (2006). Resveratrol as a chemopreventive agent: A promising molecule for fighting cancer. *Curr. Drug Targets* **7**, 423442.

29. Saiko, P., Pemberger, M., Horvath, Z., Savinc, I., Grusch, M., Handler, N, Erker, T., Jaeger, W., Fritzer-Szekeres, M., and Szekeres, T. (2008). Novel resveratrol analogs induce apoptosis and cause cell cycle arrest in HT29 human colon cancer cells: Inhibition of ribonucleotide reductase activity. *Oncol. Rep.* **19**, 16211626.

30. Juan, M. E., Wenzel, U., Daniel, H., and Planas, J. M. (2008). Resveratrol induces apoptosis through ROS-dependent mitochondria pathway in HT-29 human colorectal carcinoma cells. *J. Agric. Food Chem.* **56**, 48134818.

31. Singh, M. and Singh, N. (2009). Molecular mechanism of curcumin induced cytotoxicity in human cervical carcinoma cells. *Mol. Cell Biochem.* **325**, 107119.

32. Lilley, B. N., Gilbert, J. M., Ploegh, H. L., and Benjamin, T. L. (2006). Murine polyomavirus requires the endoplasmatic reticu-

lum protein Derlin-2 to initiate infection. *J. Virol.* **80**, 873940.

4

1. Murphy, B. R. and Webster, R. G. (1996). Orthomyxoviruses. In *Fiel's Virology*. B. N. Fields, D. M. Knipe, and P. M. Howley (Eds.). Lippincott-Raven Publishers, Philadelphia, pp. 1397–1445.

2. Fouchier, R. A., Munster, V., Wallensten, A., Bestebroer, T. M., Herfst, S., Smith, D., Rimmelzwaan, G. F., Olsen, B., and Osterhaus, A. D. (2005). Characterization of a novel influenza A virus hemagglutinin subtype (H16) obtained from black-headed gulls. *J. Virol.* **79**, 2814–2822.

3. Brown, I. H., Banks, J., Manvell, R. J., Essen, S. C., Shell, W., Slomka, M., Londt, B., and Alexander, D. J. (2006). Recent epidemiology and ecology of influenza A viruses in avian species in Europe and the Middle East. *Dev. Biol. (Basel)* **124**, 45–50.

4. Nili, H. and Asasi, K. (2002). Natural cases and an experimental study of H9N2 avian influenza in commercial broiler chickens of Iran. *Avian Pathol.* **31**, 247–252.

5. Butt, K. M., Smith, G. J., Chen, H., Zhang, L. J., Leung, Y. H., Xu, K. M., Lim, W., Webster, R. G., Yuen, K. Y., Peiris, J. S., et al. (2005). Human infection with an avian H9N2 influenza A virus in Hong Kong in 2003. *J. Clin. Microbiol.* **43**, 57605767.

6. Lin, Y. P., Shaw, M., Gregory, V., Cameron, K., Lim, W., Klimov, A., Subbarao, K., Guan, Y., Krauss, S., Shortridge, K., et al. (2000). Avian-to-human transmission of H9N2 subtype influenza A viruses: Relationship between H9N2 and H5N1 human isolates. *Proc. Natl. Acad. Sci. USA* **97**, 96549658.

7. Peiris, M., Yuen, K. Y., Leung, C. W., Chan, K. H., Ip, P. L., Lai, R. W., Orr, W. K., and Shortridge, K. F. (1999). Human infection with influenza H9N2. *Lancet* **354**, 916917.

8. Wan, H., Sorrell, E. M., Song, H., Hossain, M. J., Ramirez-Nieto, G., Monne, I., Stevens, J., Cattoli, G., Capua, I., Chen, L. M., et al. (2008). Replication and transmission of H9N2 influenza viruses in ferrets: Evaluation of pandemic potential. *PLoS ONE* **3**, e2923.

9. Nicholson, K. G., Wood, J. M., and Zambon, M. (2003). Influenza. *Lancet* **362**, 17331745.

10. Wang, C., Takeuchi, K., Pinto, L. H., and Lamb, R. A. (1993). Ion channel activity of influenza A virus M2 protein: Characterization of the amantadine block. *J. Virol* **67**, 55855594.

11. Moscona, A. (2005). Neuraminidase inhibitors for influenza. *N. Engl. J. Med.* **353**, 13631373.

12. Bright, R. A., Shay, D. K., Shu, B., Cox, N. J., and Klimov, A. I. (2006). Adamantane resistance among influenza A viruses isolated early during the 2005–2006 influenza season in the United States. *JAMA* **295**, 891894.

13. Hurt, A. C., Selleck, P., Komadina, N., Shaw, R., Brown, L., and Barr, I. G. (2007) Susceptibility of highly pathogenic A(H5N1) avian influenza viruses to the neuraminidase inhibitors and adamantanes. *Antiviral. Res.* **73**, 228–231.

14. Hayden, F. G. (2006). Respiratory viral threats. *Curr. Opin. Infect. Dis.* **19**, 169–178.

15. Chen, J. and Deng, Y. M. (2009). Influenza virus antigenic variation, host antibody production and new approach to control epidemics. *Virol. J.* **6**, 30.

16. de, Jong. M. D., Thanh, T. T., Khanh, T. H., Hien, V. M., Smith, G. J. D., Chau, N. V., Cam, B. V., Qui, P. T., Ha, D. Q., Guan, Y., et al.(2005). Oseltamivir resistance during treatment of influenza A (H5N1) infection. N. Engl. J. Med. **353**, 2667–2672.

17. Le, Q. M., Kiso, M., Someya, K., Sakai, Y. T., Nguyen, T. H., Nguyen, K. H., Pham, N. D., Ngyen, H. H., Yamada, S., Muramoto, Y., et al. (2005). Avian flu: Isolation of drug-resistant H5N1 virus. *Nature* **437**, 1108.

18. Puthavathana, P., Auewarakul, P., Charoenying, P. C., Sangsiriwut, K., Pooruk, P., Boonnak, K., Khanyok, R., Thawachsupa, P., Kijphati, R., and Sawanpanyalert, P. (2005). Molecular characterization of the complete genome of human influenza H5N1 virus isolates from Thailand. *J. Gen. Virol.* **86**, 423–433.

19. Benhar, I. (2001). Biotechnological applications of phage and cell display. *Biotechnol. Adv.* **19**, 1–33.

20. Devlin, J. J., Panganiban, L. C., and Devlin, P. E. (1990). Random peptide libraries: A source of specific protein binding molecules. *Science* **249**, 404–406.

21. Katz, B. A. (1995). Binding to protein targets of peptidic leads discovered by phage display: Crystal structures of streptavidin-bound linear and cyclic peptide ligands containing the HPQ sequence. *Biochemistry* **34**, 15421–15429.

22. Lam, K. S., Salmon, S. E., Hersh, E. M., Hruby, V. J., Kazmierski, W. M., and Knapp, R. J. (1991). A new type of synthetic peptide library for identifying ligand-binding activity. *Nature* **354**, 82–84.

23. Bair, C. L., Oppenheim, A., Trostel, A., Prag, G., and Adhya, S.(2008). A phage display system designed to detect and study protein-protein interactions. *Mol. Microbiol.* **67**, 719–728.

24. Mezo, A. R., McDonnell, K. A., Castro, A., and Fraley, C. (2008). Structure-activity relationships of a peptide inhibitor of the human FcRn:human IgG interaction. *Bioorg. Med. Chem.* **16**, 6394–6405.

25. Ramanujam, P., Tan, W. S., Nathan, S., and Yusoff, K. (2002). Novel peptides that inhibit the propagation of Newcastle disease virus. *Arch. Virol.* **147**, 981–993.

26. Song, J. M., Park, K. D., Lee, K. H., Byun, Y. H., Park, J. H., Kim, S. H., Kim, J. H., and Seong, B. L. (2007). Biological evaluation of anti-influenza viral activity of semisynthetic catechin derivatives. *Antiviral. Res.* **76**, 178–185.

27. Magdesian, M. H., Carvalho, M. M., Mendes, F. A., Saraiva, L. M., Juliano, M. A., Juliano, L., Garcia-Abreu, J., and Ferreira, S. T.(2008). Amyloid-beta binds to the extracellular cysteine-rich domain of Frizzled and inhibits Wnt/beta-catenin signaling. *J .Biol. Chem.* **283**, 9359–9368.

28. Trabocchi, A., Scarpi, D., and Guarna, A. (2008). Structural diversity of bicyclic amino acids. *Amino Acids* **341**–24.

29. Thomson, L. W., Garbee, C. I., Hibbitts, S., Brinckerhoff, C. H., Pierce, R. A., Chianese-Bullock, K. A., Deacon, D. H., Engelhard, V. H., and Slingluff, C. L. Jr. (2004). Competition among peptides in melanoma vaccines for binding to MHC molecules. *J. Immunother.* **27**, 425–431.

30. Kaushik-Basu, N., Basu, A., and Harris, D. (2008). Peptide inhibition of HIV-1: current status and future potential. *BioDrugs* **22**, 161–175.

31. Hrobowski, Y. M., Garry, R. F., and Michael, S. F. (2005). Peptide inhibitors of dengue virus and West Nile virus infectivity. *Virol. J.* **2**, 49.

32. Sulochana, K. N., and Ge, R. (2007). Developing antiangiogenic peptide drugs for angiogenesis-related diseases. *Curr. Pharm .Des.* **13**, 2074–2086.

33. Jones, J. C., Turpin, E. A., Bultmann, H., Brandt, C. R., and Schultz-Cherry, S. (2006). Inhibition of influenza virus infection by a novel antiviral peptide that targets viral attachment to cells. *J. Virol.* **80**, 11960–11967.

34. Lamb, R. A. and Krug, R. M. (1996). Orthomyxoviridae: The viruses and their replication. In *Field's Virology*. B. N. Fields, D. M. Knipe, and P. M. Howley (Eds.). Raven Publishers, Philadelphia, pp. 1353–1395.

35. Mishin, V. P., Novikov, D., Hayden, F. G., and Gubareva, L. V. (2005). Effect of hemagglutinin glycosylation on influenza virus susceptibility to neuraminidase inhibitors. *J. Virol.* **79**, 12416–12424.

36. Garten, W. and Klenk, H. D. (2008). Cleavage activation of influenza virus hemagglutinin and its role in pathogenesis. In *Avian Influenza*. H. D. Klenk, M. N. Matrosovich, and J. Stech (Eds.). Karger, Basel, pp. 156–167.

37. Bradford, M. M. (1976). A rapid and sensitive method for the quantitation of microgram quantities of protein utilizing the principle of protein-dye binding. *Anal. Biochem.* **72**, 248–254.

38. Sambrook, J., Fritsch, E. F., and Maniatis, T. (1989). Moelcular Cloning: A laboratory manual. Cold Spring Harbor Laboratories Press, New York.

39. Smith, G. P. and Scott, J. K. (1993). Libraries of peptides and proteins displayed on filamentous phage. *Methods Enzymol.* **217**, 228–257.

40. Aymard-Henry, M., Coleman, M. T., Dowdle, W. R., Laver, W. G., Schild, G. C., and Webster, R. G. (1973). Influenzavirus neuraminidase and neuraminidase-inhibition test procedures. *Bull World Health Organ.* **48**, 199–202.

41. Ausubel, F. M., Breat, R., Kingston, R. E., Moore, D. D., Seidman, J. G., Smith, J. A., et al. (1994). *Current Protocols in Molecular Biology.* Wiley, New York.

5

1. Caughey, B. and Baron, G. S. (2006). Prions and their partners in crime. *Nature* **443**, 803–810.

2. Aguzzi, A. and Heikenwalder, M. (2006). Pathogenesis of prion diseases: Current status and future outlook. *Nat. Rev. Microbiol.* **4**, 765–775.

3. Tatzelt, J. and Schätzl, H. M. (2007) Molecular basis of cerebral neurodegeneration in prion diseases. *FEB. S. J.* **274**, 606–611.

4. Saá, P., Castilla, J., and Soto, C. (2006). Ultra-efficient replication of infectious prions by automated protein misfolding cyclic amplification. *J. Biol. Chem.* **281**, 35245–35252.

5. Deleault, N. R., Harris, B. T., Rees, J. R., and Supattapone, S. (2007). Formation of native prions from minimal components *in vitro. Proc. Natl. Acad. Sci. USA* **104**, 9741–9746.

6. Aguzzi, A., Heikenwalder, M., and Polymenidou, M. (2007). Insights into prion strains and neurotoxicity. *Nat. Rev. Mol. Cell Biol.* **8**, 552–561.

7. Harris, D. A. and True, H. L. (2006). New insights into prion structure and toxicity. *Neuron* **50**, 353–357.

8. Mallucci, G., Dickinson, A., Linehan, J., Klohn, P. C., Brandner, S., et al. (2003). Depleting neuronal PrP in prion infection prevents disease and reverses spongiosis. *Science* **302**, 871–874.

9. Chesebro, B., Trifilo, M., Race, R., Meade-White, K., Teng, C., et al. (2005). Anchorless prion protein results in infectious amyloid disease without clinical scrapie. *Science* **308**, 435–439.

10. Jeffrey, M., Goodsir, C. M., Race, R. E., and Chesebro, B. (2004). Scrapie-specific neuronal lesions are independent of neuronal PrP expression. *Ann. Neurol.* **55**, 781–792.

11. Roucou, X. and LeBlanc, A. C. (2005). Cellular prion protein neuroprotective function: Implications in prion diseases. *J. Mol. Med.* **83**, 3–11.

12. Westergard, L., Christensen, H. M., and Harris, D. A. (2007). The cellular prion protein (PrP(C)): Its physiological function and role in disease. *Biochim. Biophys. Acta.* **1772**, 629–644.

13. Brown, D. R., Qin, K., Herms, J. W., Madlung, A., Manson, J., et al. (1997). The cellular prion protein binds copper *in vivo. Nature* **390**, 684–687.

14. Pauly, P. C. and Harris, D. A. (1998). Copper stimulates endocytosis of the prion protein. *J. Biol. Chem.* **273**, 33107–33110.

15. Kim, N. H., Park, S. J., Jin, J. K., Kwon, M. S., Choi, E. K., et al. (2000). Increased ferric iron content and iron-induced oxidative stress in the brains of scrapie-infected mice. *Brain Res.* **884**, 98–103.

16. Hur, K., Kim, J. I., Choi, S. I., Choi, E. K., Carp, R. I., et al. (2002). The pathogenic mechanisms of prion diseases. *Mech. Ageing Dev.* **123**, 1637–1647.

17. Basu, S., Mohan, M. L., Luo, X., Kundu, B., Kong, Q., et al. (2007). Modulation of proteinase K-resistant prion protein in cells and infectious brain homogenate by redox iron: Implications for prion replication and disease pathogenesis. *Mol. Biol. Cell* **18**, 3302–3312.

18. Moos, T. and Morgan, E. H. (2004). The metabolism of neuronal iron and its pathogenic role in neurological disease. *Ann. N. Y. Acad. Sci.* **1012**, 14–26.

19. MacKenzie, E. L., Iwasaki, K., and Tsuji, Y. (2001). Intracellular iron transport and storage: From molecular mechanisms to health implications. *Antioxid. Redox. Signal* **10**, 997–1030.

20. Petrak, J. V. and Vyoral, D. (2001). Detection of iron-containing proteins contributing to the cellular labile iron pool by a native electrophoresis metal blotting technique. *J. Inorg. Biochem.* **86**, 669–675.

21. Vyoral, D., Petrák, J., and Hradilek, A. (1998). Separation of cellular iron containing compounds by electrophoresis. *Biol. Trace. Elem. Res.* **61**, 263–275.

22. Rogers, M., Yehiely, F., Scott, M., and Prusiner, S. B. (1993). Conversion of truncated and elongated prion proteins into the scrapie isoform in cultured cells. *Proc. Natl. Acad .Sci. USA* **90**, 3182–3186.

23. Campana, V., Caputo, A., Sarnataro, D., Paladino, S., Tivodar, S., et al. (2007). Characterization of the properties and trafficking of an anchorless form of the prion protein. *J. Biol. Chem.* **282**, 22747–22756.

24. Baker, E., Richardson, D., Gross, S., and Ponka, P. (1992). Evaluation of the iron chelation potential of hydrazones of pyridoxal, salicylaldehyde and 2-hydroxy-1-naphthylaldehyde using the hepatocyte in culture. *Hepatology* **15**, 492–501.

25. Chung, M. C. (1985). A Specific iron stain for iron-binding proteins in polyacrylamide gels: Application to transferrin and lactoferrin. *Anal. Biochem.* **148**, 498–502.

26. Kascsak, R. J., Rubenstein, R., Merz, P. A., Tonna-DeMasi, M., and Fersko, R., et al. (1987). Mouse polyclonal and monoclonal antibody to scrapie-associated fibril proteins. *J. Virol.* **61**, 3688–3693.

27. Peretz, D., Williamson, R. A., Kaneko, K., Vergara, J., Leclerc, E., et al. (2001). Antibodies inhibit prion propagation and clear cell cultures of prion infectivity. *Nature* **412**, 739–743.

28. Chen, S. G., Teplow, D. B., Parchi, P., Teller, J. K., Gambetti, P., et al. (1995). Truncated forms of the human prion protein in normal brain and in prion diseases. *J. Biol. Chem.* **270**, 19173–19180.

29. Liu, X. and Theil, E. C. (2005). Ferritin as an iron concentrator and chelator target. *Ann. NY Acad. Sci.* **1054**, 136–140.

30. Ohgami, R. S., Campagna, D. R., Greer, E. L., Antiochos, B., McDonald, A., et al. (2005). Identification of a ferrireductase required for efficient transferrin-dependent iron uptake in erythroid cells. *Nat. Genet.* **37**, 1264–1269.

31. Burdo, J. R. and Connor, J. R. (2003). Brain iron uptake and homeostatic mechanisms: An overview. *Biometals* **16**, 63–75.

32. Moos, T., Rosengren, Nielsen. T., Skjørringe, T., and Morgan, E. H. (2007). Iron trafficking inside the brain. *J. Neurochem* **103**, 1730–1740.

33. Aguirre, P., Mena, N., Tapia, V., Arredondo, M., and Núñez, M. T. (2005). Iron homeostasis in neuronal cells: A role for IREG1. *B. M. C. Neurosci.* **6**, 3.

34. Brown, L. R. and Harris, D. A. (2003). Copper and zinc cause delivery of the prion protein from the plasma membrane to a subset of early endosomes and the Golgi. *J. Neurochem.* **87**, 353–363.

35. Miura, T., Sasaki, S., Toyama, A., and Takeuchi, H. (2005). Copper reduction by the octapeptide repeat region of prion protein: pH dependence and implications in cellular copper uptake. *Biochemistry* **44**, 8712–8720.

36. Shyng, S. L., Moulder, K. L. , Lesko, A., and Harris, D. A. (1995). The N-terminal domain of a glycolipid-anchored prion protein is essential for its endocytosis via clathrin-coated pits. *J. Biol. Chem.* **270**, 14793–14800.

37. Peters, P. J., Mironov, A. Jr., Peretz, D., van Donselaar, E., Leclerc, E., et al. (2003). Trafficking of prion proteins through a caveolae-mediated endosomal pathway. *J. Cell. Biol.* **162**, 703–717.

38. Mishra, R. S., Gu, Y., Bose, S., Verghese, S., Kalepu, S., et al. (2002). Cell surface accumulation of a truncated transmembrane prion protein in Gerstmann-Straussler-Scheinker disease P102L. *J. Biol. Chem.* **277**, 24554–24561.

39. Choi, C. J., Anantharam, V., Saetveit, N. J., Houk, R. S., Kanthasamy, A., et al. (2007). Normal cellular prion protein protects against manganese-induced oxidative stress and apoptotic cell death. *Toxicol. Sci.* **98**, 495–509.

40. Zivny, J. H., Gelderman, M. P., Xu, F., Piper, J., Holada, K., et al. (2008). Reduced erythroid cell and erythropoietin production in response to acute anemia in prion protein-deficient (Prnp−/−) mice. *Blood Cells Mol. Dis.* **40**, 302–307.

41. Gu, Y., Fujioka, H., Mishra, R. S., Li, R., and Singh, N. (2002). Prion peptide 106–126 modulates the aggregation of cellular

prion protein and induces the synthesis of potentially neurotoxic transmembrane PrP. *J. Biol. Chem.* **277**, 2275–2286.

42. Jin, T., Gu, Y., Zanusso, G., Sy, M., Kumar, A., et al. (2000). The chaperone protein BiP binds to a mutant prion protein and mediates its degradation by the proteasome. *J. Biol. Chem.* **275**, 38699–38704.

43. Tenopoulou, M., Kurtz, T., Doulias, P. T., Galaris, D., and Brunk, U. T. (2007). Does the calcein-AM method assay the total cellular "labile iron pool" or only a fraction of it? *Biochem. J.* **403**, 261–266.

44. Erel, O. (1998). Automated measurement of serum ferroxidase activity. *Clin. Chem.* **44**, 2313–2319.

45. Singh, A., Isaac, A. O., Luo, X., Mohan, M. L., Cohen, M. L., et al. (2009). Abnormal brain iron homeostasis in human and animal prion disorders. Plos Pathogens, (in press).

6

1. Boyle, M. P. (2007). Adult cystic fibrosis. *JAMA* **298**, 1787–93.

2. Son, M. S., Matthews, W. J. Jr., Kang. Y., Nguyen, D. T., and Hoang, T. T. (2007). *In vivo* evidence of *Pseudomonas aeruginosa* nutrient acquisition and pathogenesis in the lungs of cystic fibrosis patients. *Infect. Immun.* **75**, 5313–24.

3. Remmington, T., Jahnke, N., and Harkensee, C. (2007). Oral anti-pseudomonal antibiotics for cystic fibrosis. *Cochrane Database Syst. Rev. CD005405*

4. Lahiri, T. (2007). Approaches to the treatment of initial *Pseudomonas aeruginosa* infection in children who have cystic fibrosis. *Clin. Chest Med.* **28**, 307–18.

5. Courtney, J. M., Ennis, M., and Elborn, J. S. (2004). Cytokines and inflammatory mediators in cystic fibrosis. *J. Cyst. Fibros.* **3**, 223–31.

6. Kirov, S. M., Webb, J. S.O'May, C. Y., Reid, D. W., Woo, J. K., et al. (2007). Biofilm differentiation and dispersal in mucoid *Pseudomonas aeruginosa* isolates from patients with cystic fibrosis. *Microbiology* **153**, 3264–74.

7. Murray, T. S., Egan, M., and Kazmierczak, B. I. (2007). *Pseudomonas aeruginosa* chronic colonization in cystic fibrosis patients. *Curr. Opin. Pediatr.* **19**, 83–8.

8. Sagel, S. D., Chmiel, J. F., and Konstan, M. W. (2007). Sputum biomarkers of inflammation in cystic fibrosis lung disease. *Proc. Am. Thorac. Soc.* **4**, 406–17.

9. Elizur, A., Cannon, C. L., and Ferkol, T. W. (2008). Airway inflammation in cystic fibrosis. *Chest* **133**, 489–95.

10. Driscoll, J. A., Brody, S. L., and Kollef, M. H. (2007). The epidemiology, pathogenesis and treatment of *Pseudomonas aeruginosa* infections. *Drugs* **67**, 351–68.

11. Iredell, J. R. (2007). Optimizing antipseudomonal therapy in critical care. *Semin. Respir. Crit. Care Med.* **28**, 656–61.

12. Merlo, C. A., Boyle, M. P., Diener-West, M., Marshall, B. C., Goss, C. H., et al. (2007). Incidence and risk factors for multiple antibiotic-resistant *Pseudomonas aeruginosa* in cystic fibrosis. *Chest* **132**, 562–8.

13. Davies, J. C. and Rubin, B. K. (2007). Emerging and unusual gram-negative infections in cystic fibrosis. *Semin. Respir. Crit. Care Med.* **28**, 312–21.

14. Eisenberg, J., Pepe, M., Williams-Warren. J., Vasiliev. M., Montgomery, A. B., et al. (1997). A comparison of peak sputum tobramycin concentration in patients with cystic fibrosis using jet and ultrasonic nebulizer systems. Aerosolized Tobramycin Study Group. *Chest* **111**, 955–62.

15. Hermann, T. (2007). Aminoglycoside antibiotics: Old drugs and new therapeutic approaches. *Cell Mol. Life Sci.* **64**, 1841–52.

16. Westerman, E. M., De Boer, A. H., Le Brun, P. P., Touw, D. J., Roldaan, A. C., et al. (2007). Dry powder inhalation of colistin in cystic fibrosis patients: A single dose pilot study. *J. Cyst. Fibros.* **6**, 284–92.

17. Kaye, D. (2004). Current use for old antibacterial agents: Polymyxins, rifampin, and aminoglycosides. *Infect. Dis. Clin. North Am.* **18**, 669–89.

18. Cannella, C. A. and Wilkinson, S. T. (2006). Acute renal failure associated with inhaled tobramycin. *Am. J. Health Syst. Pharm.* **63**, 1858–61.

19. Swan, S. K. (1997). Aminoglycoside nephrotoxicity. *Semin. Nephrol.* **17**, 27–33.

20. Fernandes, B., Plummer, A., and Wildman, M. (2008). Duration of intravenous antibiotic therapy in people with cystic fibrosis. *Cochrane Database Syst. Rev.* CD006682.

21. Touw, D. J., Knox, A. J., and Smyth, A. (2007). Population pharmacokinetics of tobramycin administered thrice daily and once daily in children and adults with cystic fibrosis. *J. Cyst. Fibros.* **6**, 327–33.

22. Falagas, M. E. and Kasiakou, S. K. (2005). Colistin: The revival of polymyxins for the management of multidrug-resistant gram-negative bacterial infections. *Clin. Infect. Dis.* **40**, 1333–41.

23. Zavascki, A. P., Goldani, L. Z., Li, J., and Nation, R. L. (2007). Polymyxin B for the treatment of multidrug-resistant pathogens: A critical review. *J. Antimicrob. Chemother.* **60**, 1206–15.

24. Macfarlane, E. L., Kwasnicka, A., Ochs, M. M., and Hancock, R. E. (1999). PhoP-PhoQ homologues in *Pseudomonas aeruginosa* regulate expression of the outer-membrane protein OprH and polymyxin B resistance. *Mol. Microbiol.* **34**, 305–16.

25. Sobieszczyk, M. E., Furuya, E. Y., Hay, C. M., Pancholi, P., Della-Latta, P., et al. (2004). Combination therapy with polymyxin B for the treatment of multidrug-resistant Gram-negative respiratory tract infections. *J. Antimicrob. Chemother.* **54**, 566–9.

26. Chuchalin, A., Csiszer, E., Gyurkovics, K., Bartnicka, M. T., Sands, D., et al. (2007). A formulation of aerosolized tobramycin (Bramitob) in the treatment of patients with cystic fibrosis and *Pseudomonas aeruginosa* infection: A double-blind, placebo-controlled, multicenter study. *Paediatr. Drugs* **9(Suppl),** 121–31.

27. Govan, J, (2002). TOBI: Reducing the impact of pseudomonal infection. *Hosp. Med.* **63**, 421–5.

28. Geller, D. E., Konstan, M. W., Smith, J., Noonberg, S. B., and Conrad, C. (2007). Novel tobramycin inhalation powder in cystic fibrosis subjects: Pharmacokinetics and safety. *Pediatr. Pulmonol.* **42**, 307–13.

29. Lenoir, G., Antypkin, Y. G., Miano, A., Moretti, P., Zanda, M., et al. (2007). Efficacy, safety, and local pharmacokinetics of highly concentrated nebulized tobramycin in patients with cystic fibrosis colonized with *Pseudomonas aeruginosa. Paediatr. Drugs* **9**, 111–20.

30. Horianopoulou, M., Lambropoulos, S., Papafragas, E., and Falagas, M. E. (2005). Effect of aerosolized colistin on multidrug-resistant *Pseudomonas aeruginosa* in bronchial secretions of patients without cystic fibrosis. *J. Chemother.* **17**, 536–8.

31. Hodson, M. E., Gallagher, C. G., and Govan, J. R. (2002). A randomised clinical trial of nebulised tobramycin or colistin in cystic fibrosis. *Eur. Respir. J.* **20**, 658–64.

32. Rogan, M. P., Taggart, C. C., Greene, C. M., Murphy, P. G., O'Neill, S. J., et al. (2004). Loss of microbicidal activity and increased formation of biofilm due to decreased lactoferrin activity in patients with cystic fibrosis. *J. Infect. Dis.* **190**, 1245–53.

33. Walker, T. S., Tomlin, K. L., Worthen, G. S., Poch, K. R., Lieber, J. G., et al. (2005). Enhanced *Pseudomonas aeruginosa* biofilm development mediated by human neutrophils. *Infect. Immun.* **73**, 3693–701.

34. Ramphal, R., Lhermitte, M., Filliat, M., and Roussel, P. (1988). The binding of antipseudomonal antibiotics to macromolecules from cystic fibrosis sputum. *J. Antimicrob. Chemother.* **22**, 483–90.

35. Davis, S. D. and Bruns, W. T. (1978). Effects of sputum from patients with cystic fibrosis on the activity *in vitro* of 5 antimicrobial drugs on *Pseudomonas aeruginosa. Am. Rev. Respir. Dis.* **117**, 176–8.

36. Hunt, B. E., Weber, A., Berger, A., Ramsey, B., and Smith, A. L. (1995). Macromolecular mechanisms of sputum inhibition of tobramycin activity. *Antimicrob. Agents Chemother.* **39**, 34–9.

37. Someya, A. and Tanaka, N. (1979). Interaction of aminoglycosides and other antibiotics with actin. *J. Antibiot. (Tokyo)* **32**, 156–60.

38. Sanders, N. N., Van Rompaey, E., De Smedt, S. C., and Demeester, J. (2001). Structural alterations of gene complexes by cystic fibrosis sputum. *Am. J. Respir. Crit. Care Med.* **164**, 486–93.

39. Vasconcellos, C. A., Allen, P. G., Wohl, M. E., Drazen, J. M., Janmey, P. A., et al.

(1994). Reduction in viscosity of cystic fibrosis sputum *in vitro* by gelsolin. *Science* **263**, 969–71.

40. Kater, A., Henke, M. O., and Rubin, B. K. (2007). The role of DNA and actin polymers on the polymer structure and rheology of cystic fibrosis sputum and depolymerization by gelsolin or thymosin beta 4. *Ann. N Y Acad. Sci.* **1112**, 140–53.

41. Marshall, A. J. and Piddock, L. J. (1994). Interaction of divalent cations, quinolones and bacteria. *J. Antimicrob. Chemother.* **34**, 465–83.

42. Levy, J., Smith, A. L., Kenny, M. A., Ramsey, B., and Schoenknecht, F. D. (1983). Bioactivity of gentamicin in purulent sputum from patients with cystic fibrosis or bronchiectasis: Comparison with activity in serum. *J. Infect. Dis.* **148**, 1069–76.

43. Weiner, D.J., Bucki, R., and Janmey, P. A. (2003). The antimicrobial activity of the cathelicidin LL37 is inhibited by F-actin bundles and restored by gelsolin. *Am. J. Respir. Cell Mol. Biol.* **28**, 738–45.

44. Landry, R. M., An, D., Hupp, J. T., Singh, P. K., and Parsek, M. R. (2006). Mucin-*Pseudomonas aeruginosa* interactions promote biofilm formation and antibiotic resistance. *Mol. Microbiol.* **59**, 142–51.

45. Fabretti, F., Theilacker, C., Baldassarri, L., Kaczynski, Z., Kropec, A., et al. (2006). Alanine esters of enterococcal lipoteichoic acid play a role in biofilm formation and resistance to antimicrobial peptides. *Infect. Immun.* **74**, 4164–71.

46. Bucki, R., Sostarecz, A. G., Byfield, F. J., Savage, P. B., and Janmey, P. A. (2007). Resistance of the antibacterial agent ceragenin CSA-13 to inactivation by DNA or F-actin and its activity in cystic fibrosis sputum. *J. Antimicrob. Chemother.* 60, 535–45.

47. Lethem, M. I., James, S. L., Marriott, C., and Burke, J. F. (1990). The origin of DNA associated with mucus glycoproteins in cystic fibrosis sputum. *Eur. Respir. J.* **3**, 19–23.

48. Mozafari, M. R., Flanagan, J., Matia-Merino, L., Awati, A., Omri, A., Suntres, Z. E., et al. (2006). Recent trends in the lipid-based nanoencapsulation of antioxidants and their role in foods. *J. Sci. Food Agric.* **86**, 2038–45.

49. Mozafari, M. R., Johnson, C., Hatziantoniou, S., and Demetzos, C. (2008). Nanoliposomes and Their Applications in Food Nanotechnology. *J. Liposome Res.* **18**, 309–27.

50. Halwani,, M., Mugabe, C., Azghani, A. O., Lafrenie, R. M., Kumar, A., et al. (2007). Bactericidal efficacy of liposomal aminoglycosides against *Burkholderia cenocepacia*. *J. Antimicrob. Chemother.* **60**, 760–9.

51. Omri, A., Suntres, Z. E., and Shek, P. N. (2002). Enhanced activity of liposomal polymyxin B against *Pseudomonas aeruginosa* in a rat model of lung infection. *Biochem. Pharmacol.* **64**, 1407–13.

52. Allison, S. D. (2007). Liposomal drug delivery. *J. Infus. Nurs.* **30**, 89–95

53. Sanders, N. N., Van Rompaey, E., De Smedt, S. C., and Demeester, J. (2002). On the transport of lipoplexes through cystic fibrosis sputum. *Pharm. Res.* **19**, 451–6.

54. Meers, P., Neville, M., Malinin, V., Scotto, A. W., Sardaryan, G., et al. (2008). Biofilm penetration, triggered release and *in vivo* activity of inhaled liposomal amikacin in chronic *Pseudomonas aeruginosa* lung infections. *J. Antimicrob. Chemother.* **61**, 859–68.

55. Sanders, N. N., De Smedt, S. C., Van Rompaey, E., Simoens, P., De Baets, F., et al. (2000). Cystic fibrosis sputum: A barrier to the transport of nanospheres. *Am. J. Respir. Crit. Care Med.* **162**, 1905–11.

56. Ceri, H., Olson, M. E., Stremick, C., Read, R. R., Morck, D., et al. (1999). The Calgary Biofilm Device: New technology for rapid determination of antibiotic susceptibilities of bacterial biofilms. *J. Clin. Microbiol.* **37**, 1771–6.

57. Alipour, M., Halwani, M., Omri, A., and Suntres, Z. E. (2008). Antimicrobial effectiveness of liposomal polymyxin B against resistant Gram-negative bacterial strains. *Int. J. Pharm.* **355**, 293–8.

58. Mugabe, C., Halwani. M., Azghani, A. O., Lafrenie, R. M., and Omri, A. (2006). Mechanism of enhanced activity of liposome-entrapped aminoglycosides against resistant strains of *Pseudomonas aeruginosa*. *Antimicrob. Agents Chemother.* **50**, 2016–22.

59. Mugabe, C., Azghani, A. O., and Omri, A. (2006). Preparation and characterization of dehydration-rehydration vesicles loaded with aminoglycoside and macrolide antibiotics. *Int. J. Pharm.* **307**, 244–50.

60. Nicas, T. I. and Hancock, R. E. (1980) Outer membrane protein H1 of *Pseudomonas aeruginosa*: involvement in adaptive and mutational resistance to ethylenediaminetetraacetate, polymyxin B, and gentamicin. *J. Bacteriol.* **143**, 872–8.

61. Hancock, R. E. (2001). Cationic peptides: Effectors in innate immunity and novel antimicrobials. *Lancet. Infect. Dis.* **1**, 156–64.

62. Kharitonov, S. A. and Sjobring, U. (2007). Lipopolysaccharide challenge of humans as a model for chronic obstructive lung disease exacerbations. *Contrib. Microbiol.* **14**, 83–100.

63. Schiffelers, R., Storm, G., and Bakker-Woudenberg, I. (2001). Liposome-encapsulated aminoglycosides in pre-clinical and clinical studies. *J. Antimicrob. Chemother.* **48**, 333–44.

64. Bucki, R., Byfield, F. J., and Janmey, P. A. (2007). Release of the antimicrobial peptide LL-37 from DNA/F-actin bundles in cystic fibrosis sputum. *Eur. Respir. J.* **29**, 624–32.

65. Tang, J. X., Wen, Q., Bennett, A., Kim, B., Sheils, C. A., et al. (2005) Anionic poly(amino acid)s dissolve F-actin and DNA bundles, enhance DNase activity, and reduce the viscosity of cystic fibrosis sputum. *Am. J. Physiol. Lung Cell Mol. Physiol.* **289**, L599–605.

66. Broughton-Head, V. J., Smith, J. R., Shur, J., and Shute, J. K. (2007). Actin limits enhancement of nanoparticle diffusion through cystic fibrosis sputum by mucolytics. *Pulm. Pharmacol. Ther.* **20**, 708–17.

67. Potter, J. L., Matthews, L. W., Spector, S., and Lemm, J. (1965). Complex formation between basic antibiotics and deoxyribonucleic acid in human pulmonary secretions. *Pediatrics* **36**, 714–20.

68. Davies, M., Stewart-Tull, D. E., and Jackson, D. M. (1978). The binding of lipopolysaccharide from *Escherichia coli* to mammalian cell membranes and its effect on liposomes. *Biochim. Biophys. Acta.* **508**, 260–76.

69. Bataillon, V., Lhermitte, M., Lafitte, J. J., Pommery, J., and Roussel, P. (1992). The binding of amikacin to macromolecules from the sputum of patients suffering from respiratory diseases. *J. Antimicrob. Chemother.* **29**, 499–508.

70. Stern, M., Caplen, N. J., Browning, J. E., Griesenbach, U., Sorgi, F., et al. (1998). The effect of mucolytic agents on gene transfer across a CF sputum barrier *in vitro*. *Gene Ther.* **5**, 91–8.

7

1. Bruyn, G. A. and van Furth, R. (1991). Pneumococcal polysaccharide vaccines: Indications, efficacy and recommendations. *Eur. J. Clin. Microbiol. Infect. Dis.* **10**(11), 897–910.

2. Ryan, M. W. and Antonelli, P. J. (2000). Pneumococcal antibiotic resistance and rates of meningitis in children. *Laryngoscope* **110**(6), 961–964.

3. Cutts, F. T., Zaman, S. M., Enwere, G., Jaffar, S., Levine, O. S., Okoko, J. B., Oluwalana, C., Vaughan, A., Obaro, S. K., Leach, A., et al. (2005). Efficacy of nine-valent pneumococcal conjugate vaccine against pneumonia and invasive pneumococcal disease in The Gambia: Randomised, double-blind, placebo-controlled trial. *Lancet* **365**(9465), 1139–1146.

4. Swiatlo, E., Champlin, F. R., Holman, S. C., Wilson, W. W., Watt, J. M. (2002). Contribution of choline-binding proteins to cell surface properties of *Streptococcus pneumoniae*. *Infect. Immun.* **70**(1), 412–415.

5. Sandgren, A., Albiger, B., Orihuela, C. J., Tuomanen, E., Normark, S., and Henriques-Normark, B. (2005). Virulence in mice of pneumococcal clonal types with known invasive disease potential in humans. *J. Infect. Dis.* **192**(5), 791–800.

6. Liang, X. and Ji, Y. (2007). Comparative analysis of staphylococcal adhesion and internalization by epithelial cells. *Methods Mol. Biol.* **391**, 145–151.

7. Howden, B. P., Stinear, T. P., Allen, D. L., Johnson, P. D., Ward, P. B., and Davies, J. K. (2008). Genomics Reveals a Point Mutation in the Two-Component Sensor Gene

graS that Leads to Vancomycin-Intermediate Resistance in Clinical Staphylococcus aureus. *Antimicrob. Agents Chemother.* **52**(10), 3755–3762.

8. Frederick, J. R., Rogers, E. A., and Marconi, R. T. (2008). Analysis of a growth-phase-regulated two-component regulatory system in the periodontal pathogen Treponema denticola. *J. Bacteriol.* **190**(18), 6162–6169.

9. Bush, K. and Macielag, M. (2000). New approaches in the treatment of bacterial infections. *Curr. Opin. Chem. Biol.* **4**(4):433–439.

10. Martin, P. K., Li, T., Sun, D., Biek, D. P., and Schmid, M. B. (1999). Role in cell permeability of an essential two-component system in Staphylococcus aureus. *J. Bacteriol.* **181**(12), 3666–3673.

11. Watanabe, T., Hashimoto, Y., Yamamoto, K., Hirao, K., Ishihama, A., Hino, M., and Utsumi, R. (2003). Isolation and characterization of inhibitors of the essential histidine kinase, YycG in *Bacillus subtilis* and Staphylococcus aureus. *J. Antibiot. (Tokyo)* **56**(12), 1045–1052.

12. Fabret, C. and Hoch, J. A. (1998). A two-component signal transduction system essential for growth of *Bacillus subtilis*: Implications for anti-infective therapy. *J. Bacteriol.* **180**(23), 6375–6383.

13. Hancock, L. and Perego, M. (2002). Two-component signal transduction in Enterococcus faecalis. *J. Bacteriol.* **184**(21), 5819–5825.

14. Barrett, J. F. and Hoch, J. A. (1998). Two-component signal transduction as a target for microbial anti-infective therapy. *Antimicrob. Agents Chemother.* **42**(7), 1529–1536.

15. Macielag, M. J. and Goldschmidt, R. (2000). Inhibitors of bacterial two-component signalling systems. *Expert Opin. Investig. Drugs* **9**(10), 2351–2369.

16. Matsushita, M. and Janda, K. D. (2002) Histidine kinases as targets for new antimicrobial agents. *Bioorg. Med. Chem.* **10**(4), 855–867.

17. Stock, A. M., Robinson, V. L., and Goudreau, P. N. (2000). Two-component signal transduction. *Annu. Rev. Biochem.* **69**, 183–215.

18. Wagner, C., Saizieu Ad, A., Schonfeld, H. J., Kamber, M., Lange, R., Thompson, C. J., Page, M. G. (2002). Genetic analysis and functional characterization of the *Streptococcus pneumoniae* vic operon. *Infect. Immun.* **70**(11), 6121–6128.

19. Echenique, J. R. and Trombe, M. C. (2001). Competence repression under oxygen limitation through the two-component MicAB signal-transducing system in *Streptococcus pneumoniae* and involvement of the PAS domain of MicB. *J. Bacteriol.* **183**(15), 4599–4608.

20. Throup, J. P., Koretke, K. K., Bryant, A. P., Ingraham, K. A., Chalker, A. F., Ge, Y., Marra, A., Wallis, N. G., Brown, J. R., Holmes, D. J., et al. (2000). A genomic analysis of two-component signal transduction in *Streptococcus pneumoniae*. *Mol. Microbiol.* **35**(3), 566–576.

21. Ng, W. L., Tsui, H. C., and Winkler, M. E. (2005). Regulation of the pspA virulence factor and essential pcsB murein biosynthetic genes by the phosphorylated VicR (YycF) response regulator in *Streptococcus pneumoniae*. *J. Bacteriol.* **187**(21), 7444–7459.

22. Riboldi-Tunnicliffe, A., Trombe, M. C., Bent, C. J., Isaacs, N. W., Mitchell, T. J. (2004). Crystallization and preliminary crystallographic studies of the D59A mutant of MicA, a YycF response-regulator homologue from *Streptococcus pneumoniae*. *Acta Crystallogr. D Biol. Crystallogr.* **60**(Pt 5), 950–951.

23. Bent, C. J., Isaacs, N. W., Mitchell, T. J., and Riboldi-Tunnicliffe, A. (2004). Crystal structure of the response regulator 02 receiver domain, the essential YycF two-component system of *Streptococcus pneumoniae* in both complexed and native states. *J. Bacteriol.* **186**(9), 2872–2879.

24. Paterson, G. K., Blue, C. E., and Mitchell, T. J. (2006). Role of two-component systems in the virulence of *Streptococcus pneumoniae*. *J. Med. Microbiol.* **55**(Pt 4), 355–363.

25. Kadioglu, A., Echenique, J., Manco, S., Trombe, M. C., and Andrew, P. W. (2003). The MicAB two-component signaling system is involved in virulence of *Streptococc-*

cus pneumoniae. Infect. Immun. 71(11), 6676–6679.

26. Andries, K., Verhasselt, P., Guillemont, J., Gohlmann, H. W., Neefs, J. M., Winkler, H., Van Gestel, J., Timmerman, P., Zhu, M., Lee, E., et al. (2005). A diarylquinoline drug active on the ATP synthase of Mycobacterium tuberculosis. Science 307(5707), 223–227.

27. Kim, D. and Forst, S. (2001). Genomic analysis of the histidine kinase family in bacteria and archaea. Microbiology 147(Pt 5), 1197–1212.

28. Marina, A., Waldburger, C. D., and Hendrickson, W. A. (2005). Structure of the entire cytoplasmic portion of a sensor histidine-kinase protein. Embo. J. 24(24), 4247–4259.

29. Zhang, K. Y. and Eisenberg, D. (1994). The three-dimensional profile method using residue preference as a continuous function of residue environment. Protein Sci. 3(4), 687–695.

30. Ewing, T. J., Makino, S., Skillman, A. G., and Kuntz, I. D. (2001). DOCK 4.0: search strategies for automated molecular docking of flexible molecule databases. J. Comput. Aided Mol. Des. 15(5), 411–428.

31. Kuntz, I. D. (1992). Structure-based strategies for drug design and discovery. Science 257(5073), 1078–1082.

32. Morris, G. M., Goodsell, D. S., Halliday, R. S., Huey, R., Hart, W. E., Belew, R. K., and Olson, A. J. (1998). Automated docking using Lamarckian genetic algorithm and an empirical binding free energy function. J. Comp. Chem. 19, 1639–1662.

33. Ng, W. L., Robertson, G. T., Kazmierczak, K. M., Zhao, J., Gilmour, R., and Winkler, M. E. (2003). Constitutive expression of PcsB suppresses the requirement for the essential VicR (YycF) response regulator in Streptococcus pneumoniae R6. Mol. Microbiol. 50(5), 1647–1663.

34. Lange, R., Wagner, C., de Saizieu, A., Flint, N., Molnos, J., Stieger, M., Caspers, P., Kamber, M., Keck, W., and Amrein, K. E. (1999). Domain organization and molecular characterization of 13 two-component systems identified by genome sequencing of

Streptococcus pneumoniae. Gene 237(1), 223–234.

35. Mohedano, M. L., Overweg, K., de la Fuente, A., Reuter, M., Altabe, S., Mulholland, F., de Mendoza, D., Lopez, P., and Wells, J. M. (2005). Evidence that the essential response regulator YycF in Streptococcus pneumoniae modulates expression of fatty acid biosynthesis genes and alters membrane composition. J. Bacteriol. 187(7), 2357–2367.

36. Qin, Z., Zhang, J., Xu, B., Chen, L., Wu, Y., Yang, X., Shen, X., Molin, S., Danchin, A., Jiang, H., et al. (2006). Structure-based discovery of inhibitors of the YycG histidine kinase: new chemical leads to combat Staphylococcus epidermidis infections. BMC Microbiol. 6, 96–114.

8

1. Kates, M. (1992). Archaebacterial lipids: Structure, biosynthesis and function. In The Archaebacteria: Biochemistry and Biotechnology, Volume 58. M. J. Danson, D. W. Hough, and G. G. Lunt (Eds.). Portland Press and Chapel Hill, London, pp. 51–77.

2. Sprott, G. D., Tolson, D. L., and Patel, G. B. (1997). Archaeosomes as novel antigen delivery systems. FEMS Microbiol. Lett. 154, 17–22.

3. Kates, M. (1993). Biology of halophilic bacteria, Part II. Membrane lipids of extreme halophiles: Biosynthesis, function and evolutionary significance. Experientia 49, 1027–1036.

4. Sprott, G. D., Dicaire, C. J., Fleming, L, P, and Patel, G. B. (1996). Stability of liposomes prepared from archaeobacterial lipids and phosphatidylcholine mixtures. Cells Mater. 6, 143–155.

5. Kitano, T., Onoue, T., and Yamauchi, K. (2003). Archaeal lipids forming a low energy-surface on air-water interface. Chem. Phys. Lipids 126, 225–232.

6. Yamauchi, K., Onoue, Y., Tsujimoto, T., and Kinoshita, M. (1997). Archaebacterial lipids: High surface activity of polyisoprenoid surfactants in water. Coll. Surf. B: Bioint. 10, 35–39.

7. Mathai, J. C., Sprott, G. D., and Zeidel, M. L. (2001). Molecular mechanisms of water and solute transport across archaebacterial lipid membranes. *J. Biol. Chem.* **276**, 27266–27271.

8. Vossenberg, J. L., Driessen, A. J., Grant, W. D., and Konings, W. N. (1999). Lipid membranes from halophilic and alkali-halophilic Archaea have a low H+ and Na+ permeability at high salt concentration. *Extremophiles* **3**, 253–257.

9. Eckburg, P. B., Lepp, P. W., and Relman, D. A. (2003). Archaea and their potential role in human disease. *Infect. Imm.* **71**, 591–596.

10. Brennan, F. R. and Dougan, G. (2005). Non-clinical safety evaluation of novel vaccines and adjuvants: New products, new strategies. *Vaccine* **23**, 3210–3222.

11. Lofthouse, S. (2002). Immunological aspects of controlled antigen delivery. *Adv. Drug Deliv. Rev.* **54**, 863–870.

12. Patel, G. B., Omri, A., Deschatelets, L., and Sprott, G. D. (2002). Safety of archaeosome adjuvants evaluated in a mouse model. *J. Liposome Res.* **12**, 353–372.

13. Omri, A., Agnew, B. J., and Patel, G. B. (2003). Short-term repeated-dose toxicity profile of archaeosomes administered to mice via intravenous and oral routes. *Int. J. Toxicol.* **22**, 9–23.

14. Friede, M. and Aguado, M. T. (2005). Need for new vaccine formulations and potential of particulate antigen and DNA delivery systems. *Adv. Drug Deliv. Rev.* **57**, 325–331.

15. Lizama, C., Monteoliva-Sanchez, M., Suarez-Garcia, A., Rosello-Mora, R., Aguilera, M., Campos, V., and Ramos-Cormenzana, A. (2002). *Halorubrum tebenquichense* sp. nov., a novel halophilic archaeon isolated from the Atacama Saltern, Chile. *Int. J. Syst. Evol. Microbiol.* **52**, 149–155.

16. Stackebrandt, E. and Goebel, B. M. (1994). Taxonomic note: A place for DNA-DNA reassociation and 16S rRNA sequence analysis in the present species definition in bacteriology. *Int. J. Syst. Bacteriol.* **44**, 846–849.

17. Wayne, L. G., Brenner, D. J., Colwell, R. R., Grimont, P. A. D., Kandler, O., Krichevsky, M. I., Moore, L. H., Moore, W. E. C., Murray, R. G. E., Stackebrandt, E., Starr, M. P., and Trüper, H. G. (1987). Report of the ad hoc committee on reconciliation of approaches to bacterial systematics. *Int. J. Syst. Bacteriol.* **37**, 463–464.

18. Berg, D. E., Akopyants, N. S., and Kersulyte, D. (1994). Fingerprinting microbial genomes using the RAPD or AP-PCR. *Methods Mol. Cell Biol.* **5**, 13–24.

19. Lopalco, P., Lobasso, S., Babudri, F. A., and Corcelli, A. (2004). Osmotic shock stimulates de novo synthesis of two cardiolipins in an extreme halophilic archaeon. *J. Lipid Res.* **45**, 194–201.

20. Daleke, D. L., Hong, K., and Papahadjopoulos, D. (1990). Endocytosis of liposomes by macrophages: Binding, acidification and leakage of liposomes monitored by a new fluorescence assay. *Biochim. Biophys. Acta* **1024**, 352–366.

21. Straubinger, R. M., Papahadjopoulos, D., and Hong, K. (1990). Endocytosis and intracellular fate of liposomes using pyranine as a probe. *Biochemistry* **29**, 4929–4939.

22. Oren, A. and Ventosa, A. (1996). A proposal for the transfer of *Halorubrobacterium distributum* and *Halorubrobacterium coriense* to the genus Halorubrum as *Halorubrum distributum* comb. nov. and *Halorubrum coriense* comb. nov., respectively. *Int. J. Syst. Bacteriol.* **46**, 1180.

23. Kharroub, K., Quesada, T., Ferrer, R., Fuentes, S., Aguilera, M., Boulahrouf, A., Ramos-Cormenzana, A., and Monteoliva-Sanchez, M. (2006). *Halorubrum ezzemoulense* sp. nov., a *Halophilic archaeon* isolated from *Ezzemoul sabkha*, Algeria. *Int. J. Syst. Evol. Microbiol.* **56**, 583–588.

24. Kamekura, M., Dyall-Smith, M. L., Upasani, V., Ventosa, A., and Kates, M. (1997). Diversity of alkaliphilic halobacteria: Proposals for transfer of *Natronobacterium vacuolatum*, *Natronobacterium magadii*, and *Natronobacterium pharaonis* to Halorubrum, Natrialba, and Natronomonas gen. nov., respectively, as *Halorubrum vacuolatum* comb. nov., *Natrialba magadii* comb. nov., and *Natronomonas pharaonis* comb. nov., respectively. *Int. J. Syst. Bacteriol.* **47**, 853–857.

25. McGenity, T. J. and Grant, W. D. (1995). Transfer of *Halobacterium trapanicum* NRC 34021 and Halobacterium gen nov, or

Halorubrum saccharorubrum comb. nov, and *Halorubrum sodomense* com. Nov. *Halorubrum trapanicum* comb nov, and *Halorubrum lacusprofundi* comb nov and *Halorubrum lacusprofundi* comb nov. *Syst. Appl. Microbiol.* **18**, 237–243.

26. Kamekura, M., Seno, Y., and Dyall-Smith, M. (1996). Halolysin R4, a serine protein-ase from the halophilic archaeon *Haloferax mediterranei*; gene cloning, expression and structural studies. *Biochim. Biophys. Acta* **1294**, 159–167.

27. Franzmann, P. D. (1988). *Halobacterium lacusprofundi* sp. nov., a halophilic bacte-rium isolated from Deep Lake, Antarctica. *Syst. Appl. Microbiol.* **11**, 20–27.

28. Tomlinson, G. A. and Hochstein, L. I. (1976). *Halobacterium saccharovorum* sp. nov., a carbohydrate-metabolizing, ex-tremely halophilic bacterium. *Can. J. Mi-crobiol.* **22**, 587–591.

29. Fan, H., Xue, Y., Ma, Y., Ventosa, A., and Grant, W. D. (2004). *Halorubrum tibetense* sp. nov., a novel haloalkaliphilic archaeon from Lake Zabuye in Tibet, China. *Int. J. Syst. Evol. Microbiol.* **54**, 1213–1216.

30. Kamekura, M. and Kates, M. (1999). Struc-tural diversity of membrane lipids in mem-bers of Halobacteriaceae. *Biosci. Biotech-nol. Biochem.* **63**, 969–972.

31. Kates, M., Moldoveanu, N., and Stewart, L. C. (1993). On the revised structure of the major phospholipid of *Halobacterium salinarium*. *Biochim. Biophys. Acta* **1169**, 46–53.

32. Kates, M. (1993). Membrane lipids of ar-chaea. In *The biochemistry of Archaea (Ar-chaebacteria)*. M. Kates, D. J. Kushner, and A. T. Matheson (Eds.). Elsevier, Amster-dam, pp. 261–295.

33. Kates, M. (1996). Structural analysis of phospholipids and glycolipids in extremely halophilic archaebacteria. *J. Microbiol. Methods* **25**, 113–128.

34. Noguchi, Y., Hayashi, A., Tsujimoto, K., Miyabayashi, K., Mizukami, T., Naito, Y., and Ohashi, M. (2004). Composition analy-sis of polar lipids in halobacteria with mass spectrometry. *J. Mass Spectrm. Soc. Jpn.* **52**, 307–316.

35. Corcelli, A. and Lobasso, S. (2006). Charac-terization of lipids of halophilic Archaea. In *Methods in Microbiology—Extremophiles*. Oren AFRaA. (Ed.). Elsevier, Amsterdam, pp. 585–613.

36. Corcelli, A., Lattanzio, V. M., Mascolo, G., Bardudri, F., Oren, A., and Kates, M. (2004). Novel sulfonolipid in the extremely halophilic bacterium salinibacter rubber. *Appl. Environ. Microbiol.* **70**, 6678–6685.

37. Tolson, D. L., Latta, R. K., Patel, G. B., and Sprott, G. D. (1996). Uptake of archaeo-bacterial liposomes by phagocytic cells. *J. Liposome Res.* **6**, 755–776.

38. Krishnan, L., Sad, S., Patel, G. B., and Sprott, G. D. (2001). The potent adjuvant activity of archaeosomes correlates to the recruitment and activation of macrophages and dendritic cells *in vivo*. *J. Immunol.* **166**, 1885–1893.

39. Krishnan, L., Sad, S., Patel, G. B., and Sprott, G. D. (2000). Archaeosomes induce long-term CD8+ cytotoxic T cell respponse to entrapped soluble protein by the exog-enous cytosolic pathway, in the absence of CD4+ T cell help. *J. Immunol.* **165**, 5177–5185.

40. Krishnan, L. and Sprott, G. D. (2008). Ar-chaeosome adjuvants: Immunological ca-pabilities and mechanism (s) of action. *Vac-cine* **26**, 2043–2055.

41. Dudani, R., Chapdelaine, Y., van Faassen, H., Smith, D. K., Shen, H., Krishnan, L., and Sad, S. (2002). Preexisting inflamma-tion due to Mycobacterium bovis BCG in-fection differentially modulates T-cell prim-ing against a replicating or nonreplicating immunogen. *Infect. Immun.* **70**, 1957–1964.

42. Sprott, G. D., Dicaire, C. J., Gurnani, K., Deschatelets, L. A., and Krishnan, L. (2004). Liposome adjuvants prepared from the total polar lipids of *Haloferax volcanii*, Planococcus spp. and *Bacillus firmus* differ in ability to elicit and sustain immune re-sponses. *Vaccine* **22**, 2154–2162.

43. Krishnan, L., Dicaire, C. J., Patel, G. B., and Sprott, G. D. (2000). Archaeosome vaccine adjuvants induce strong humoral, cell-me-diated, and memory responses: Comparison to conventional liposomes and alum. *Infect. Immun.* **68**, 54–63.

44. Krishnan, L. and Sprott, G. D. (2003) Archaeosomes as self adjuvanting delivery systems for cancer vaccines. *J. Drug Target* **11**, 515–524.

45. Sprott, G. D., Dicaire, C. J., Côté, J. P., and Whitfield, D. M. (2008). Adjuvant potential of archaeal synthetic glycolipid mimetics critically depends on the glyco head group structure. *Glycobiology* **18**, 559–565.

46. Whitfield, D. M., Eichler, E. V., and Sprott, G. D. (2008). Synthesis of archaeal glycolipid adjuvants—What is the optimum number of sugars? *Carbohydrate Res.* **343**, 2349–2360.

47. Sprott, G. D., Côté, J. P., and Jarrell, H. C. (2009). Glycosidase-induced fusion of isoprenoid gentiobiosyl lipid membranes at acidic pH. *Glycobiology* **19**, 267–276.

48. Ihara, K., Watanabe, S., and Tamura, T. (1997). *Haloarcula argentinensis* sp. nov. and *Haloarcula mukohataei* sp. nov., two new extremely halophilic archaea collected in Argentina. *Int. J. Syst. Bacteriol.* **47**, 73–77.

49. Oren, A., Ventosa, A., and Grant, W. D. (1997). Proposed minimal standards for description of new taxa in the order Halobacteriales. *Int. J. Syst. Bacteriol.* **47**, 233–238.

50. Dussault, H. P. (1955). An improved technique for staining red halophilic bacteria. *J. Bacteriol.* **70**, 484–485.

51. Cashion, P., Hodler-Franklin, M. A., McCully, J., and Franklin, M. (1977). A rapid method for base ratio determination of bacterial DNA. *Anal. Biochem.* **81**, 461–466.

52. Marmur, J. and Doty, P. (1962). Determination of the base composition of deoxyribonucleic acid from its thermal denaturation temperature. *J. Mol. Biol.* **5**, 109–118.

53. Ferragut, C. and Leclerc, H. (1976). A comparative study of different methods of determining the Tm of bacterial DNA. *Ann. Microbiol. (Paris)* **127A**, 223–235.

54. Sanger, F., Nicklen, S., and Coulson, A. R. (1977). DNA sequencing with chain-terminating inhibitors. *Proc. Natl. Acad. Sci. USA* **74**, 5463–5467.

55. Thompson, J. D., Higgins, D. G., and Gibson, T. J. (1994). CLUSTAL W: Improving the sensitivity of progressive multiple sequence alignment through sequence weighting, position-specific gap penalties and weight matrix choice. *Nucleic Acids Res.* **22**, 4673–4680.

56. Thompson, J. D., Gibson, T. J., Plewniak, F., Jeanmougin, F., and Higgins, D. G. (1997). The CLUSTAL_X windows interface: Flexible strategies for multiple sequence alignment aided by quality analysis tools. *Nucleic Acids Res.* **25**, 4876–4882.

57. Felsenstein, J. (1985). Confidence limits on phylogenies: An approach using the bootstrap. *Evolution* **39**, 783–791.

58. De Ley, J., Cattoir, H., and Reynaerts, A. (1970). The quantitative measurement of DNA hybridization from renaturation rates. *Eur. J. Biochem.* **12**, 133–142.

59. Huss, V. A. R., Festl, H., and Schleifer, K. H. (1983). Studies on the spectrophotometric determination of DNA hybridization from renaturation rates. *Syst. Appl. Microbiol.* **4**, 184–192.

60. Kates, M. K. S. (1995). Protocol 5: Isoprenoids and polar lipids of extreme halophiles. In *Archaea. A Laboratory Manual. Halophiles*. E. M. DSaF (Ed.). Cold Spring Harbor Laboratory Press, New York, pp. 35–53.

61. Bötcher, C. J. F., van Gent, C. M., and Pries, C. (1961). A rapid and sensitive sub-micro phosphorus determination. *Anal. Chim. Acta* **24**, 203–204.

62. Stewart, J. C. (1980). Colorimetric determination of phospholipids with ammonium ferrothiocyanate. *Anal. Biochem.* **104**, 10–14.

63. Bradford, M. M. (1976). A rapid and sensitive method for the quantitation of microgram quantities of protein utilizing the principle of protein-dye binding. *Anal. Biochem.* **72**, 248–254.

64. Jost, L. M., Kirkwood, J. M., and Whiteside, T. L. (1992). Improved short- and long-term XTT-based colorimetric cellular cytotoxicity assay for melanoma and other tumor cells. *J. Immunol. Methods* **147**, 153–165.

9

1. Coates, J. D., Chakraborty, R., Lack, J. G., O'Connor, S. M., Cole, K. A., Bender, K. S., and Achenbach, L. A. (2001). Anaerobic benzene oxidation coupled to nitrate reduc-

tion in pure culture by two strains of *Dechloromonas*. *Nature* **411**(6841), 10391043.

2. Chakraborty, R., O'Connor, S. M., Chan, E., and Coates, J. D. (2005). Anaerobic degradation of benzene, toluene, ethylbenzene, and xylene compounds by Dechloromonas strain RCB. *Appl. Environ. Microbiol.* **71**(12), 86498655.

3. Thrash, J. C., Van Trump, J. I., Weber, K. A., Miller, E., Achenbach, L. A., and Coates, J. D. (2007). Electrochemical stimulation of microbial perchlorate reduction. *Environ. Sci. Technol.* **41**(5), 17401746.

4. Kasai, Y., Takahata, Y., Manefield, M., and Watanabe, K. (2006). RNA-based stable isotope probing and isolation of anaerobic benzene-degrading bacteria from gasoline-contaminated groundwater. *Appl. Environ. Microbiol.* **72**(5), 35863592.

5. Chakraborty, R. and Coates, J.D. (2005). Hydroxylation and carboxylation—two crucial steps of anaerobic benzene degradation by *Dechloromonas* strain RCB. *Appl. Environ. Microbiol.* **71**(9), 54275432.

6. Beller, H. R. and Spormann, A. M. (1998). Analysis of the novel benzylsuccinate synthase reaction for anaerobic toluene activation based on structural studies of the product. *J. Bacteriol.* **180**(20), 54545457.

7. Edwards, A., Voss, H., Rice, P., Civitello. A., Stegemann, J., Schwager, C., Zimmermann, J., Erfle, H., Caskey, C. T., and Ansorge, W. (1990). Automated DNA sequencing of the human HPRT locus. *Genomics* **6**(4), 593608.

8. Ewing, B., Hillier, L., Wendl, M. C., and Green, P. (1998). Base-calling of automated sequencer traces using phred. I. Accuracy assessment. *Genome Res.* **8**(3), 175185.

9. Ewing, B. and Green, P. (1998). Base-calling of automated sequencer traces using phred. II. Error probabilities. *Genome Res.* **8**(3), 186194.

10. Gordon, D., Abajian, C., and Green, P. (1998). Consed: A graphical tool for sequence finishing. *Genome Res.* **8**(3), 195202.

11. Badger, J. H. and Olsen, G. J. (1999). CRITICA: Coding region identification tool invoking comparative analysis. *Mol. Biol. Evol.* **16**(4), 512524.

12. Delcher, A. L., Harmon, D., Kasif, S., White, O., and Salzberg, S. L. (1999). Improved microbial gene identification with GLIMMER. *Nucleic Acids Res.* **27**(23), 46364641.

13. Alm, E. J., Huang, K. H., Price, M. N., Koche, R. P., Keller, K., Dubchak, I. L., and Arkin, A. P. (2005). The MicrobesOnline Web site for comparative genomics. *Genome Res.* **15**(7), 10151022.

14. Chakraborty, R. and Coates, J. D. (2004). Anaerobic degradation of monoaromatic hydrocarbons. *Appl. Microbiol. Biotechnol.* **64**(4), 437446.

15. Egland, P. G., Pelletier, D. A., Dispensa, M., Gibson, J., and Harwood, C. S. (1997). A cluster of bacterial genes for anaerobic benzene ring biodegradation. *Proc. Natl. Acad. Sci. USA* **94**(12), 64846489.

16. Heider, J. and Fuchs, G. (1997). Anaerobic metabolism of aromatic compounds. *Eur. J. Biochem.* **243**(3), 577596.

17. Boll, M., Fuchs, G., and Heider, J. (2002). Anaerobic oxidation of aromatic compounds and hydrocarbons. *Curr. Opin. Chem. Biol.* **6**(5), 604611.

18. Storm, C. E. and Sonnhammer, E. L. (2003). Comprehensive analysis of orthologous protein domains using the HOPS database. *Genome Res.* **13**(10), 23532362.

19. Eisen, J. A. (1998). Phylogenomics: Improving functional predictions for uncharacterized genes by evolutionary analysis. *Genome Res.* **8**(3), 163167.

20. Kuhner, S., Wohlbrand, L., Fritz, I., Wruck, W., Hultschig, C., Hufnagel, P., Kube, M., Reinhardt, R., and Rabus, R. (2005). Substrate-dependent regulation of anaerobic degradation pathways for toluene and ethylbenzene in a denitrifying bacterium, strain EbN1. *J. Bacteriol.* **187**(4), 14931503.

21. Rabus, R., Kube, M., Heider, J., Beck, A., Heitmann, K., Widdel, F., and Reinhardt, R. (2005). The genome sequence of an anaerobic aromatic-degrading denitrifying bacterium, strain EbN1. *Arch. Microbiol.* **183**(1), 2736.

22. Kube, M., Heider, J., Amann, J., Hufnagel, P., Kuhner, S., Beck, A., Reinhardt, R., and Rabus, R. (2004). Genes involved in the anaerobic degradation of toluene in a denitri-

fying bacterium, strain EbN1. *Arch. Microbiol.* **181**(3), 182194.

23. Kane, S. R., Beller, H. R., Legler, T. C, and Anderson, R. T. (2002). Biochemical and genetic evidence of benzylsuccinate synthase in toluene-degrading, ferric iron-reducing *Geobacter metallireducens*. *Biodegradation* **13**(2), 149154.

24. Bender, K. S., Shang, C., Chakraborty, R., Belchik, S. M., Coates, J. D., and Achenbach, L. A. (2005). Identification, characterization, and classification of genes encoding perchlorate reductase. *J. Bacteriol.* **187**(15), 50905096.

25. Rabus, R., Kube, M., Beck, A., Widdel, F., and Reinhardt, R. (2002). Genes involved in the anaerobic degradation of ethylbenzene in a denitrifying bacterium, strain EbN1. *Arch. Microbiol.* **178**(6), 506516.

26. Weelink, S. A., Tan, N. C., ten Broeke, H., Kieboom, C., van Doesburg, W., Langenhoff, A. A., Gerritse, J., Junca, H., and Stams, A. J. (2008). Isolation and characterization of *Alicycliphilus denitrificans* strain BC, which grows on benzene with chlorate as the electron acceptor. *Appl. Environ. Microbiol.* **74**(21), 66726681.

27. Sibley, M. H. and Raleigh, E. A. (2004). Cassette-like variation of restriction enzyme genes in *Escherichia coli* C and relatives. *Nucleic Acids Res.* **32**(2), 522534.

28. Laurie, A. D. and Lloyd-Jones, G. (1999). The phn genes of *Burkholderia* sp. strain RP007 constitute a divergent gene cluster for polycyclic aromatic hydrocarbon catabolism. *J. Bacteriol.* **181**(2), 531540.

29. Pelletier, D. A. and Harwood, C. S. (2000). 2-Hydroxycyclohexanecarboxyl coenzyme A dehydrogenase, an enzyme characteristic of the anaerobic benzoate degradation pathway used by *Rhodopseudomonas palustris*. *J. Bacteriol.* **182**(10), 27532760.

30. Arai, H., Ohishi, T., Chang, M. Y., and Kudo, T. (2000). Arrangement and regulation of the genes for meta-pathway enzymes required for degradation of phenol in *Comamonas testosteroni* TA441. *Microbiology* **146**(Pt 7), 17071715.

31. Arai, H., Yamamoto, T., Ohishi, T., Shimizu, T., Nakata, T., and Kudo, T. (1999). Genetic organization and characteristics of the 3-(3-hydroxyphenyl)propionic acid degradation pathway of *Comamonas testosteroni* TA441. *Microbiology* **145**(Pt 10), 28132820.

32. Hofer, B., Eltis, L. D., Dowling, D. N., and Timmis, K. N. (1993). Genetic analysis of a Pseudomonas locus encoding a pathway for biphenyl/polychlorinated biphenyl degradation. *Gene.* **130**(1), 47–55.

33. Ward, N. C., Croft, K. D., Puddey, I. B., and Hodgson, J. M. (2004). Supplementation with grape seed polyphenols results in increased urinary excretion of 3-hydroxyphenylpropionic acid, an important metabolite of proanthocyanidins in humans. *J. Agric. Food Chem.* **52**(17), 55455549.

34. Ma YaH, S. D. (2000). The catechol 2,3-dioxygenase gene and toluene monooxygenase genes from *Burkholderia* sp. AA1, an isolate capable of degrading aliphatic hydrocarbons and toluene. *J. Ind. Microbiol. Biotechnol.* **25**, 127131.

35. Kitayama, A., Kawakami, Y., and Nagamune, T. (1996). Gene organization and low regiospecificity in aromatic-ring hydroxylati on of a benzene monooxygenase of *Pseudomonas aeruginosa* JI104. *J. Ferment. Bioeng.* **82**, 421425.

36. Tao, Y., Fishman, A., Bentley, W. E., and Wood, T. K. (2004). Oxidation of benzene to phenol, catechol, and 1,2,3-trihydroxybenzene by toluene 4-monooxygenase of Pseudomonas mendocina KR1 and toluene 3-monooxygenase of *Ralstonia pickettii* PKO1. *Appl. Environ. Microbiol.* **70**(7), 38143820.

37. Zhang, H., Luo, H., and Kamagata, Y. (2003). Characterization of the phenol hydroxylase from *Burkholderia kururiensis* KP23 involved in trichloroethylene degradation by gene cloning and disruption. *Microbes. Environ.* **18**, 167173.

38. Canada, K. A., Iwashita, S., Shim, H., and Wood, T. K. (2002). Directed evolution of toluene ortho-monooxygenase for enhanced 1-naphthol synthesis and chlorinated ethene degradation. *J. Bacteriol.* **184**(2), 344349.

39. Mizuno, T., Kaneko, T., and Tabata, S. (1996). Compilation of all genes encoding bacterial two-component signal transducers in the genome of the cyanobacterium, *Syn-*

echocystis sp. strain PCC 6803. *DNA Res.* 3(6), 407414.

40. Galperin, M. Y. (2006). Structural classification of bacterial response regulators: Diversity of output domains and domain combinations. *J. Bacteriol.* 188(12), 41694182.

41. Ryjenkov, D. A., Tarutina, M., Moskvin, O. V., and Gomelsky, M. (2005). Cyclic diguanylate is a ubiquitous signaling molecule in bacteria: Insights into biochemistry of the GGDEF protein domain. *J. Bacteriol.* 187(5), 17921798.

42. Mendez-Ortiz, M. M., Hyodo, M., Hayakawa, Y., and Membrillo-Hernandez, J. (2006). Genome-wide transcriptional profile of *Escherichia coli* in response to high levels of the second messenger 3',5'-cyclic diguanylic acid. *J. Biol. Chem.* 281(12), 80908099.

43. Wandersman, C. and Delepelaire, P. (1990). TolC, an Escherichia coli outer membrane protein required for hemolysin secretion. *Proc. Natl. Acad. Sci. USA* 87(12), 47764780.

44. Richarme, G. and Caldas, T. D. (1997). Chaperone properties of the bacterial periplasmic substrate-binding proteins. *J. Biol. Chem.* 272(25), 1560715612.

45. Binet, R., Letoffe, S., Ghigo, J. M., Delepelaire, P., and Wandersman, C. (1997). Protein secretion by gram-negative bacterial ABC exporters—A review. *Gene.* 192(1), 711.

46. Omori, K. and Idei, A. (2003). Gram-negative bacterial ATP-binding cassette protein exporter family and diverse secretory proteins. *J. Biosci. Bioeng.* 95(1), 112.

47. Pahel, G. and Tyler, B. (1979). A new glnA-linked regulatory gene for glutamine synthetase in Escherichia coli. *Proc. Natl. Acad. Sci. USA* 76(9), 45444548.

48. Meibom, K. L., Li, X. B., Nielsen, A. T., Wu, C. Y., Roseman, S., and Schoolnik, G. K. (2004). The *Vibrio cholerae* chitin utilization program. *Proc. Natl. Acad. Sci. USA* 101(8), 25242529.

49. Tominaga, A., Lan, R., and Reeves, P. R. (2005). Evolutionary changes of the flhDC flagellar master operon in Shigella strains. *J. Bacteriol.* 187(12), 4295–4302.

50. Hueck, C. J. (1998). Type III protein secretion systems in bacterial pathogens of animals and plants. *Microbiol. Mol. Biol. Rev.* 62(2), 379433.

51. Marsh, J. W. and Taylor, R. K. (1999). Genetic and transcriptional analyses of the *Vibrio cholerae* mannose-sensitive hemagglutinin type 4 pilus gene locus. *J. Bacteriol.* 181(4), 11101117.

52. Mougous, J. D., Cuff, M. E., Raunser, S., Shen, A., Zhou, M., Gifford, C. A., Goodman, A. L., Joachimiak, G., Ordonez, C. L., Lory, S., et al. (2006). A virulence locus of *Pseudomonas aeruginosa* encodes a protein secretion apparatus. *Science* 312(5779), 15261530.

53. Pukatzki, S., Ma, A. T., Sturtevant, D., Krastins, B., Sarracino, D., Nelson, W. C., Heidelberg, J. F., and Mekalanos, J. J. (2006). Identification of a conserved bacterial protein secretion system in *Vibrio cholerae* using the Dictyostelium host model system. *Proc. Natl. Acad. Sci. USA* 103(5), 15281533.

54. Parsons, D. A. and Heffron, F. (2005). SciS, an icmF homolog in *Salmonella enterica* serovar Typhimurium, limits intracellular replication and decreases virulence. *Infect. Immun.* 73(7), 43384345.

55. Bingle, L. E., Bailey, C. M., and Pallen, M. J. (2008). Type VI secretion: A beginner's guide. *Curr. Opin. Microbiol.* 11(1), 38.

56. Watnick, P. and Kolter, R. (2000). Biofilm, city of microbes. *J. Bacteriol.* 182(10), 26752679.

57. Barraud, N., Hassett, D. J., Hwang, S. H., Rice, S. A., Kjelleberg, S., Webb, J. S. (2006). Involvement of nitric oxide in biofilm dispersal of *Pseudomonas aeruginosa*. *J. Bacteriol.* 188(21), 73447353.

58. Federle, M. J. and Bassler, B. L. (2003). Interspecies communication in bacteria. *J. Clin. Invest.* 112(9), 12911299.

59. Fuqua, C., Parsek, M. R. and Greenberg, E. P. (2001). Regulation of gene expression by cell-to-cell communication: Acyl-homoserine lactone quorum sensing. *Annu. Rev. Genet.* 35, 439468.

60. Withers, H., Swift, S., and Williams, P. (2001). Quorum sensing as an integral component of gene regulatory networks in

Gram-negative bacteria. *Curr. Opin. Microbiol.* **4**(2), 186193.

61. Huang, J. J., Han, J. I., Zhang, L. H., and Leadbetter, J. R. (2003). Utilization of acyl-homoserine lactone quorum signals for growth by a soil pseudomonad and *Pseudomonas aeruginosa* PAO1. *Appl. Environ. Microbiol.* **69**(10), 59415949.

62. Cha, C., Gao, P., Chen, Y. C., Shaw, P. D., and Farrand, S. K. (1998). Production of acyl-homoserine lactone quorum-sensing signals by gram-negative plant-associated bacteria. *Mol. Plant. Microbe. Interact.* **11**(11), 11191129.

63. Thakor, N., Trivedi, U., and Patel, K. C. (2005). Biosynthesis of medium chain length poly(3-hydroxyalkanoates) (mcl-PHAs) by *Comamonas testosteroni* during cultivation on vegetable oils. *Bioresour. Technol.* **96**(17), 18431850.

64. Potter, M., Madkour, M. H., Mayer, F., and Steinbuchel, A. (2002). Regulation of phasin expression and polyhydroxyalkanoate (PHA) granule formation in *Ralstonia eutropha* H16. *Microbiology* **148**(Pt 8), 24132426.

65. Delledonne, M., Porcari, R., and Fogher, C. (1990). Nucleotide sequence of the nodG gene of *Azospirillum brasilense*. *Nucleic. Acids. Res.* **18**(21), 6435.

66. Trautwein, K., Kuhner, S., Wohlbrand, L., Halder, T., Kuchta, K., Steinbuchel, A., and Rabus, R. (2008). Solvent stress response of the denitrifying bacterium "Aromatoleum aromaticum" strain EbN1. *Appl. Environ. Microbiol.* **74**(8), 22672274.

67. Zago, A., Chugani, S., and Chakrabarty, A. M. (1999). Cloning and characterization of polyphosphate kinase and exopolyphosphatase genes from Pseudomonas aeruginosa 8830. *Appl. Environ. Microbiol.* **65**(5), 20652071.

68. Rao, N. N. and Kornberg, A. (1996). Inorganic polyphosphate supports resistance and survival of stationary-phase *Escherichia coli*. *J. Bacteriol.* **178**(5), 13941400.

69. Reusch, R. N. and Sadoff, H. L. (1988). Putative structure and functions of a poly-beta-hydroxybutyrate/calcium polyphosphate channel in bacterial plasma membranes. *Proc. Natl. Acad. Sci. USA* **85**(12), 41764180.

70. Santaella, C., Schue, M., Berge, O., Heulin, T., and Achouak, W. (2008). The exopolysaccharide of *Rhizobium* sp. YAS34 is not necessary for biofilm formation on *Arabidopsis thaliana* and *Brassica napus* roots but contributes to root colonization. *Environ. Microbiol.* **10**(8), 21502163.

71. Haft, D. H., Paulsen, I. T., Ward, N., and Selengut, J. D. (2006). Exopolysaccharide-associated protein sorting in environmental organisms: The PEP-CTERM/EpsH system. Application of a novel phylogenetic profiling heuristic. *BMC Biol.* **4**, 29.

72. Cabello, P., Pino, C., Olmo-Mira, M. F., Castillo, F., Roldan, M. D., and Moreno-Vivian, C. (2004). Hydroxylamine assimilation by *Rhodobacter capsulatus* E1F1 requirement of the hcp gene (hybrid cluster protein) located in the nitrate assimilation nas gene region for hydroxylamine reduction. *J. Biol. Chem.* **279**(44), 4548545494.

73. Anjum, M. F., Stevanin, T. M., Read, R. C., and Moir, J. W. (2002). Nitric oxide metabolism in *Neisseria meningitidis*. *J. Bacteriol.* **184**(11), 29872993.

74. Lin, J. T., Goldman, B. S., and Stewart, V. (1994). The nasFEDCBA operon for nitrate and nitrite assimilation in *Klebsiella pneumoniae* M5al. *J. Bacteriol.* **176**(9), 25512559.

75. Gutierrez, J. C., Ramos, F., Ortner, L., and Tortolero, M. (1995). nasST, two genes involved in the induction of the assimilatory nitrite-nitrate reductase operon (nasAB) of *Azotobacter vinelandii*. *Mol. Microbiol.* **18**(3), 579591.

76. Ogawa, K., Akagawa, E., Yamane, K., Sun, Z. W., LaCelle, M., Zuber, P., and Nakano, M. M. (1995). The nasB operon and nasA gene are required for nitrate/nitrite assimilation in *Bacillus subtilis*. *J. Bacteriol.* **177**(5), 14091413.

77. Allen, A. E., Booth, M. G., Frischer, M. E., Verity, P. G., Zehr, J. P., and Zani, S. (2001). Diversity and detection of nitrate assimilation genes in marine bacteria. *Appl. Environ. Microbiol.* **67**(11), 53435348.

78. Siddiqui, R. A., Warnecke-Eberz, U., Hengsberger, A., Schneider, B., Kostka,

S., and Friedrich, B. (1993). Structure and function of a periplasmic nitrate reductase in *Alcaligenes eutrophus* H16. *J. Bacteriol.* **175**(18), 58675876.

79. Turner, S. M., Moir, J. W., Griffiths, L., Overton, T. W., Smith, H., and Cole, J. A. (2005). Mutational and biochemical analysis of cytochrome c', a nitric oxide-binding lipoprotein important for adaptation of *Neisseria gonorrhoeae* to oxygen-limited growth. *Biochem. J.* **388**(Pt 2), 545553.

80. Baek, S. H. and Shapleigh, J. P. (2005). Expression of nitrite and nitric oxide reductases in free-living and plant-associated *Agrobacterium tumefaciens* C58 cells. *Appl. Environ. Microbiol.* **71**(8), 44274436.

81. Waller, A. S., Cox, E. E., and Edwards, E. A. (2004). Perchlorate-reducing microorganisms isolated from contaminated sites. *Environ. Microbiol.* **6**(5), 517527.

82. Kanamori, T., Kanou, N., Atomi, H., and Imanaka, T. (2004). Enzymatic characterization of a prokaryotic urea carboxylase. *J. Bacteriol.* **186**(9), 25322539.

83. Simon, J., Einsle, O., Kroneck, P. M., and Zumft, W. G. (2004). The unprecedented nos gene cluster of *Wolinella succinogenes* encodes a novel respiratory electron transfer pathway to cytochrome c nitrous oxide reductase. *FEBS. Lett.* **569**(1–3), 712.

84. Yoon, S. S., Hennigan, R. F., Hilliard, G. M., Ochsner, U. A., Parvatiyar, K., Kamani, M. C., Allen, H. L., DeKievit, T. R., Gardner, P. R., Schwab, U., et al. (2002). *Pseudomonas aeruginosa* anaerobic respiration in biofilms: Relationships to cystic fibrosis pathogenesis. *Dev. Cell.* **3**(4), 593603.

85. Baar, C., Eppinger, M., Raddatz, G., Simon, J., Lanz, C., Klimmek, O., Nandakumar, R., Gross, R., Rosinus, A., Keller, H., et al. (2003). Complete genome sequence and analysis of *Wolinella succinogenes*. *Proc. Natl. Acad. Sci. USA* **100**(20), 1169011695.

86. Schmehl, M., Jahn, A., Meyer zu Vilsendorf, A., Hennecke, S., Masepohl, B., Schuppler, M., Marxer, M., Oelze, J., and Klipp, W. (1993). Identification of a new class of nitrogen fixation genes in *Rhodobacter capsulatus*: A putative membrane complex involved in electron transport to nitrogenase. *Mol. Gen. Genet.* **241**(5–6), 602615.

87. Baginsky, C., Brito, B., Imperial, J., Palacios, J. M., and Ruiz-Argueso, T. (2002). Diversity and evolution of hydrogenase systems in rhizobia. *Appl. Environ. Microbiol.* **68**(10), 49154924.

88. Menon, A. L., Mortenson, L. E., and Robson, R. L. (1992). Nucleotide sequences and genetic analysis of hydrogen oxidation (hox) genes in *Azotobacter vinelandii*. *J. Bacteriol.* **174**(14), 45494557.

89. Durmowicz, M. C. and Maier, R. J. (1997). Roles of HoxX and HoxA in biosynthesis of hydrogenase in *Bradyrhizobium japonicum*. *J. Bacteriol.* **179**(11), 36763682.

90. Pawlowski, K., Klosse, U., and de Bruijn, F. J. (1991). Characterization of a novel *Azorhizobium caulinodans* ORS571 two-component regulatory system, NtrY/NtrX, involved in nitrogen fixation and metabolism. *Mol. Gen. Genet.* **231**(1), 124138.

91. Robinson, J. J., Stein, J. L., and Cavanaugh, C. M. (1998). Cloning and sequencing of a form II ribulose-1,5-biphosphate carboxylase/oxygenase from the bacterial symbiont of the hydrothermal vent tubeworm *Riftia pachyptila*. *J. Bacteriol.* **180**(6), 15961599.

92. Oda, Y., Wanders, W., Huisman, L. A., Meijer, W. G., Gottschal, J. C., and Forney, L. J. (2002). Genotypic and phenotypic diversity within species of purple nonsulfur bacteria isolated from aquatic sediments. *Appl. Environ. Microbiol.* **68**(7), 34673477.

93. Gibson, J. L. and Tabita, F. R (1997). Analysis of the cbbXYZ operon in *Rhodobacter sphaeroides*. *J. Bacteriol.* **179**(3), 663669.

94. Thony-Meyer, L., Beck, C., Preisig, O., and Hennecke, H. (1994). The ccoNOQP gene cluster codes for a cb-type cytochrome oxidase that functions in aerobic respiration of *Rhodobacter capsulatus*. *Mol. Microbiol.* **14**(4), 705716.

95. Friedrich, C. G., Bardischewsky, F., Rother, D., Quentmeier, A., and Fischer, J. (2005). Prokaryotic sulfur oxidation. *Curr. Opin. Microbiol.* **8**(3), 253259.

96. Kappler, U., Friedrich, C. G., Truper, H. G., and Dahl, C. (2001). Evidence for two pathways of thiosulfate oxidation in *Starkeya novella* (formerly *Thiobacillus novellus*). *Arch. Microbiol.* **175**(2), 102111.

10

1. Timmis, K. N. (2002). *Pseudomonas putida*: A cosmopolitan opportunist par excellence. *Environ. Microbiol.* **4**, 779781.

2. dos Santos, V. A. P. M., Heim, S., Moore, E. R. B., Stratz, M., and Timmis, K. N. (2004). Insights into the genomic basis of niche specificity of *Pseudomonas putida* KT2440. *Environ. Microbiol.* **6**, 1264–1286.

3. Moore, E. R. B., Tindall, B. J., Martins dos Santos, V. A. P., Pieper, D. H., Ramos, J. L., et al. (2006). Nonmedical: Pseudomonas. In *The Prokaryotes: A Handbook on the Biology of Bacteria.* M. Dworkin, S. Falkow, E. Rosenberg, K. Schleifer, and E. Stackebrandt (Eds.). Springer, New York, pp. 646–703.

4. Mosqueda, G., Ramos-Gonzalez, M. I., and Ramos, J. L. (1999). Toluene metabolism by the solvent-tolerant *Pseudomonas putida* DOT-T1 strain, and its role in solvent impermeabilization. *Gene* **232**, 69–76.

5. de Bont, J. A. M. (1998). Solvent-tolerant bacteria in biocatalysis. *Trends Biotechnol.* **16**, 493–499.

6. Wierckx, N. J. P., Ballerstedt, H., de Bont, J. A. M., and Wery, J. (2005). Engineering of solvent-tolerant *Pseudomonas putida* S12 for bioproduction of phenol from glucose. *Appl. Environ. Microbiol.* **71**, 8221–8227.

7. Nijkamp, K., van Luijk, N., de Bont, J. A. M., and Wery, J. (2005). The solvent-tolerant *Pseudomonas putida* S12 as host for the production of cinnamic acid from glucose. *Appl. Microbiol. Biotechnol.* **69**, 170–177.

8. Choi, W. J., Lee, E. Y., Cho, M. H., and Choi, C. Y. (1997). Enhanced production of cis,cis-muconate in a cell-recycle bioreactor. *J. Ferment. Bioeng.* **84**, 70–76.

9. Ramos-Gonzalez, M. I., Ben-Bassat, A., Campos, M. J., and Ramos, J. L. (2003) Genetic engineering of a highly solvent-tolerant *Pseudomonas putida* strain for biotransformation of toluene to p-hydroxybenzoate. *Appl. Environ. Microbiol.* **69**, 5120–5127.

10. Verhoef, S., Ruijssenaars, H. J., de Bont, J. A. M., and Wery, J. (2007). Bioproduction of p-hydroxybenzoate from renewable feedstock by solvent-tolerant *Pseudomonas putida* S12. *J. Biotechnol.* **132**, 49–56.

11. Nijkamp, K., Westerhof, R. G. M. , Ballerstedt, H., de Bont, J. A. M., and Wery, J. (2007). Optimization of the solvent-tolerant *Pseudomonas putida* S12 as host for the production of p-coumarate from glucose. *Appl. Microbiol. Biotechnol.* **74**, 617–624.

12. Stephan, S., Heinzle, E., Wenzel, S. C., Krug, D., Muller, R., et al. (2006). Metabolic physiology of *Pseudomonas putida* for heterologous production of myxochromide. *Process Biochem.* **41**, 2146–2152.

13. Schmid, A., Dordick, J. S., Hauer, B., Kiener, A., Wubbolts, M., et al. (2001). Industrial biocatalysis today and tomorrow. *Nature* **409**, 258–268.

14. Nelson, K. E., Weinel, C., Paulsen, I. T., Dodson, R. J., Hilbert, H., et al. (2002). Complete genome sequence and comparative analysis of the metabolically versatile *Pseudomonas putida* KT2440. *Environ Microbiol* **4**, 799–808.

15. Wackett, L. P. (2003). *Pseudomonas putida*—A versatile biocatalyst. *Nat. Biotechnol.* **21**, 136–138.

16. Jimenez, J. I., Minambres, B., Garcia, J. L., and Diaz, E. (2002). Genomic analysis of the aromatic catabolic pathways from *Pseudomonas putida* KT2440. *Environ. Microbiol.* **4**, 824–841.

17. Huijberts, G. N. M., and Eggink, G. (1996). Production of poly(3-hydroxyalkanoates) by *Pseudomonas putida* KT2442 in continuous cultures. *Appl. Microbiol. Biotechnol.* **46**, 233–239.

18. Steinbüchel, A. and Hein, S. (2001). Biochemical and molecular basis of microbial synthesis of polyhydroxyalkanoates in microorganisms. *Adv. Biochem. Eng. Biotechnol.* **71**, 81–123.

19. Price, N. D., Reed, J. L., and Palsson, B. O. (2004). Genome-scale models of microbial cells: Evaluating the consequences of constraints. *Nat. Rev. Microbiol.* **2**, 886–897.

20. Reed, J. L. and Palsson, B. O. (2003). Thirteen years of building constraint-based *in silico* models of *Escherichia coli. J. Bacteriol.* **185**, 2692–2699.

21. Papin, J. A., Price, N. D., Wiback, S. J., Fell, D. A., and Palsson, B. O. (2003). Metabolic pathways in the post-genome era. *Trends. Biochem. Sci.* **28**, 250–258.

22. Varma, A. and Palsson, B. O. (1994). Stoi-chiometric flux balance models quantitatively predict growth and metabolic by-product secretion in wild-type *Escherichia coli* W3110. *Appl. Environ. Microbiol.* **60**, 3724–3731.

23. Covert, M. W., Knight, E. M., Reed, J. L., Herrgard, M. J., and Palsson, B. O. (2004). Integrating high-throughput and computational data elucidates bacterial networks. *Nature* **429**, 92–96.

24. Price, N. D., Papin, J. A., Schilling, C. H., and Palsson, B. O. (2003). Genome-scale microbial *in silico* models: The constraints-based approach. *Trends Biotechnol.* **21**, 162–169.

25. Joyce, A. R. and Palsson, B. O. (2007). Toward whole cell modeling and simulation: Comprehensive functional genomics through the constraint-based approach. *Prog. Drug. Res.* **64**, 267–309.

26. Lee, K. H., Park, J. H., Kim, T. Y., Kim, H. U., and Lee, S. Y. (2007). Systems metabolic engineering of *Escherichia coli* for L-threonine production. *Mol. Syst. Biol.* **3**, 149.

27. Pharkya, P., Burgard, A. P., and Maranas, C. D. (2004). OptStrain: A computational framework for redesign of microbial production systems. *Genome Res.* **14**, 2367–2376.

28. Burgard, A. P., Pharkya, P., and Maranas, C. D. (2003). OptKnock: A bilevel programming framework for identifying gene knockout strategies for microbial strain optimization. *Biotechnol. Bioeng.* **84**, 647–657.

29. Pharkya, P., Burgard, A. P. and Maranas, C. D. (2003). Exploring the overproduction of amino acids using the bilevel optimization framework OptKnock. *Biotechnol. Bioeng.* **84**, 887–899.

30. Papin, J. A., Stelling, J., Price, N. D., Klamt, S., Schuster, S., et al. (2004). Comparison of network-based pathway analysis methods. *Trends Biotechnol.* **22**, 400–405.

31. Bochner, B. R., Gadzinski, P., and Panomitros, E. (2001) Phenotype microarrays for high-throughput phenotypic testing and assay of gene function. *Genome. Res.* **11**, 1246–1255.

32. Fischer, E., Zamboni, N., and Sauer, U. (2004). High-throughput metabolic flux analysis based on gas chromatography–mass spectrometry derived ^{13}C constraints. *Anal. Biochem.* **325**, 308–316.

33. Reed, J. L., Vo, T. D., Schilling, C. H., and Palsson, B. O. (2003). An expanded genome-scale model of *Escherichia coli* K-12 (iJR904 GSM/GPR). *Genome. Biol.* **4**, R54.

34. Osterman, A., and Overbeek, R. (2003). Missing genes in metabolic pathways: A comparative genomics approach. *Curr. Opin. Chem. Biol.* **7**, 238–251.

35. Kanehisa, M. and Goto, S. (2000). KEGG: Kyoto Encyclopedia of Genes and Genomes. *Nucleic Acids Res.* **28**, 27–30.

36. Revelles, O., Espinosa-Urgel, M., Fuhrer, T., Sauer, U., and Ramos, J. L. (2005). Multiple and interconnected pathways for L-lysine catabolism in Pseudomonas putida KT2440. *J. Bacteriol.* **187**, 7500–7510.

37. Duetz, W. A., Marques, S., Wind, B., Ramos, J. L., and van Andel, J. G. (1996). Catabolite repression of the toluene degradation pathway in *Pseudomonas putida* harboring pWWO under various conditions of nutrient limitation in chemostat culture. *Appl. Environ. Microbiol.* **62**, 601–606.

38. Pfeiffer, T. and Schuster, S. (2005). Game-theoretical approaches to studying the evolution of biochemical systems. *Trends. Biochem. Sci.* **30**, 20–25.

39. Schuster, S., Pfeiffer, T., and Fell, D. A. (2008). Is maximization of molar yield in metabolic networks favoured by evolution? *J. Theor. Biol.* **252**, 497–504.

40. Pramanik, J. and Keasling, J. D. (1998). Effect of *Escherichia coli* biomass composition on central metabolic fluxes predicted by a stoichiometric model. *Biotechnol. Bioeng.* **60**, 230–238.

41. Mahadevan, R. and Schilling, C. H. (2003). The effects of alternate optimal solutions in constraint-based genome-scale metabolic models. *Metab. Eng.* **5**, 264–276.

42. Russell, J. B. and Cook, G. M. (1995). Energetics of bacterial growth: Balance of anabolic and catabolic reactions. *Microbiol. Rev.* **59**, 48–62.

43. Hempfling, W. P. and Mainzer, S. E. (1975). Effects of varying carbon source limiting

growth on yield and maintenance characteristics of *Escherichia coli* in continuous culture. *J. Bacteriol.* **123**, 1076–1087.

44. Mainzer, S. E. and Hempfling, W. P. (1976). Effects of growth temperature on yield and maintenance during glucose-limited continuous culture of *Escherichia coli. J. Bacteriol.* **126**, 251–256.

45. Isken, S., Derks, A., Wolffs, P. F. G., and de Bont, J. A. M. (1999). Effect of organic solvents on the yield of solvent-tolerant *Pseudomonas putida* S12. *Appl. Environ. Microbiol.* **65**, 2631–2635.

46. Fieschko, J. and Humphrey. A. E. (1984). Statistical analysis in the estimation of maintenance and true growth yield coefficients. *Biotechnol. Bioeng.* **26**, 394–396.

47. Bratbak, G. (1985). Bacterial biovolume and biomass estimations. *Appl. Environ. Microbiol.* **49**, 1488–1493.

48. Burgard, A. P., Nikolaev, E. V., Schilling, C. H., and Maranas, C. D. (2004). Flux coupling analysis of genome-scale metabolic network reconstructions. *Genome. Res.* **14**, 301–312.

49. Fuhrer, T., Fischer, E., and Sauer, U. (2005). Experimental identification and quantification of glucose metabolism in seven bacterial species. *J. Bacteriol.* **187**, 1581–1590.

50. del Castillo, T., Ramos, J. L., Rodriguez-Herva, J. J., Fuhrer, T., Sauer, U., et al. (2007). Convergent peripheral pathways catalyze initial glucose catabolism in *Pseudomonas putida*: Genomic and flux analysis. *J. Bacteriol.* **189**, 5142–5152.

51. Cozzone, A. J. (1998). Regulation of acetate metabolism by protein phosphorylation in enteric bacteria. *Annu. Rev. Microbiol.* **52**, 127–164.

52. Teusink, B., Wiersma, A., Molenaar, D., Francke, C., de Vos, W. M., et al. (2006). Analysis of growth of *Lactobacillus plantarum* WCFS1 on a complex medium using a genome-scale metabolic model. *J. Biol. Chem.* **281**, 40041–40048.

53. Schuetz, R., Kuepfer, L., and Sauer, U. (2007). Systematic evaluation of objective functions for predicting intracellular fluxes in *Escherichia coli. Mol. Syst. Biol.* **3**, 119.

54. Pal, C., Papp, B., Lercher, M. J., Csermely, P., Oliver, S. G., et al. (2006). Chance and necessity in the evolution of minimal metabolic networks. *Nature* **440**, 667–670.

55. Jensen, P. R. and Michelsen, O. (1992). Carbon and energy metabolism of atp mutants of *Escherichia coli. J. Bacteriol.* **174**, 7635–7641.

56. von Meyenburg, K., Jorgensen, B. B., Nielsen, J., and Hansen, F. G. (1982). Promoters of the atp operon coding for the membrane-bound ATP synthase of *Escherichia coli* mapped by Tn10 insertion mutations. *Mol. Gen. Genet.* **188**, 240–248.

57. Kornberg, H. L. (1966). Role and control of glyoxylate cycle in *Escherichia coli. Biochem. J.* **99**, 1–11.

58. Fischer, E. and Sauer, U. (2005). Large-scale *in vivo* flux analysis shows rigidity and suboptimal performance of *Bacillus subtilis* metabolism. *Nat. Genet.* **37**, 636–640.

59. Oh, Y. K., Palsson, B. O., Park, S. M., Schilling, C. H., and Mahadevan, R. (2007). Genome-scale reconstruction of metabolic network in *Bacillus subtilis* based on high-throughput phenotyping and gene essentiality data. *J. Biol. Chem.* **282**, 28791–28799.

60. Oberhardt, M. A., Puchalka, J., Fryer, K. E., dos Santos, V. A. P. M., and Papin, J. A. (2008). Genome-scale metabolic network analysis of the opportunistic pathogen *Pseudomonas aeruginosa* PAO1. *J. Bacteriol.* **190**, 2790–2803.

61. Steinbuchel, A. (2001). Perspectives for biotechnological production and utilization of biopolymers: Metabolic engineering of polyhydroxyalkanoate biosynthesis pathways as a successful example. *Macromol. Biosci.* **1**, 1–24.

62. Giavaresi, G., Tschon, M., Borsari, V., Daly, J. H., Liggat, J. J., et al. (2004). New polymers for drug delivery systems in orthopaedics: *In vivo* biocompatibility evaluation. *Biomed. Pharmacother.* **58**, 411–417.

63. van der Walle, G. A. M., de Koning, G. J. M., Weusthuis, R. A., and Eggink, G. (2001). Properties, modifications and applications of biopolyesters. *Adv. Biochem. Eng. Biotechnol.* **71**, 263–291.

64. Klinke, S., Dauner, M., Scott, G., Kessler, B., and Witholt, B. (2000). Inactivation of isocitrate lyase leads to increased production of medium-chain-length poly(3-hy-

droxyalkanoates) in *Pseudomonas putida*. *Appl. Environ. Microbiol.* **66**, 909–913.

65. Patil, K. R., Akesson, M., and Nielsen, J. (2004). Use of genome-scale microbial models for metabolic engineering. *Curr. Opin. Biotechnol.* **15**, 64–69.

66. Jamshidi, N. and Palsson, B. O. (2008). Formulating genome-scale kinetic models in the post-genome era. *Mol. Syst. Biol.* **4**, 171.

67. Joyce, A. R. and Palsson, B. O. (2006). The model organism as a system: Integrating 'omics' data sets. *Nat. Rev. Mol. Cell. Biol.* **7**, 198–210.

68. Seker, S., Beyenal, H., Salih, B., and Tanyolac, A. (1997). Multi-substrate growth kinetics of *Pseudomonas putida* for phenol removal. *Appl. Microbiol. Biotechnol.* **47**, 610–614.

69. Kumar, A., Kumar, S., and Kumar, S. (2005). Biodegradation kinetics of phenol and catechol using *Pseudomonas putida* MTCC 1194. *Biochem. Eng. J.* **22**, 151–159.

70. Wang, S. J. and Loh, K. C. (2001). Biotransformation kinetics of *Pseudomonas putida* for cometabolism of phenol and 4-chlorophenol in the presence of sodium glutamate. *Biodegradation*. **12**, 189–199.

71. Abuhamed, T., Bayraktar, E., Mehmetoglu, T., and Mehmetoglu, U. (2004). Kinetics model for growth of *Pseudomonas putida* F1 during benzene, toluene and phenol biodegradation. *Process Biochem.* **39**, 983–988.

72. Edwards, J. S. and Palsson, B. O. (2000). Metabolic flux balance analysis and the *in silico* analysis of *Escherichia coli* K-12 gene deletions. *BMC Bioinformatics* **1**, 1.

73. Vanrolleghem, P. A. and Heijnen, J. J. (1998). A structured approach for selection among candidate metabolic network models and estimation of unknown stoichiometric coefficients. *Biotechnol. Bioeng.* **58**, 133–138.

74. Palsson, B. O. (2006). *Systems Biology: Properties of Reconstructed Networks*. Cambridge University Press, New York.

75. Varma, A. and Palsson, B. O. (1993). Metabolic capabilities of *Escherichia-Coli* .1. Synthesis of biosynthetic precursors and cofactors. *J. Theor. Biol.* **165**, 477–502.

76. Lee, J. M., Gianchandani, E. P., and Papin, J. A. (2006). Flux balance analysis in the era of metabolomics. *Brief Bioinform.* **7**, 140–150.

77. Edwards, J. S. and Palsson, B. O. (2000). The *Escherichia coli* MG1655 *in silico* metabolic genotype: Its definition, characteristics, and capabilities. *Proc. Natl. Acad. Sci. USA* **97**, 5528–5533.

78. Schuster, S., Fell, D. A., and Dandekar, T. (2000). A general definition of metabolic pathways useful for systematic organization and analysis of complex metabolic networks. *Nature Biotechnology* **18**, 326–332.

79. Reed, J. L. and Palsson, B. O. (2004). Genome-scale *in silico* models of *E-coli* have multiple equivalent phenotypic states: Assessment of correlated reaction subsets that comprise network states. *Genome. Res.* **14**, 1797–1805.

80. Bonarius, H. P. J., Schmid, G., and Tramper, J. (1997). Flux analysis of underdetermined metabolic networks: The quest for the missing constraints. *Trends Biotechnol.* **15**, 308–314.

81. Kanehisa, M., Goto, S., Hattori, M., Aoki-Kinoshita, K. F., Itoh, M., et al. (2006). From genomics to chemical genomics: New developments in KEGG. *Nucleic. Acids. Res.* **34**, D354–D357.

82. Winsor, G. L., Lo, R., Sui, S. J. H., Ung, K. S. E., Huang, S. S., et al. (2005). Pseudomonas aeruginosa Genome Database and PseudoCAP: Facilitating community-based, continually updated, genome annotation. *Nucleic. Acids Res.* **33**, D338–D343.

83. Schomburg, I., Chang, A., and Schomburg, D. (2002). BRENDA, enzyme data and metabolic information. *Nucleic Acids Res.* **30**, 47–49.

84. Goryshin, I. Y. and Reznikoff, W. S. (1998). Tn5 *in vitro* transposition. *J. Biol. Chem.* **273**, 7367–7374.

85. Caetano-Anolles, G. (1993). Amplifying DNA with arbitrary oligonucleotide primers. *Genome. Res. 3*, 85.

86. O'Toole, G. A. and Kolter R (1998). Initiation of biofilm formation in Pseudomonas fluorescens WCS365 proceeds via multiple, convergent signalling pathways: A genetic analysis. *Mol. Microbiol.* **28**, 449–461.

87. Sanger, F., Nicklen, S., and Coulson, A. R. (1977). DNA sequencing with chain-terminating inhibitors. *Proc. Natl. Acad. Sci. USA* **74**, 5463–5467.

88. Hoschle, B., Gnau, V., and Jendrossek, D. (2005). Methylcrotonyl-CoA and geranyl-CoA carboxylases are involved in leucine/isovalerate utilization (Liu) and acyclic terpene utilization (Atu), and are encoded by liuB/liuD and atuC/atuF, in *Pseudomonas aeruginosa*. *Microbiology* **151**, 3649–3656.

11

1. Trimbur, D. E., Gutshall, K. R., Prema, P., and Brenchley, J. E. (1994). Characterization of a psychrotrophic *Arthrobacter* gene and its cold-active β-galactosidase. *Appl. Environ. Microbiol.* **60**, 45444552.

2. Gutshall, K. R., Trimbur, D. E., Kasmir, J. J., and Brenchley, J. E. (1995). Analysis of a novel gene and β-galactosidase isozyme from a psychrotrophic *Arthrobacter* isolate. *J. Bacteriol.* **177**, 19811988.

3. Coombs, J. M. and Brenchley, J. E. (1999). Biochemical and phylogenetic analyses of a cold-active β-galactosidase from the lactic acid bacterium *Carnobacterium piscicola* BA. *Appl. Environ. Microbiol.* **65**, 54435450.

4. Sheridan, P. P. and Brenchley, J. E. (2000). Characterization of a salt-tolerant family 42 beta-galactosidase from a psychrophilic antarctic *Planococcus* isolate. *Appl. Environ. Microbiol.* **66**, 24382444.

5. Hoyoux, A., Jennes, I., Dubois, P., Genicot, S., Dubail, F., François, J. M., Baise, E., Feller, G., and Gerday, C. (2001). Cold-adapted beta-galactosidase from the Antarctic psychrophile *Pseudoalteromonas haloplanktis*. *Appl. Environ. Microbiol.* **67**, 15291535.

6. Fernandes, S., Geueke, B., Delgado, O., Coleman, J., and Hatti-Kaul, R. (2002). Beta-galactosidase from a cold-adapted bacterium: Purification, characterization and application for lactose hydrolysis. *Appl. Microbiol. Biotechnol.* **58**, 313321.

7. Karasová-Lipovová, P., Strnad, H., Spiwok, V., Malá, S., Králová, B., and Russell, N. J. (2003). The cloning, purification and characterisation of a cold-active β-galactosidase from the psychrotolerant Antarctic bacterium *Arthrobacter* sp. C2-2. *Enzyme Microb. Technol.* **33**, 836844.

8. Coker, J. A., Sheridan, P. P., Loveland-Curtze, J., Gutshall, K. R., Auman, A. J., and Brenchley, J. E. (2003). Biochemical characterization of a β-galactosidase with a low temperature optimum obtained from an Antarctic *Arthrobacter* isolate. *J. Bacteriol.* **185**, 54735482.

9. Nakagawa, T., Fujimoto, Y., Ikehata, R., Miyaji, T., and Tomizuka, N. (2006). Purification and molecular characterization of cold-active beta-galactosidase from *Arthrobacter psychrolactophilus* strain F2. *Appl. Microbiol. Biotechnol.* **72**, 720725.

10. Turkiewicz, M., Kur, J., Białkowska, A., Cieśliński, H., Kalinowska, H., and Bielecki, S. (2003). Antarctic marine bacterium *Pseudoalteromonas* sp. 22b as a source of cold-adapted beta-galactosidase. *Biomol. Eng.* **20**, 317324.

11. Cieśliński, H., Kur, J., Białkowska, A., Baran, I., Makowski, K., and Turkiewicz, M. (2005). Cloning, expression, and purification of a recombinant cold-adapted beta-galactosidase from antarctic bacterium *Pseudoalteromonas* sp. 22b. *Protein Expr. Purif.* **39**, 2734.

12. Skalova, T., Dohnalek, J., Spiwok, V., Lipovova, P., Vondrackova, E., Petrokova, H., Duskova, J., Strnad, H., Kralova, B., and Hasek, J. (2005). Cold-active beta-galactosidase from *Arthrobacter* sp. C2-2 forms compact 660 kDa hexamers: Crystal structure at 1.9A resolution. *J. Mol. Biol.* **353**, 282294.

13. Nakagawa, T., Ikehata, R., Myoda, T., Miyaji, T., and Tomizuka, N. (2007). Overexpression and functional analysis of cold-active β-galactosidase from *Arthrobacter psychrolactophilus* strain F2. *Protein Expr. Purif.* **54**, 295299.

14. Hu, J. M., Li, H., Cao, L. X., Wu, P. C., Zhang, C. T., Sang, S. L., Zhang, X. Y., Chen, M. J., Lu, J. Q., and Liu, Y. H. (2007). Molecular cloning and characterization of the gene encoding cold-active beta-galactosidase from a psychrotrophic and halo-

tolerant *Planococcus* sp. L4. *J. Agric. Food Chem.* **55**, 22172224.

15. Kumar, V., Ramakrishnan, S., Teeri, T. T., Knowles, J. K. C., and Hartley, B. S. (1992). *Saccharomyces cerevisiae* cells secreting an *Aspergillus niger* β-galactosidase grow on whey permeate. *Bio./Technol.* **10**, 8285.

16. Ramakrishnan, S. and Hartley, B. S. (1993). Fermentation of lactose by yeast cells secreting recombinant fungal lactase. *Appl. Environ. Microbiol.* **59**, 42304235.

17. Domingues, L., Onnela, M.-L., Teixeira, J. A., Lima, N., and Penttilä, M. (2000). Construction of a flocculent brewer's yeast strain secreting *Aspergillus niger* β-galactosidase. *Appl. Microbiol. Biotechnol.* **54**, 97103.

18. Domingues, L., Teixeira, J. A., Penttilä, M., and Lima, N. (2002). Construction of a flocculent *Saccharomyces cerevisiae* strain secreting high levels of *Aspergillus niger* β-galactosidase. *Appl. Microbiol. Biotechnol.* **58**, 645650.

19. Domingues, L., Lima, N., and Teixeira, J. A. (2005). *Aspergillus niger* β-galactosidase production by yeast in a continuous high cell density reactor. *Process Biochem.* **40**, 11511154.

20. Becerra, M., Cerdán, E., and González Siso, M. I. (1997). Heterologous *Kluyveromyces lactis* β-galactosidase production and release by *Saccharomyces cerevisiae* osmotic-remedial thermosensitive autolytic mutants. *Biochim. Biophys. Acta.* **1335**, 235241.

21. Becerra, M., Rodriguez-Belmonte, E., Cerdán, M. E., and González Siso, M. I. (2004). Engineered autolytic yeast strains secreting *Kluyveromyces lactis* β-galactosidase for production of heterologous proteins in lactose media. *J. Biotechnol.* **109**, 131137.

22. Rodríguez, A. P., Leiro, R. F., Trillo, M. C., Cerdán, M. E., González Siso, M. I., and Becerra, M. (2006). Secretion and properties of a hybrid *Kluyveromyces lactisAspergillus niger* β-galactosidase. *Microb. Cell Fact.* **5**, 41.

23. Yuan, T., Yang, P., Wang, Y., Meng, K., Luo, H., Zhang, W., Wu, N., Fan, Y., and Yao, B. (2008). Heterologous expression of a gene encoding a thermostable β-galactosidase

form *Alicyclobacillus acidocaldaris. Biotechnol. Lett.* **30**, 343348.

24. Rodríguez, A. P., Leiro, R. F., Trillo, M. C., Cerdán, M. E., Siso, M. I. G., and Becerra, M. (2006). Secretion and properties of a hybrid *Kluyveromyces lactis*—Aspergillus niger beta-galactosidase. *Microb. Cell Fact.* **5**, 41.

25. Rubio-Texeira, M. (2006). Endless versatility in the biotechnological applications of *Kluyveromyces LAC* genes. *Biotechnol. Adv.* **24**, 212225.

26. Rubio-Texeira, M., Arévalo-Rodríguez, M., Lequerica, J. L., and Polaina, J. (2001). Lactose utilization by *Saccharomyces cerevisiae* strains expressing *Kluyveromyces lactis LAC* genes. *J. Biotechnol.* **84**, 97106.

27. Rubio-Texeira, M., Castrillo, J. I., Adam, A. C., Ugalde, U. O., and Polaina, J. (1998). Highly efficient assimilation of lactose by a metabolically engineered strain of *Saccharomyces cerevisiae. Yeast* **14**, 827837.

28. Guimarães, P. M., Teixeira, J. A., and Domingues, L. (2008). Fermentation of high concentrations of lactose to ethanol by engineered flocculent *Saccharomyces cerevisiae. Biotechnol. Lett.* **30**, 19538.

29. Wanarska, M., Hildebrandt, P., and Kur, J. (2007). A freeze-thaw method for disintegration of *Escherichia coli* cells producing T7 lysozyme used in pBAD expression systems. *Acta. Biochim. Pol.* **54**, 671672.

12

1. Pomeranz, L. E., Reynolds, A. E., and Hengartner, C. J. (2005). Molecular biology of pseudorabies virus: Impact on neurovirology and veterinary medicine. *Microbiol. Mol. Biol. Rev.* **69**(3), 462500.

2. Roizman, B. and Pellett, P. E. (2001). The family Herpesviridae: A brief introduction. In *Fields virology,* Volume 2 and 4th edition. D. M. Knipe and P. M. Howley (Eds.). Lippincott Williams & Wilkins, Pa, Philadelphia, pp. 23812397.

3. Taddeo, B., Esclatine, A., and Roizman, B. (2002). The patterns of accumulation of cellular RNAs in cells infected with a wild-type and a mutant herpes simplex virus 1

lacking the virion host shutoff gene. *Proc. Nat. Acad. Sci. USA* **99**(26), 1703117036.

4. Jones, J. O. and Arvin, A. M. (2003). Microarray analysis of host cell gene transcription in response to varicella-zoster virus infection of human T cells and fibroblasts *in vitro* and SCIDhu skin xenografts *in vivo*. *J. Virol.* **77**(2), 12681280.

5. Ray, N. and Enquist, L. W. (2004). Transcriptional response of a common permissive cell type to infection by two diverse alphaherpesviruses. *J. Virol.* **78**(7), 34893501.

6. Paulus, C., Sollars, P. J., Pickard, G. E., and Enquist, L. W. (2006). Transcriptome signature of virulent and attenuated pseudorabies virus-infected rodent brain. *J. Virol.* **80**(4), 17731786.

7. Poletto, R., Siegford, J. M., Steibel, J. P., Coussens, P. M., and Zanella, A. J. (2006). Investigation of changes in global gene expression in the frontal cortex of early-weaned and socially isolated piglets using microarray and quantitative real-time RT-PCR. *Brain Res.* **1068**(1), 715.

8. Brittle, E. E., Reynolds, A. E., and Enquist, L. W. (2004). Two modes of pseudorabies virus neuroinvasion and lethality in mice. *J. virol.* **78**(23), 1295112963.

9. Allison, D. B., Cui, X., Page, G. P., and Sabripour, M. (2006). Microarray data analysis: From disarray to consolidation and consensus. *Nat. Rev.* **7**(1), 5565.

10. Petalidis, L., Bhattacharyya, S., Morris, G. A., Collins, V. P., Freeman, T. C., and Lyons, P. A. (2003). Global amplification of mRNA by template-switching PCR: Linearity and application to microarray analysis. *Nucl. Acids Res.* **31**(22), e142.

11. Sykacek, P., Furlong, R. A., and Micklem, G. (2005). A friendly statistics package for microarray analysis. *Bioinformatics (Oxford, England)* **21**(21), 40694070.

12. Wernisch, L., Kendall, S. L., Soneji, S., Wietzorrek, A., Parish, T., Hinds, J., Butcher, P. D., and Stoker, N. G. (2003). Analysis of whole-genome microarray replicates using mixed models. *Bioinformatics (Oxford, England)* **19**(1), 5361.

13. Altschul, S. F., Gish, W., Miller, W., Myers, E. W., and Lipman, D. J. (1990). Basic local alignment search tool. *J. Mol. Biol.* **215**(3), 403410.

14. Khatri, P., Draghici, S., Ostermeier, G. C., and Krawetz, S. A. (2002). Profiling gene expression using onto-express. *Genomics* **79**(2), 266270.

15. Draghici, S., Khatri, P., Martins, R. P., Ostermeier, G. C., and Krawetz, S. A. (2003). Global functional profiling of gene expression. *Genomics* **81**(2), 98104.

16. Livak, K. J. and Schmittgen, T. D. (2001). Analysis of relative gene expression data using real-time quantitative PCR and the 2(-Delta Delta C(T)) Method. *Methods (San Diego, Calif)* **25**(4), 402408.

17. Lee, C., Bachand, A., Murtaugh, M. P., and Yoo, D. (2004). Differential host cell gene expression regulated by the porcine reproductive and respiratory syndrome virus GP4 and GP5 glycoproteins. *Vet. Immunol. Immuno.* **102**(3), 189198.

18. Nau, G. J., Richmond, J. F., Schlesinger, A., Jennings, E. G., Lander, E. S., and Young, R. A. (2002). Human macrophage activation programs induced by bacterial pathogens. *Proc. Nat. Acad. Sci. USA* **99**(3), 15031508.

19. Chan, V. L. (2003). Bacterial genomes and infectious diseases. *Ped. Res.* **54**(1), 17.

20. Shah, G., Azizian, M., Bruch, D., Mehta, R., and Kittur, D. (2004). Cross-species comparison of gene expression between human and porcine tissue, using single microarray platformPreliminary results. *Clin. Transplant.* **18**(Suppl. 12), 7680.

21. McEwen, B. S., Biron, C. A., Brunson, K. W., Bulloch, K., Chambers, W. H., Dhabhar, F. S., Goldfarb, R. H., Kitson, R. P., Miller, A. H., Spencer, R. L., et al. (1997). The role of adrenocorticoids as modulators of immune function in health and disease: Neural, endocrine and immune interactions. *Brain Res. Brain Res. Rev.* **23**(1–2), 79133.

22. Rassnick, S., Enquist, L. W., Sved, A. F., and Card, J. P. (1998). Pseudorabies virus-induced leukocyte trafficking into the rat central nervous system. *J. Virol.* **72**(11), 91819191.

23. Campadelli-Fiume, G., Cocchi, F., Menotti, L., and Lopez, M. (2000). The novel receptors that mediate the entry of herpes simplex

viruses and animal alphaherpesviruses into cells. *Rev. Med. Virol.* **10**(5), 305319.

24. Spear, P. G., Eisenberg, R. J., and Cohen, G. H. (2000). Three classes of cell surface receptors for alphaherpesvirus entry. *Virology* **275**(1), 18.

25. Aravalli, R. N., Hu, S., Rowen, T. N., Gekker, G., and Lokensgard, J. R. (2006). Differential apoptotic signaling in primary glial cells infected with herpes simplex virus 1. *J. Neurovirol.* **12**(6), 501510.

26. Higaki, S., Deai, T., Fukuda, M., and Shimomura, Y. (2004). Microarray analysis in the HSV-1 latently infected mouse trigeminal ganglion. *Cornea* **23** (Suppl. 8), S4247.

27. Flori, L., Rogel-Gaillard, C., Cochet, M., Lemonnier, G., Hugot, K., Chardon, P., Robin, S., and Lefevre, F. (2008). Transcriptomic analysis of the dialogue between Pseudorabies virus and porcine epithelial cells during infection. *BMC genom.* **9**, 123.

28. Reiner, G., Melchinger, E., Kramarova, M., Pfaff, E., Buttner, M., Saalmuller, A., and Geldermann, H. (2002). Detection of quantitative trait loci for resistance/susceptibility to pseudorabies virus in swine. *J. Gen. Virol.* **83**(Pt. 1), 167172.

29. Patarca, R., Freeman, G. J., Singh, R. P., Wei, F. Y., Durfee, T., Blattner, F., Regnier, D. C., Kozak, C. A., Mock, B. A., Morse, H. C., et al. (1989). Structural and functional studies of the early T lymphocyte activation 1 (Eta-1) gene. Definition of a novel T cell-dependent response associated with genetic resistance to bacterial infection. *J. Exp. Med.* **170**(1), 145161.

30. Lebedev, A. A., Krause, M. H., Isidro, A. L., Vagin, A. A., Orlova, E. V., Turner, J., Dodson, E. J., Tavares, P., and Antson, A. A. (2007). Structural framework for DNA translocation via the viral portal protein. *EMBO J.* **26**(7), 19841994.

13

1. Diaz, P. I., Chalmers, N. I., Rickard, A. H., Kong, C., Milburn, C. L., Palmer, R. J. Jr., and Kolenbrander, P. E. (2006). Molecular characterization of subject-specific oral microflora during initial colonization of enamel. *Appl. Environ. Microbiol.* **72**, 28372848.

2. Rosan, B. and Lamont, R. J. (2000). Dental plaque formation. *Microbes. Infect.* **2**, 15991607.

3. Ximenez-Fyvie, L. A., Haffajee, A. D., and Socransky, S. S. (2000). Comparison of the microbiota of supra- and subgingival plaque in health and periodontitis. *J. Clin. Periodontol.* **27**, 648657.

4. Socransky, S. S., Haffajee, A. D., Ximenez-Fyvie, L. A., Feres, M., and Mager, D. (1900). Ecological considerations in the treatment of *Actinobacillus actinomycetemcomitans* and *Porphyromonas gingivalis* periodontal infections. *Periodontol* 2000, **20**, 341362.

5. Kolenbrander, P. E., Andersen, R. N., Blehert, D. S., Egland, P. G., Foster, J. S., and Palmer, R. J. Jr. (2002). Communication among oral bacteria. *Microbiol. Mol. Biol. Rev.* **66**, 486505.

6. Kolenbrander, P. E., Palmer, R. J. Jr., Rickard, A. H., Jakubovics, N. S., Chalmers, N. I., and Diaz, P. I. (2006). Bacterial interactions and successions during plaque development. *Periodontol* 2000 **42**, 4779.

7. Marsh, P. D. (2006). Dental plaque as a biofilm and a microbial community—Implications for health and disease. *BMC Oral Health* **6** (Suppl. 1), S14.

8. Jenkinson, H. F. and Lamont, R. J. (2005). Oral microbial communities in sickness and in health. *Trends Microbiol.* **13**, 589595.

9. Whiteley, M., Bangera, M. G., Bumgarner, R. E., Parsek, M. R., Teitzel, G. M., Lory, S., and Greenberg, E. P. (2001). Gene expression in *Pseudomonas aeruginosa* biofilms. *Nature* **413**, 860864.

10. Stoodley, P., Sauer, K., Davies, D. G., and Costerton, J. W. (2002). Biofilms as complex differentiated communities. *Annu. Rev. Microbiol.* **56**, 187209.

11. Jakubovics, N. S., Gill, S. R., Iobst, S. E., Vickerman, M. M., and Kolenbrander, P. E. (2008). Regulation of gene expression in a mixed-genus community: Stabilized arginine biosynthesis in *Streptococcus gordonii* by coaggregation with *Actinomyces naeslundii*. *J. Bacteriol.* **190**, 36463657.

12. Simionato, M. R., Tucker, C. M., Kuboniwa, M., Lamont, G., Demuth, D. R., Tribble, G. D., and Lamont, R. J. (2006). *Porphyromo-*

nas gingivalis genes involved in community development with *Streptococcus gordonii*. *Infect. Immun.* **74**, 64196428.

13. Ang, C. S., Veith, P. D., Dashper, S. G., and Reynolds, E. C. (2008). Application of 16O/18O reverse proteolytic labeling to determine the effect of biofilm culture on the cell envelope proteome of *Porphyromonas gingivalis* W50. *Proteomics* **8**, 16451660.

14. Aas, J. A., Paster, B. J., Stokes, L. N., Olsen, I., and Dewhirst, F. E. (2005). Defining the normal bacterial flora of the oral cavity. *J. Clin. Microbiol.* **43**, 57215732.

15. Kuboniwa, M., Tribble, G. D., James, C. E., Kilic, A. O., Tao, L., Herzberg, M. C., Shizukuishi, S., and Lamont, R. J. (2006). *Streptococcus gordonii* utilizes several distinct gene functions to recruit *Porphyromonas gingivalis* into a mixed community. *Mol. Microbiol.* **60**, 121139.

16. Lamont, R. J., El-Sabaeny, A., Park, Y., Cook, G. S., Costerton, J. W., and Demuth, D. R. (2002). Role of the *Streptococcus gordonii* SspB protein in the development of *Porphyromonas gingivalis* biofilms on streptococcal substrates. *Microbiology* **148**, 16271636.

17. Capestany, C.A., Tribble, G. D., Maeda, K., Demuth, D. R., and Lamont, R. J. (2008). Role of the Clp system in stress tolerance, biofilm formation, and intracellular invasion in *Porphyromonas gingivalis*. *J. Bacteriol.* **190**, 14361446.

18. Slots, J. and Gibbons, R. J. (1978). Attachment of *Bacteroides melaninogenicus* subsp. asaccharolyticus to oral surfaces and its possible role in colonization of the mouth and of periodontal pockets. *Infect. Immun.* **19**, 254264.

19. Bradshaw, D. J., Marsh, P. D., Watson, G. K., and Allison, C. (1998). Role of *Fusobacterium nucleatum* and coaggregation in anaerobe survival in planktonic and biofilm oral microbial communities during aeration. *Infect. Immun.* **66**, 47294732.

20. Yao, E. S., Lamont, R. J., Leu, S. P., and Weinberg, A. (1996). Interbacterial binding among strains of pathogenic and commensal oral bacterial species. *Oral Microbiol. Immunol.* **11**, 3541.

21. Foster, J. S. and Kolenbrander, P. E. (2004). Development of a multispecies oral bacterial community in a saliva-conditioned flow cell. *Appl. Environ. Microbiol.* **70**, 43404348.

22. Ebersole, J. L., Feuille, F., Kesavalu, L., and Holt, S. C. (1997). Host modulation of tissue destruction caused by periodontopathogens: Effects on a mixed microbial infection composed of *Porphyromonas gingivalis* and *Fusobacterium nucleatum*. *Microb. Pathog.* **23**, 2332.

23. Saito, A., Inagaki, S., Kimizuka, R., Okuda, K., Hosaka, Y., Nakagawa, T., and Ishihara, K. (2008). *Fusobacterium nucleatum* enhances invasion of human gingival epithelial and aortic endothelial cells by *Porphyromonas gingivalis*. *FEMS Immunol. Med. Microbiol.* **54**, 349355.

24. Storey, J. D. and Tibshirani, R. (2003). Statistical significance for genomewide studies. *Proc. Natl. Acad. Sci. USA* **100**, 94409445.

25. Benjamini, Y. and Yekutieli, D. (2005). Quantitative trait Loci analysis using the false discovery rate. *Genetics* **171**, 783790.

26. Storey Research Group, Qvalue Retrieved from Retrieved from [http://genomics.princeton.edu/storeylab/qvalue/].

27. Xia, Q., Hendrickson, E. L., Wang, T., Lamont, R. J., Leigh, J. A., and Hackett, M. (2007). Protein abundance ratios for global studies of prokaryotes. *Proteomics* **7**, 29042919.

28. Knudsen, S. (2004). *Guide to analysis of DNA microarray data*. Wiley-Liss, Hoboken NJ, pp. 3355.

29. Hendrickson, E. L., Lamont, R. J., and Hackett, M. (2008). Tools for interpreting large-scale protein profiling in microbiology. *J. Dent. Res.* **87**, 10041015.

30. Cleveland, W. S. (1981). A program for smoothing scatterplots by robust locally weighted regression. *American Statistician* **35**, 54.

31. Naito, M., Hirakawa, H., Yamashita, A., Ohara, N., Shoji, M., Yukitake, H., Nakayama, K., Toh, H., Yoshimura, F., Kuhara, S., et al. (2008). Determination of the genome sequence of *Porphyromonas gingivalis* strain ATCC 33277 and genomic comparison with strain W83 revealed extensive ge-

nome rearrangements in *P. gingivalis*. *DNA Res.* **15**, 215225.

32. Xia, Q., Wang, T., Park, Y., Lamont, R. J., and Hackett, M. (2007). Differential quantitative proteomics of *Porphyromonas gingivalis* by linear ion trap mass spectrometry: non-label methods comparison, q-values and LOWESS curve fitting. *Int. J. Mass Spectrom.* **259**, 105116.

33. Xia, Q., Wang, T., Taub, F., Park, Y., Capestany, C. A., Lamont, R. J., and Hackett, M. (2007). Quantitative proteomics of intracellular *Porphyromonas gingivalis*. *Proteomics* **7**, 43234337.

34. Eng, J. K., McCormack, A. L., and Yates, J. R. (1994). An approach to correlate tandem mass-spectral data of peptides with amino-acid-sequences in a protein database. *J. Am. Soc. Mass Spectrom.* **5**, 976989.

35. Chiu, S. W., Chen, S. Y., and Wong, H. C. (2008). Localization and expression of MreB in *Vibrio parahaemolyticus* under different stresses. *Appl. Environ. Microbiol.* **74**, 70167022.

36. Nomura, M., Gourse, R., and Baughman, G. (1984). Regulation of the synthesis of ribosomes and ribosomal components. *Annu. Rev. Biochem.* **53**, 75117.

37. Schenk, G., Duggleby, R. G., and Nixon, P. F. (1998). Properties and functions of the thiamin diphosphate dependent enzyme transketolase. *Int. J. Biochem. Cell Biol.* **30**, 12971318.

38. Roper, J. M., Raux, E., Brindley, A. A., Schubert, H. L., Gharbia, S. E., Shah, H. N., and Warren, M. J. (2000). The enigma of cobalamin (Vitamin B12) biosynthesis in *Porphyromonas gingivalis*. Identification and characterization of a functional corrin pathway. *J. Biol. Chem.* **275**, 4031640323.

39. Grenier, D. (1992). Nutritional interactions between two suspected periodonto-pathogens, *Treponema denticola* and *Porphyromonas gingivalis*. *Infect. Immun.* **60**, 52985301.

40. Nelson, K. E., Fleischmann, R. D., DeBoy, R. T., Paulsen, I. T., Fouts, D. E., Eisen, J. A., Daugherty, S. C., Dodson, R. J., Durkin, A. S., Gwinn, M., et al. (2003). Complete genome sequence of the oral pathogenic bacterium *Porphyromonas gingivalis* strain W83. *J. Bacteriol.* **185**, 55915601.

41. Volkert, M. R. and Landini, P. (2001). Transcriptional responses to DNA damage. *Curr. Opin. Microbiol.* **4**, 178185.

42. Lewis, J. P., Plata, K., Yu, F., Rosato, A., and Anaya, C. (2006). Transcriptional organization, regulation and role of the *Porphyromonas gingivalis* W83 hmu haemin-uptake locus. *Microbiology* **152**, 33673382.

43. Leveille, S., Caza, M., Johnson, J. R., Clabots, C., Sabri, M., and Dozois, C. M. (2006). Iha from an *Escherichia coli* urinary tract infection outbreak clonal group A strain is expressed *in vivo* in the mouse urinary tract and functions as a catecholate siderophore receptor. *Infect. Immun.* **74**, 34273436.

44. Merritt, J., Kreth, J., Shi, W., and Qi, F. (2005). LuxS controls bacteriocin production in *Streptococcus mutans* through a novel regulatory component. *Mol. Microbiol.* **57**, 960969.

45. Washburn, M. P., Ulaszek, R., Deciu, C., Schieltz, D. M., and Yates, J. R. 3rd (2002). Analysis of quantitative proteomic data generated via multidimensional protein identification technology. *Anal. Chem.* **74**, 16501657.

46. Washburn, M. P., Wolters, D., and Yates, J. R. 3rd (2001). Large-scale analysis of the yeast proteome by multidimensional protein identification technology. *Nat. Biotechnol.* **19**, 242247.

47. *Porphyromonas gingivalis W83 Genome Page*. Retrieved from [http://cmr.jcvi.org/tigr-scripts/CMR/GenomePage.cgi?org=gpg].

48. *Streptococcus gordonii Challis NCTC7868 Genome Page*. Retrieved from [http://cmr.jcvi.org/tigr-scripts/CMR/GenomePage.cgi?org=gsg].

49. *Fusobacterium nucleatum ATCC 25586 Genome Page*. Retrieved from [http://cmr.jcvi.org/cgi-bin/CMR/GenomePage.cgi?org=ntfn01].

50. *Mammalian Gene Collection*. Retrieved from [http://mgc.nci.nih.gov].

51. Peng, J., Elias, J. E., Thoreen, C. C., Licklider, L. J., and Gygi, S. P. (2003). Evaluation of multidimensional chromatography

coupled with tandem mass spectrometry (LC/LC-MS/MS) for large-scale protein analysis: The yeast proteome. *J. Proteome. Res.* **2**, 4350.

52. Elias, J. E., Gibbons, F. D., King, O. D., Roth, F. P., and Gygi, S. P. (2004). Intensity-based protein identification by machine learning from a library of tandem mass spectra. *Nat. Biotechnol.* **22**, 214219.

53. Tabb, D. L., McDonald, W. H., and Yates, J. R. 3rd (2002). DTASelect and Contrast: Tools for assembling and comparing protein identifications from shotgun proteomics. *J. Proteome. Res.* **1**, 2126.

54. Liu, H., Sadygov, R. G., and Yates, J. R. 3rd (2004). A model for random sampling and estimation of relative protein abundance in shotgun proteomics. *Anal. Chem.* **76**, 41934201.

55. Bosch, G., Skovran, E., Xia, Q., Wang, T., Taub, F., Miller, J. A., Lidstrom, M. E., and Hackett, M. (2008). Comprehensive proteomics of *Methylobacterium extorquens* AM1 metabolism under single carbon and nonmethylotrophic conditions. *Proteomics* **8**, 34943505.

56. Sokal, R. R. and Rohlf, F. J. (1995). *Biometry, the Principles and Practice of Statistics in Biological Research*. WH Freeman, New York, pp. 715724.

57. Huang da, W., Sherman, B. T., Tan, Q., Kir, J., Liu, D., Bryant, D., Guo, Y., Stephens, R., Baseler, M. W., Lane, H. C., et al. (2007). DAVID Bioinformatics Resources: Expanded annotation database and novel algorithms to better extract biology from large gene lists. *Nucleic. Acids Res.* **35**, W169175.

58. Aslanidis, C. and de Jong, P. J. (1990). Ligation-independent cloning of PCR products (LIC-PCR). *Nucleic. Acids Res.* **18**, 60696074.

59. Wu, J., Lin, X., and Xie, H. (2009). Regulation of hemin binding proteins by a novel transcriptional activator in *Porphyromonas gingivalis. J. Bacteriol.* **191**, 115122.

60. O'Toole, G. A. and Kolter, R. (1998). Initiation of biofilm formation in *Pseudomonas fluorescens* WCS365 proceeds via multiple, convergent signalling pathways: A genetic analysis. *Mol. Microbiol.* **28**, 449461.

61. Capestany, C. A., Kuboniwa, M., Jung, I. Y., Park, Y., Tribble, G. D., and Lamont, R. J. (2006). Role of the *Porphyromonas gingivalis* InlJ protein in homotypic and heterotypic biofilm development. *Infect. Immun.* **74**, 30023005.

Index